Coupled-Oscillator Based Active-Array Antennas

JPL DEEP-SPACE COMMUNICATIONS AND NAVIGATION SERIES

The Deep-Space Communications and Navigation Systems
Center of Excellence
Jet Propulsion Laboratory
California Institute of Technology

Joseph H. Yuen, Editor-in-Chief

Published Titles in this Series

Radiometric Tracking Techniques for Deep-Space Navigation
C. L. Thornton and J. S. Border

Formulation for Observed and Computed Values of Deep Space Network Data Types for Navigation
Theodore D. Moyer

Bandwidth-Efficient Digital Modulation with Application to Deep-Space Communications
Marvin K. Simon

Large Antennas of the Deep Space Network
William A. Imbriale

Antenna Arraying Techniques in the Deep Space Network
David H. Rogstad, Alexander Mileant, and Timothy T. Pham

Radio Occultations Using Earth Satellites: A Wave Theory Treatment
William G. Melbourne

Deep Space Optical Communications
Hamid Hemmati

Spaceborne Antennas for Planetary Exploration
William A. Imbriale, Editor

Autonomous Software-Defined Radio Receivers for Deep Space Applications
Jon Hamkins and Marvin K. Simon, Editors

Low-Noise Systems in the Deep Space Network
Macgregor S. Reid, Editor

Coupled-Oscillator Based Active-Array Antennas
Ronald J. Pogorzelski and Apostolos Georgiadis

Coupled-Oscillator Based Active-Array Antennas

Ronald J. Pogorzelski
Jet Propulsion Laboratory
California Institute of Technology

Apostolos Georgiadis
Centre Tecnològic de Telecomunicacions de Catalunya
Castelldefels, Barcelona, Spain

WILEY
A JOHN WILEY & SONS, INC., PUBLICATION

Published by John Wiley & Sons, Inc., Hoboken, New Jersey
Published simultaneously in Canada

Library of Congress Cataloging-in-Publication Data:

Pogorzelski, Ronald J.
Coupled-oscillator based active-array antennas / Ronald J. Pogorzelski and Apostolos Georgiadis.
pages cm. — (JPL deep-space communications and navigation series ; 11)
ISBN 978-1-118-23529-4
1. Antenna arrays. 2. Electric networks, Active. 3. Nonlinear oscillators. 4. Coupled mode theory. I. Georgiadis, Apostolos. II. Title.
TK7871.67.A77P64 2012
621.382'4—dc23 2011045388

Printed in the United States of America.

10 9 8 7 6 5 4 3 2 1

We dedicate this book to our wives,

Barbara and Ana,

who sustained us in this endeavor.

Contents

Foreword

The Deep Space Communications and Navigation Systems Center of Excellence (DESCANSO) was established in 1998 by the National Aeronautics and Space Administration (NASA) at the California Institute of Technology's Jet Propulsion Laboratory (JPL). DESCANSO is chartered to harness and promote excellence and innovation to meet the communications and navigation needs of future deep-space exploration.

DESCANSO's vision is to achieve continuous communications and precise navigation—anytime, anywhere. In support of that vision, DESCANSO aims to seek out and advocate new concepts, systems, and technologies; foster key technical talents; and sponsor seminars, workshops, and symposia to facilitate interaction and idea exchange.

The Deep Space Communications and Navigation Series, authored by scientists and engineers with many years of experience in their respective fields, lays a foundation for innovation by communicating state-of-the-art knowledge in key technologies. The series also captures fundamental principles and practices developed during decades of deep-space exploration at JPL. In addition, it celebrates successes and imparts lessons learned. Finally, the series will serve to guide a new generation of scientists and engineers.

Joseph H. Yuen
DESCANSO Leader

Preface

This book is a compilation of research results obtained primarily over the past two decades in the application of groups of oscillators coupled in various configurations to the excitation of phased-array antennas. Much of the work was carried out at the Jet Propulsion Laboratory of the California Institute of Technology under contract with the National Aeronautics and Space Administration (NASA) building on the early work at the University of Massachusetts, Cornell University, and the University of California, Santa Barbara. More recent work at several institutions in Spain and especially at the Centre Tecnologic de Telecomunicacions de Catalunya (CTTC), as well as at a variety of institutions across Europe and Asia, is also described. A motivation for much of this work was the promise of a method of providing beam agility at electronic speed that is simpler than the conventional method of using a phase shifter at each element or module and controlling these phase shifters in a coordinated manner. More generally, however, the effort has focused on the integration of transmitter, receiver, and antenna including the beam-steering function in a single planar package.

The intended audience for the book comprises primarily designers of phased-array antennas and the associated electronics, but the book may also be of interest to those who may, through understanding the principles presented, envision other innovative applications of oscillator arrays such as distribution of timing signals and phase locking in general. In the same way, graduate students may find inspiration for research work leading to theses or dissertations based on extending the work described here.

With regard to the references, as a general rule we have used peer-reviewed archival journal articles and not conference presentations in the interest of ease of access. We have, however, made a few exceptions in this regard in cases of very recent work that, as far as we know, has not yet appeared in the peer-reviewed literature and in one case for the use of figures with proper attribution. We have endeavored to present a comprehensive treatment of the work in this field to date but recognize that we cannot be sure that we are aware of everyone in the world with interest in and contributions to this fascinating area of research. We, therefore, extend apologies to any who feel their work has been slighted in any way. Be assured it was unintentional.

The book begins with a note concerning the early use of coupled oscillators in the field of mathematical biology wherein researchers used them as an artifice in representing the behavior of neurons in what is known as a central pattern generator in a manner amenable to mathematical analysis. The application to phased-array antennas owes its origin primarily to Karl Stephan at the University of Massachusetts [1–3] and to Richard C. Compton at Cornell and his student, Robert A. York [4–7]. However, the modern emphasis on the study of the dynamics of such arrays was inspired by the interest of James W. Mink of the U. S. Army Research Office [8] in spatial power combining at millimeter wave frequencies. Thus, the presentation continues with a discussion of the utility of oscillator arrays in phased-array antennas and a detailed discussion of the mathematical analysis of the dynamic behavior of such arrays. The mathematics is at a level that should be easily accessible to graduate students in the physical sciences. Advanced calculus, linear algebra, complex variables, and Laplace transforms are the primary tools.

The treatment is arranged in two passes. On the first pass in Part I, we formulate the analysis in the simplest possible manner while retaining the essence of the dynamic behavior, the so-called phase model. Most of the results are based on a linearization of the equations valid for small inter-oscillator phase differences. This permits introduction of the key features of array behavior with a minimum of complexity. We then describe a number of experimental demonstrations of this approach to phased-array beam agility and validation of the approximate theoretical results in Part II. In Part III, we return for a second pass at the analysis, this time including a more sophisticated theoretical description of the oscillators permitting detailed study of the impact of their nonlinear properties. Much of the contemporary research in this area is focused on these properties and their potential utility in modern physical array implementations with many and varied applications. In Part III the presentation of experimental work is integrated with the theoretical as appropriate.

In preparing material for this book, a number of sign errors, typographical errors, and, in rare cases, errors of substance were uncovered in the references.

Every effort has been made to correct these so that where the book differs from the literature, it is the book version that is correct.

Ronald J. Pogorzelski and Apostolos Georgiadis
Pasadena, California and Castelldefels, Barcelona, Spain
June 2011

Acknowledgments

The work of R. Pogorzelski reported here was carried out at the Jet Propulsion Laboratory (JPL), California Institute of Technology under contract with the National Aeronautics and Space Administration (NASA) with additional funding from the U.S. Ballistic Missile Defense Organization (BMDO). Dr. Pogorzelski gratefully acknowledges support from the NASA/JPL Office of the Chief Scientist and Chief Technologist for the writing of this book. He further wishes to acknowledge the contribution of his co-workers at JPL as represented by their co-authorship of many of the references included here. In addition, he thanks Dr. Vahraz Jamnejad of JPL for helpful discussions concerning causality and coupling delay and Mr. Robert J. Beckon of JPL for his help with the cover graphic. Many of the results described here were either obtained or checked using Mathematica™ by Wolfram Research, Inc. (Reference herein to any specific commercial product, process, or service by trade name, trademark, manufacturer, or otherwise does not constitute or imply its endorsement by the United States Government or the Jet Propulsion Laboratory, California Institute of Technology.)

The work of A. Georgiadis has been supported by the Juan de la Cierva Program 2004, the Torres Quevedo Grant PTQ-06-02-0555, project TEC2008-02685/TEC on Novel Architectures for Reconfigurable Reflectarrays and Phased Array Antennas (NARRA) of the Ministry of Science and Innovation Spain, and the European COST Action IC0803 RF/Microwave Communication Subsystems for Emerging Wireless Technologies (RFCSET).

Dr. Georgiadis would like to especially acknowledge Dr. Ana Collado for her invaluable contribution in every aspect of the results presented in Part III of this book. Additionally, he would like to acknowledge Dr. Konstantinos Slavakis for his contribution associated with beamforming and optimization and for long discussions related to convex optimization. Finally, he would like to acknowledge Dr. Maurizio Bozzi and Francesco Giuppi from the University of Pavia and Selva Via from CTTC for their contribution in the recent development of coupled oscillator systems using substrate integrated waveguide (SIW) technology.

Lastly, both Dr. Pogorzelski and Dr. Georgiadis would like to acknowledge the tireless efforts of Mr. Roger V. Carlson of JPL in obtaining the permissions to reprint items from the literature and in editing the manuscript to conform to the format required by the publisher.

Ronald J. Pogorzelski
Apostolos Georgiadis

Authors

Ronald J. Pogorzelski received his BSEE and MSEE degrees from Wayne State University, Detroit, Michigan in 1964 and 1965, respectively, and his PhD degree in electrical engineering and physics from the California Institute of Technology, Pasadena, in 1970, where he studied under Professor Charles H. Papas.

From 1969 to 1973, he was Assistant Professor of Engineering at the University of California, Los Angeles, where his research dealt with relativistic solution of Maxwell's equations. From 1973 to 1977, he was Associate Professor of Electrical Engineering at the University of Mississippi. There his research interests encompassed analytical and computational aspects of electromagnetic radiation and scattering. In 1977 he joined TRW as a senior staff engineer and remained there until 1990, serving as a subproject manager in the Communications and Antenna Laboratory and as a department manager and the Manager of the Senior Analytical Staff in the Electromagnetic Applications Center. From 1981 to 1990 he was also on the faculty of the University of Southern California, first as a part-time instructor and then as an adjunct full professor. In 1990, he joined General Research Corporation as Director of the Engineering Research Group in Santa Barbara, California. Since 1993, he had been with the Jet Propulsion Laboratory as Supervisor of the Spacecraft Antenna Research Group until 2010. From 1999 to 2002 he was a lecturer in electrical engineering at Caltech. In June 2001 he was appointed a JPL Senior Research Scientist. He retired from JPL in May 2010 and is currently Senior Research Scientist Emeritus at JPL.

Dr. Pogorzelski's work has resulted in more than 100 technical publications and presentations. In 1980, he was the recipient of the R. W. P. King Award of the

IEEE Antennas and Propagation Society for a paper on propagation in underground tunnels. Over the years he has served on a number of symposium committees and has chaired a number of symposium sessions. Notably, he was Vice Chairman of the Steering Committee for the 1981 IEEE AP-S Symposium in Los Angeles, California and Technical Program Chair for the corresponding symposium held in Newport Beach, California in 1995. From 1980 to 1986 he was an associate editor of the *IEEE Transactions on Antennas and Propagation* and from 1986 to 1989 he served as its editor. From 1989 to 1990 he served as Secretary/Treasurer of the Los Angeles Chapter of the IEEE Antennas and Propagation Society, was a member of the Society Administrative Committee from 1989 to 2000, served as Vice President of the Society Administrative Committee in 1992, and was its 1993 president. From 1989 to 1992 he was a member of the Society's IEEE Press Liaison Committee. He has also represented IEEE Division IV on the Technical Activities Board Publication Products Council, Periodicals Council, and New Technology Directions Committee. In 1995 he also served as a member of a blue ribbon panel evaluating the U.S. Army's Team Antenna Program in helicopter antennas. He served for ten years as a program evaluator for the Accreditation Board for Engineering and Technology (now ABET, Inc.). Dr. Pogorzelski is a member of Tau Beta Pi, Eta Kappa Nu, and Sigma Xi Honor Societies and has been elected a full member of U.S. National Committee of the Union Radio Scientifique Internationale (USNC/URSI) Commissions A, B, and D. He is a past chair of U.S. Commission B. In 1984 he was appointed an Academy Research Council Representative to the XXIst General Assembly of URSI in Florence, Italy, and in 1999 he was appointed a U.S. Participant in the XXVIth General Assembly of URSI in Toronto, Canada, and similarly in the XXVIIth General Assembly of URSI in Maastricht, the Netherlands in 2002. He has been a member of the Technical Activities Committee of U.S. Commission B and has also served on its Membership Committee from 1988 to 2002, serving as Committee Chair from 1993 to 2002. He was appointed to a two-year term as Member at Large of the U.S. National Committee of URSI in 1996 and again in 1999. Dr. Pogorzelski is an IEEE Third Millennium Medalist and a Fellow of the IEEE.

Apostolos Georgiadis was born in Thessaloniki, Greece. He received his B.S. degree in physics and M.S. degree in telecommunications from the Aristotle University of Thessaloniki, Greece, in 1993 and 1996, respectively. He received his Ph.D. degree in electrical engineering from the University of Massachusetts at Amherst in 2002.

In 1995, he spent a semester with Radio Antenna Communications (R.A.C.), Milan Italy, where he was involved with Yagi antennas for UHF applications. In 2000, he spent three months with Telaxis Communications, South Deerfield,

Massachusetts, where he assisted in the design and testing of a pillbox antenna for local multipoint distribution service (LMDS) applications. In 2002, he joined Global Communications Devices (GCD), North Andover, Massachusetts, where he was a systems engineer involved with CMOS transceivers for wireless network applications. In June 2003, he was with Bermai Inc., Minnetonka, Minnesota, where he was an RF/analog systems architect. In 2005, he joined the University of Cantabria, Santander, Spain as a researcher. While with the University of Cantabria, he collaborated with Advanced Communications Research and Development, S.A. (ACORDE S.A.), Santander, Spain, in the design of integrated CMOS voltage-controlled oscillators (VCOs) for ultra-wideband (UWB) applications. Since 2007, he has been a senior research associate at Centre Tecnològic de Telecomunicacions de Catalunya (CTTC), Barcelona, Spain, in the area of communications subsystems where he is involved in active antennas and antenna arrays and more recently with radio-frequency identification (RFID) technology and energy harvesting.

Dr. Georgiadis is an IEEE senior member. He was the recipient of a 1996 Fulbright Scholarship for graduate studies with the University of Massachusetts at Amherst; the 1997 and 1998 Outstanding Teaching Assistant Award presented by the University of Massachusetts at Amherst; the 1999, 2000 Eugene M. Isenberg Award presented by the Isenberg School of Management, University of Massachusetts at Amherst; and the 2004 Juan de la Cierva Fellowship presented by the Spanish Ministry of Education and Science. He is involved in a number of technical program committees and serves as a reviewer for several journals, including *IEEE Transactions on Antennas and Propagation* and *IEEE Transactions on Microwave Theory and Techniques*. He was the co-recipient of the EUCAP 2010 Best Student Paper Award and the ACES 2010 2nd Best Student Paper Award. He is the Chairman of COST Action IC0803, RF/Microwave communication subsystems for emerging wireless technologies (RFCSET), and he is the Coordinator of the Marie Curie Industry-Academia Pathways and Partnerships project Symbiotic Wireless Autonomous Powered system (SWAP).

Part I: Theory and Analysis

Chapter 1
Introduction—Oscillators and Synchronization

Oscillation is among the simplest of dynamic behaviors to describe mathematically and has thus been conveniently used in modeling a wide variety of physical phenomena ranging from mechanical vibration to quantum mechanical behavior and even neurological systems. Certainly not the least of these is the area of electronic circuits. Many years ago, van der Pol created his classical model of an oscillator including the nonlinear saturation effects that determine the amplitude of the steady-state oscillation [9]. Soon afterward, Adler provided a simple theory of what is now known as injection locking, and coupled oscillators became a valuable design resource for the electronics engineer and the antenna designer [10]. Moreover, circuit theorists were able to apply these principles to long chains and closed rings of coupled oscillators to model biological behaviors such as intestinal and colorectal myoelectrical activity in humans [11, 12].

1.1 Early Work in Mathematical Biology and Electronic Circuits

Biologists, in trying to understand how neurons coordinate the movements of animals, have defined what is known as a "central pattern generator," or "CPG" for short. A CPG in this context is a group of neurons that produce rhythmic or periodic signals without sensory input. Biologists have found that CPGs are conveniently modeled mathematically if treated as a set of oscillators that are coupled to each other, most often using nearest-neighbor coupling but sometimes using more elaborate coupling schemes. Taking this viewpoint and performing the subsequent mathematical analysis has enabled biologists to fruitfully study the manner in which vertebrates (such as the lamprey) coordinate their muscles in locomotion (swimming) and how bipeds (such as you and I) do so in walking or running. The muscles are controlled by signals from a CPG [13, 14]. Electronics engineers have also found oscillators to be useful but more as a component of a man-made system than as a model of a naturally occurring one as in biology. Legend has it that the first electronic oscillator was made by accident in trying to construct an amplifier and encountering unwanted feedback that produced oscillatory behavior. In any case, to deliberately make an oscillator, one starts with an amplifier and provides a feedback path that puts some of the amplifier output into its input whence it is amplified and again returned to the input, and so on. The feedback signal is arranged to arrive at the input in-phase with the preexisting signal at that point so the feedback is regenerative. Thus, the amplitude of the circulating signal would continue to increase indefinitely. However, the amplification or gain of practical amplifiers decreases as the signal amplitude increases. Thus, an equilibrium is quickly reached where the amplitude is just right so the amplifier gain balances the losses in the loop. Then the oscillation amplitude stops increasing and becomes constant. This equilibrium occurs at a particular frequency of oscillation depending on the frequency response of the amplifier and the phase characteristics of the feedback path. Thus, the amplitude and frequency become stable and constant. These can be controlled by changing the circuit component values.

Before long it was realized that an oscillator could also be controlled by injecting a signal from outside the circuit into the feedback loop. This, in a sense, adds energy to the circuit at the injection frequency, making it easier for the circuit to sustain oscillation at that frequency. Therefore, if the injected signal is strong enough, the oscillator will oscillate, not at its natural or free running frequency but, rather, at the injection signal frequency, and the oscillator is said to be "injection locked." If the injection signal comes from another oscillator similar to the one being injected and the coupling is bidirectional, the pair is said to be "mutually injection locked."

If many oscillators are mutually injection locked by providing signal paths between them, mutual coupling paths, they can be made to oscillate as a synchronized ensemble. The ensemble properties of such a system are both interesting and useful, and it is this aspect that so intrigued the mathematical biologists. However, some years ago, it was noted by antenna design engineers that these ensemble properties may be exploited in providing driving signals for phased-array antennas. This is because the phases of the oscillators in a coupled group are coordinated and form useful distributions across the oscillator array.

These phase distributions will be discussed in great detail in the remainder of this book, but, for now, we only note that in, for example, using a linear array of mutually injection locked oscillators coupled to nearest neighbors, one may create linear phase progressions across the array by merely changing the free-running frequencies of the end oscillators of the array anti-symmetrically; that is, one up in frequency and the other down by the same amount. Such a linear distribution of signal phases, when used to excite the elements of a linear array of radiating antenna elements, produces a radiated beam whose direction depends on the phase slope. This slope is determined by the amount by which the free-running frequencies of the end oscillators are changed. Electronic oscillators can be designed so that their free-running frequencies are determined by the bias applied to a varactor in the circuit. These are called voltage-controlled oscillators or "VCOs." So we have now described an antenna wherein the beam direction is controlled by a DC bias voltage, a very convenient and useful arrangement that is, in large part, the subject of this book.

1.2 van der Pol's Model

Although having published some related earlier results, in the fall of 1934, Balthasar van der Pol, of the Natuurkuedig Laboratorium der N. V. Philips' Gloeilampenfabricken in Eindhoven, published in the *Proceedings of the Institute of Radio Engineers*, what has become a classic paper on his analyses of the nonlinear behavior of triode vacuum-tube based electronic oscillators [9]. The beauty of his work lies in the fact that he included in his model only the degree of complexity necessary to produce the important phenomena observed. Thus, his mathematical description remained reasonably tractable, permitting detailed analytical, and more recently computational, study of all the salient behaviors of such circuits.

An important aspect that was missing from the earlier, linear treatments was that of gain saturation. Recall that it is this saturation of the gain that produces a stable steady-state amplitude of oscillation. van der Pol included this as a negative damping of his oscillator which depends quadratically on the oscillation amplitude and becomes positive for sufficiently large amplitude. He also allowed for a driving signal with a frequency different from the resonant frequency of the oscillator. The inclusion of these two features in his model will enable us to use it to describe in this book both the steady-state and the transient behavior of coupled oscillator arrays.

Consider the oscillator of Fig. 1-1 and let Y_L be a resonant parallel combination of an inductor, a capacitor, and a resistor. Application of Kirchhoff's current law to the node at the top of Y_L, using phasors with $e^{j\omega t}$ time dependence, yields

$$j\omega I_D + \left(\frac{1}{L} + \frac{j\omega}{R} - \omega^2 C\right)V = 0 \tag{1.2-1}$$

Now, van der Pol recognized that the active device current, i_d, would be a nonlinear function of the node voltage and modeled that nonlinear function in the time domain as

$$i_D(t) = -\varepsilon\left(g_1 v(t) - g_3 v^3(t)\right) \tag{1.2-2}$$

using the constants ε, g_1, and g_3 for consistency with Section 7.5 where the van der Pol model is revisited in the context of circuit parameter extraction. Thus we have that

$$\frac{d}{dt} i_D(t) = -\varepsilon g_1 \frac{d}{dt} v(t) + 3\varepsilon g_3 v^2(t) \frac{d}{dt} v(t) \tag{1.2-3}$$

or in phasor notation,

$$j\omega I_D = -j\omega\varepsilon\left(g_1 - 3g_3 V^2\right)V \tag{1.2-4}$$

where capital letters denote phasors. Substituting this into Eq. (1.2-1) yields

$$-j\omega\varepsilon\left(g_1 - 3g_3 V^2\right)V + \left(\frac{1}{L} + \frac{j\omega}{R} - \omega^2 C\right)V = 0 \tag{1.2-5}$$

which may be rewritten in the form

$$\left[-j\omega\varepsilon\left(g_1 - 3g_3 V^2\right) + \frac{1}{L} + \frac{j\omega}{R} - \omega^2 C\right]V = j\omega Y V = 0 \tag{1.2-6}$$

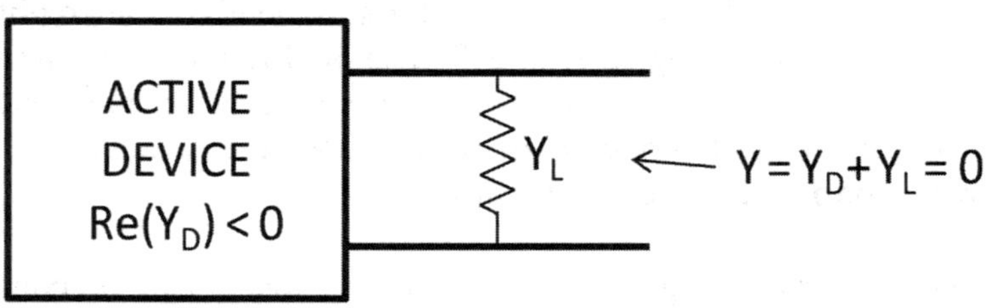

Fig. 1-1. An oscillator as a negative admittance.

where

$$Y=-\varepsilon\left(g_1-3g_3V^2\right)+\frac{1}{j\omega L}+\frac{1}{R}+j\omega C \tag{1.2-7}$$

Now, expanding this admittance in a Taylor series about the resonant frequency,

$$\omega_0=\frac{1}{\sqrt{LC}} \tag{1.2-8}$$

results in

$$\begin{aligned} Y&=-\varepsilon\left(g_1-3g_3V^2\right)+\frac{1}{j\omega L}+\frac{1}{R}+j\omega C \\ &\approx-\varepsilon\left(g_1-3g_3V^2\right)+\frac{2jQ}{\omega_0 R}(\omega-\omega_0) \end{aligned} \tag{1.2-9}$$

where

$$Q=\omega_0 RC \tag{1.2-10}$$

is the traditional quality factor of the oscillator. Use of this expression for the admittance is how we will introduce the van der Pol model into our analysis of an injection locked oscillator below.

1.3 Injection Locking (Adler's Formalism) and Its Spectra (Locked and Unlocked)

To analytically describe the injection locking phenomenon, an oscillator can be viewed as an admittance with a negative real part connected to a resonant load admittance with a positive real part as shown in Fig. 1-1. Using this representation, we proceed now to develop a differential equation for the dynamic behavior of the phase of the oscillation.

The voltage across the load admittance can be written in time-varying phasor form as

$$V=A(t)e^{j\theta(t)} \tag{1.3-1}$$

where,

$$\theta(t) = \omega_0 t + \varphi(t) \tag{1.3-2}$$

Note that V may also be written

$$V = e^{j[\theta(t) - j \ln A(t)]} \tag{1.3-3}$$

Kurokawa [15] suggested that the time derivative of this phasor be written in the form

$$\frac{dV}{dt} = j\left[\omega_0 + \frac{d\varphi}{dt} - j\frac{d}{dt}\ln A\right]V \tag{1.3-4}$$

and that the quantity in brackets be identified as the "instantaneous frequency," ω_{inst}. That is,

$$\frac{dV}{dt} = j\omega_{inst}V \tag{1.3-5}$$

where

$$\omega_{inst} = \left[\omega_0 + \frac{d\varphi}{dt} - j\frac{d}{dt}\ln A\right] \tag{1.3-6}$$

The negative admittance of the device, Y_D, is a function of both the frequency and the amplitude of the oscillating voltage across it. The oscillator operates at the frequency and amplitude that makes this negative admittance equal to the negative of the load admittance, Y_L, so that the total admittance is zero. Following Chang, Shapiro, and York [16], we may expand the admittance in a Taylor series about this operating point in the form

$$Y(\omega_{inst}, A) = Y_L + Y_D(\omega_0, A_0) + (\omega_{inst} - \omega_0)\left.\frac{\partial Y}{\partial \omega}\right|_{\omega_0} + \cdots \tag{1.3-7}$$

where we have neglected the amplitude dependence of Y_D. Multiplying by V, we obtain Kirchhoff's current law at the top node of Fig. 1-1.

$$\begin{aligned} &Y(\omega_{inst}, A)V = \\ &\quad Y_L V + Y_D(\omega_0, A_0)V + (\omega_{inst} - \omega_0)\left.\frac{\partial Y}{\partial \omega}\right|_{\omega_0} V + \cdots = 0 \end{aligned} \tag{1.3-8}$$

In steady state, the oscillator will oscillate with frequency ω_0 and amplitude A_0 making the derivative term zero. Then the load current cancels the oscillator current for a total of zero current exiting the node. However, if a signal is injected at the node from an external source, this equilibrium is changed to

$$I_{inj} + Y(\omega_{inst}, A)V = \\ I_{inj} + Y_L V + Y_D(\omega_0, A_0)V + (\omega_{inst} - \omega_0)\left.\frac{\partial Y}{\partial \omega}\right|_{\omega_0} V + \quad \cdots \quad = 0 \tag{1.3-9}$$

Inserting Eq. (1.3-6) for the instantaneous frequency results in

$$I_{inj} + Y_D(\omega_0, A_0)V + Y_L V + \left[\frac{d\varphi}{dt} - j\frac{d}{dt}\ln A\right]\left.\frac{\partial Y}{\partial \omega}\right|_{\omega_0} V = 0 \tag{1.3-10}$$

or

$$\left[\frac{d\varphi}{dt} - j\frac{d}{dt}\ln A\right] + \frac{Y(\omega_0, A_0)}{\left.\frac{\partial Y}{\partial \omega}\right|_{\omega_0}} + \frac{I_{inj}}{\left.\frac{\partial Y}{\partial \omega}\right|_{\omega_0} V} = 0 \tag{1.3-11}$$

We will now substitute the negative admittance appropriate to the van der Pol oscillator model and analyze the oscillator assuming that a current, I_{inj}, is injected.

Recall that near ω_0 van der Pol's model gives

$$Y = -\varepsilon\left(g_1 - 3g_3 V^2\right) + \frac{2jQ}{\omega_0 R_{osc}}(\omega - \omega_0) \tag{1.3-12}$$

so that

$$\left.\frac{\partial Y}{\partial \omega}\right|_{\omega_0} = \frac{2jQ}{\omega_0 R_{osc}} \tag{1.3-13}$$

Taking the real part of (1.3-11) using (1.3-13) yields

$$\frac{d\varphi}{dt}+\mathrm{Re}\left(\frac{I_{inj}}{\dfrac{2jQ}{\omega_0 R_{osc}}V}\right)=0 \tag{1.3-14}$$

Letting $V_{inj}=R_{osc}I_{inj}$, we obtain

$$\frac{d\varphi}{dt}+\frac{\omega_0}{2Q}\mathrm{Im}\left(\frac{V_{inj}}{V}\right)=0 \tag{1.3-15}$$

Using phasor notation for the injection signal $V_{inj}=A_{inj}e^{j\theta_{inj}}$, and using Eq. (1.3-2), we obtain

$$\frac{d\theta}{dt}=\omega_0+\frac{\omega_0}{2Q}\frac{A_{inj}}{A}\mathrm{Im}\,e^{j\left(\theta_{inj}-\theta\right)}=\omega_0+\frac{\omega_0}{2Q}\frac{A_{inj}}{A}\sin\left(\theta_{inj}-\theta\right) \tag{1.3-16}$$

Defining $\dfrac{\omega_0}{2Q}\dfrac{A_{inj}}{A}=\Delta\omega_{lock}$, the so-called "locking range," we have

$$\frac{d\theta}{dt}=\omega_0+\Delta\omega_{lock}\sin\left(\theta_{inj}-\theta\right) \tag{1.3-17}$$

known as Adler's equation [10]. Taking the imaginary part of Eq. (1.3-11) leads in the same manner to a differential equation for the amplitude dynamics, but treatment of that aspect will be postponed until Chapter 7, which deals with nonlinear analysis of oscillator arrays. For clarity and simplicity in the initial description of the array properties, the amplitude variation will be assumed negligible. If you are particularly interested, however, you may wish to consult Nogi et al. [17], Meadows et al. [18] , and Seetharam et al. [19], which discuss some aspects of amplitude behavior.

Although the differential equation given by Eq. (1.3-17) is first order, it is nonlinear. Remarkably, however, it can nevertheless be solved analytically. Once the solution is obtained, it can be used to describe the dynamic behavior of the locking process and, very interestingly, the spectrum of the oscillations under both locked and unlocked conditions. We begin by solving Eq. (1.3-17) and then proceed to exhibit the spectral properties of the solution.

First, we define

$$\psi = \theta - \theta_{inj} = \left(\varphi - \varphi_{inj}\right) + \left(\omega_0 - \omega_{inj}\right) t \tag{1.3-18}$$

so that Eq. (1.3-17) may be written

$$\frac{d\psi}{dt} = -\Delta\omega_{lock}\left(\sin\psi + \frac{\Delta\omega_{inj}}{\Delta\omega_{lock}}\right) \tag{1.3-19}$$

where $\Delta\omega_{inj} = \omega_{inj} - \omega_0$. Now defining $K = \dfrac{\Delta\omega_{inj}}{\Delta\omega_{lock}}$ and $\tau = \Delta\omega_{lock} t$, we have the deceptively simple looking differential equation

$$\frac{d\psi}{d\tau} = -\left(\sin\psi + K\right) \tag{1.3-20}$$

Integrating from an initial time, τ_0, to an arbitrary subsequent time, τ,

$$\int_{\psi(\tau_0)}^{\psi(\tau)} \frac{d\psi}{\left(\sin\psi + K\right)} = -\int_{\tau_0}^{\tau} d\tau \tag{1.3-21}$$

we arrive at

$$\tau = \tau_0 - \int_{\psi(\tau_0)}^{\psi(\tau)} \frac{d\psi}{\left(\sin\psi + K\right)} \tag{1.3-22}$$

and it remains to carry out the integration. Using the substitution

$$u = \tan\left(\frac{\psi}{2}\right) \tag{1.3-23}$$

the integral may be cast in the form

$$\frac{1}{K}\int_{u_0}^{u} \frac{2du}{u^2 + \dfrac{2u}{K} + 1} \tag{1.3-24}$$

where

$$u_0 = \tan\left(\frac{\psi(\tau_0)}{2}\right) \tag{1.3-25}$$

By factoring the denominator of the integrand and expanding in partial fractions, the integral, Eq. (1.3-24), can be expressed in terms of the natural logarithm function in the form

$$\frac{1}{K}\int_{u_0}^{u}\frac{2du}{u^2+\frac{2u}{K}+1}=\frac{1}{\sqrt{1-K^2}}\ln\left(\frac{u-u_2}{u-u_1}\right)\Bigg|_{u_0}^{u} \tag{1.3-26}$$

where u_1 and u_2 are the roots of the quadratic in the denominator of the integrand. That is, Eq. (1.3-22) becomes

$$\int_{\psi(\tau_0)}^{\psi(\tau)}\frac{d\psi}{(\sin\psi+K)}=$$

$$\frac{1}{\sqrt{1-K^2}}\ln\left(\frac{K\tan\left(\frac{\psi}{2}\right)+\left(1-\sqrt{1-K^2}\right)}{K\tan\left(\frac{\psi}{2}\right)+\left(1+\sqrt{1-K^2}\right)}\right)\Bigg|_{\psi(\tau_0)}^{\psi(\tau)}=\tau_0-\tau \tag{1.3-27}$$

Recall that the natural logarithm function is related to the inverse hyperbolic tangent function by

$$\ln\left(\frac{1+x}{1-x}\right)=2\tanh^{-1}(x) \tag{1.3-28}$$

if $0\le x^2<1$. Upon using Eq. (1.3-28) in Eq. (1.3-27), we obtain

$$\tau=\tau_0+\frac{2}{\sqrt{1-K^2}}\tanh^{-1}\left(\frac{\sqrt{1-K^2}}{K\tan\left(\frac{\psi}{2}\right)+1}\right)\Bigg|_{\psi(\tau_0)}^{\psi(\tau)} \tag{1.3-29}$$

Provided that $K^2<1$. This condition is equivalent to

$$\left|\Delta\omega_{inj}\right|<\left|\Delta\omega_{lock}\right| \tag{1.3-30}$$

which means that the injection signal frequency is within one locking range of the free-running frequency of the oscillator corresponding to the so-called "locked" condition. If $K^2 \geq 1$, the oscillator is said to be "unlocked" and the solution given by Eq. (1.3-27) becomes

$$\tau = \tau_0 - \frac{2}{\sqrt{K^2-1}} \tan^{-1}\left(\frac{\sqrt{K^2-1}}{K \tan\left(\frac{\psi}{2}\right)+1} \right)\Bigg|_{\psi(\tau_0)}^{\psi(\tau)} \tag{1.3-31}$$

Now, rewriting Eqs. (1.3-29) and (1.3-31) explicitly evaluated at the limits and rearranging a bit results in

$$\frac{1}{2}\sqrt{1-K^2}\,(\tau-\tau_0) = \tanh^{-1}\left(\frac{\sqrt{1-K^2}}{K \tan\left(\frac{\psi(\tau)}{2}\right)+1} \right) - \tanh^{-1}\left(\frac{\sqrt{1-K^2}}{K \tan\left(\frac{\psi(\tau_0)}{2}\right)+1} \right) \tag{1.3-32}$$

and

$$-\frac{1}{2}\sqrt{K^2-1}\,(\tau-\tau_0) = \tan^{-1}\left(\frac{\sqrt{K^2-1}}{K \tan\left(\frac{\psi(\tau)}{2}\right)+1} \right) - \tan^{-1}\left(\frac{\sqrt{K^2-1}}{K \tan\left(\frac{\psi(\tau_0)}{2}\right)+1} \right) \tag{1.3-33}$$

We now make use of the following pair of identities.

$$\tanh^{-1}(x) - \tanh^{-1}(x_0) = \tanh^{-1}\left(\frac{x-x_0}{1-xx_0} \right) \tag{1.3-34}$$

$$\tan^{-1}(x) - \tan^{-1}(x_0) = \tan^{-1}\left(\frac{x-x_0}{1+xx_0} \right) \tag{1.3-35}$$

Applying these to Eqs. (1.3-32) and (1.3-33), respectively, we obtain

$$\tanh\left[\frac{1}{2}\sqrt{1-K^2}\,(\tau-\tau_0)\right]= \frac{-\left[\tan\left(\frac{\psi(\tau)}{2}\right)-\tan\left(\frac{\psi(\tau_0)}{2}\right)\right]}{K+K\tan\left(\frac{\psi(\tau)}{2}\right)\tan\left(\frac{\psi(\tau_0)}{2}\right)+\tan\left(\frac{\psi(\tau)}{2}\right)+\tan\left(\frac{\psi(\tau_0)}{2}\right)} \tag{1.3-36}$$

$$\tan\left[\frac{1}{2}\sqrt{1-K^2}\,(\tau-\tau_0)\right]= \frac{\left[\tan\left(\frac{\psi(\tau)}{2}\right)-\tan\left(\frac{\psi(\tau_0)}{2}\right)\right]}{K+K\tan\left(\frac{\psi(\tau)}{2}\right)\tan\left(\frac{\psi(\tau_0)}{2}\right)+\tan\left(\frac{\psi(\tau)}{2}\right)+\tan\left(\frac{\psi(\tau_0)}{2}\right)} \tag{1.3-37}$$

These equations may now be solved for $\psi(\tau)$. The results are

$$\psi(\tau)= 2\tan^{-1}\left\{\frac{\tan\left(\frac{\psi(\tau_0)}{2}\right)-\tanh\left[\frac{1}{2}\sqrt{1-K^2}\,(\tau-\tau_0)\right]\left[K+\tan\left(\frac{\psi(\tau_0)}{2}\right)\right]}{1+\tanh\left[\frac{1}{2}\sqrt{1-K^2}\,(\tau-\tau_0)\right]\left[1+K\tan\left(\frac{\psi(\tau_0)}{2}\right)\right]}\right\} \tag{1.3-38}$$

$$\psi(\tau)= 2\tan^{-1}\left\{\frac{\tan\left(\frac{\psi(\tau_0)}{2}\right)+\tan\left[\frac{1}{2}\sqrt{K^2-1}\,(\tau-\tau_0)\right]\left[K+\tan\left(\frac{\psi(\tau_0)}{2}\right)\right]}{1-\tan\left[\frac{1}{2}\sqrt{K^2-1}\,(\tau-\tau_0)\right]\left[1+K\tan\left(\frac{\psi(\tau_0)}{2}\right)\right]}\right\} \tag{1.3-39}$$

These represent the exact analytic solution of Eq. (1.3-20) giving the dynamic behavior of the phase of an externally injection locked oscillator for all time subsequent to τ_0. While they are actually the same solution, Eq. (1.3-38) is conveniently applied when $K^2 < 1$, and Eq. (1.3-39) is conveniently applied when $K^2 > 1$. When $K^2 = 1$, Eqs. (1.3-38) and (1.3-39) are identical.

We will now proceed to study the spectral properties of this solution. It will be expedient to return to the logarithmic representation in Eq. (1.3-27). For the locked condition we have

$$\ln\left\{\left(\frac{K\tan\left(\frac{\psi(\tau)}{2}\right)+\left(1-\sqrt{1-K^2}\right)}{K\tan\left(\frac{\psi(\tau)}{2}\right)+\left(1+\sqrt{1-K^2}\right)}\right)\left(\frac{K\tan\left(\frac{\psi(\tau_0)}{2}\right)+\left(1+\sqrt{1-K^2}\right)}{K\tan\left(\frac{\psi(\tau_0)}{2}\right)+\left(1-\sqrt{1-K^2}\right)}\right)\right\}= \quad (1.3\text{-}40)$$

$$(\tau_0-\tau)\sqrt{1-K^2}$$

Exponentiating both sides yields

$$\left\{\left(\frac{K\tan\left(\frac{\psi(\tau)}{2}\right)+\left(1-\sqrt{1-K^2}\right)}{K\tan\left(\frac{\psi(\tau)}{2}\right)+\left(1+\sqrt{1-K^2}\right)}\right)\left(\frac{K\tan\left(\frac{\psi(\tau_0)}{2}\right)+\left(1+\sqrt{1-K^2}\right)}{K\tan\left(\frac{\psi(\tau_0)}{2}\right)+\left(1-\sqrt{1-K^2}\right)}\right)\right\}= \quad (1.3\text{-}41)$$

$$e^{-(\tau-\tau_0)\sqrt{1-K^2}}$$

For simplicity of notation, the second factor in the curly brackets, being a constant that depends on the initial conditions, will be defined to be $1/C_0$. Thus,

$$\left(\frac{K\tan\left(\frac{\psi(\tau)}{2}\right)+\left(1-\sqrt{1-K^2}\right)}{K\tan\left(\frac{\psi(\tau)}{2}\right)+\left(1+\sqrt{1-K^2}\right)}\right)=C_0e^{-(\tau-\tau_0)\sqrt{1-K^2}} \quad (1.3\text{-}42)$$

Now solving for $\psi(\tau)$, we obtain

$$\psi(\tau)=2\tan^{-1}\left[\frac{\sqrt{1-K^2}}{K}\left(\frac{1+C_0e^{-(\tau-\tau_0)\sqrt{1-K^2}}}{1-C_0e^{-(\tau-\tau_0)\sqrt{1-K^2}}}\right)-\frac{1}{K}\right] \quad (1.3\text{-}43)$$

Recall that

$$\tan^{-1}(x) = \frac{j}{2}\ln\left(\frac{j+x}{j-x}\right) \quad (1.3\text{-}44)$$

So that Eq. (1.3-43) may be written in the form

$$\psi(\tau) = j\ln\left\{\frac{j + \dfrac{\sqrt{1-K^2}}{K}\left(\dfrac{1+C_0 e^{-(\tau-\tau_0)\sqrt{1-K^2}}}{1-C_0 e^{-(\tau-\tau_0)\sqrt{1-K^2}}}\right) - \dfrac{1}{K}}{j - \dfrac{\sqrt{1-K^2}}{K}\left(\dfrac{1+C_0 e^{-(\tau-\tau_0)\sqrt{1-K^2}}}{1-C_0 e^{-(\tau-\tau_0)\sqrt{1-K^2}}}\right) + \dfrac{1}{K}}\right\} \quad (1.3\text{-}45)$$

Again exponentiating both sides, we obtain

$$e^{j\psi(\tau)} = \left\{\frac{j - \dfrac{\sqrt{1-K^2}}{K}\left(\dfrac{1+C_0 e^{-(\tau-\tau_0)\sqrt{1-K^2}}}{1-C_0 e^{-(\tau-\tau_0)\sqrt{1-K^2}}}\right) + \dfrac{1}{K}}{j + \dfrac{\sqrt{1-K^2}}{K}\left(\dfrac{1+C_0 e^{-(\tau-\tau_0)\sqrt{1-K^2}}}{1-C_0 e^{-(\tau-\tau_0)\sqrt{1-K^2}}}\right) - \dfrac{1}{K}}\right\} \quad (1.3\text{-}46)$$

This can be rearranged as

$$e^{j\psi(\tau)} = \left\{\frac{\left(jK+1-\sqrt{1-K^2}\right) - \left(jK+1+\sqrt{1-K^2}\right)C_0 e^{-(\tau-\tau_0)\sqrt{1-K^2}}}{\left(jK-1+\sqrt{1-K^2}\right) - \left(jK-1-\sqrt{1-K^2}\right)C_0 e^{-(\tau-\tau_0)\sqrt{1-K^2}}}\right\} \quad (1.3\text{-}47)$$

Equation (1.3-47) gives the dynamic behavior of the oscillator voltage as the phase evolves from $\psi(\tau_0)$ to $\psi(\tau)$. This behavior is exponential, not oscillatory, and the steady-state value of the phase at infinite time is $-\sin^{-1}(K)$. Returning to Eq. (1.3-1) and using Eq. (1.3-18), we find that the oscillator voltage in steady state is

$$V_{ss} = A(t)e^{j\theta(t)} = Ae^{j(\psi+\theta_{inj})} = Ae^{j(-\sin^{-1}(K)+\varphi_{inj}+\omega_{inj}t)} \quad (1.3\text{-}48)$$

Thus, the spectrum is a single line at frequency ω_{inj} and there is a steady-state phase difference between the oscillator signal and the injection signal of $\sin^{-1}(K)$.

Suppose we allow K to become larger than unity in magnitude. In such a case, the injection signal frequency lies outside the locking range around the free running frequency and the oscillator will be in the "unlocked" condition described by Eq. (1.3-39). Now, however, the spectral properties of the solution become more interesting. We follow an approach suggested by Armand [20]. In this situation, Eq. (1.3-47) becomes

$$e^{j\psi(\tau)} = \left\{ \frac{\left(jK+1-j\sqrt{K^2-1}\right)-\left(jK+1+j\sqrt{K^2-1}\right)C_0 e^{-j\sqrt{K^2-1}(\tau-\tau_0)}}{\left(jK-1+j\sqrt{K^2-1}\right)-\left(jK-1-j\sqrt{K^2-1}\right)C_0 e^{-j\sqrt{K^2-1}(\tau-\tau_0)}} \right\} \tag{1.3-49}$$

or

$$e^{j\psi(\tau)} = \left\{ \frac{A_1 - A_2 C_0 e^{-jT}}{B_1 - B_2 C_0 e^{-jT}} \right\} \tag{1.3-50}$$

where

$$\begin{aligned} A_1 &= jK+1-j\sqrt{K^2-1} \\ A_2 &= jK+1+j\sqrt{K^2-1} \\ B_1 &= jK-1+j\sqrt{K^2-1} \\ B_2 &= jK-1-j\sqrt{K^2-1} \end{aligned} \tag{1.3-51}$$

$$T = \sqrt{K^2-1}\left(\tau-\tau_0\right) \tag{1.3-52}$$

and

$$C_0 = \left(\frac{K \tan\left(\frac{\psi(\tau_0)}{2}\right) + \left(1 - j\sqrt{K^2 - 1}\right)}{K \tan\left(\frac{\psi(\tau_0)}{2}\right) + \left(1 + j\sqrt{K^2 - 1}\right)} \right) \tag{1.3-53}$$

Expanding Eq. (1.3-49) in a geometric series yields

$$e^{j\psi(\tau)} = \frac{A_1}{B_1} + \left\{ \frac{A_1}{B_1} - \frac{A_2}{B_2} \right\} \sum_{n=1}^{\infty} \left(\frac{B_2}{B_1} C_0 \right)^n e^{-jnT} \tag{1.3-54}$$

Now, the magnitude of the common ratio of the series is

$$\left| \frac{B_2}{B_1} C_0 \right|$$

$$= \left| \frac{jK - 1 - j\sqrt{K^2 - 1}}{jK - 1 + j\sqrt{K^2 - 1}} \right| \left| \frac{K \tan\left(\frac{\psi(\tau_0)}{2}\right) + \left(1 - j\sqrt{K^2 - 1}\right)}{K \tan\left(\frac{\psi(\tau_0)}{2}\right) + \left(1 + j\sqrt{K^2 - 1}\right)} \right| \tag{1.3-55}$$

$$= \sqrt{\frac{1 + \left(K - \sqrt{K^2 - 1}\right)^2}{1 + \left(K + \sqrt{K^2 - 1}\right)^2}}$$

This is less than unity for positive K, and the series converges for all T. If, on the other hand, K is negative, we instead expand the reciprocal of Eq. (1.3-49),

$$e^{-j\psi(\tau)} = \left\{ \frac{B_1 - B_2 C_0 e^{-jT}}{A_1 - A_2 C_0 e^{-jT}} \right\} = \frac{B_1}{A_1} + \left\{ \frac{B_1}{A_1} - \frac{B_2}{A_2} \right\} \sum_{n=1}^{\infty} \left(\frac{A_2}{A_1} C_0 \right)^n e^{-jnT} \tag{1.3-56}$$

and the magnitude of the common ratio is

$$\left|\frac{A_2}{A_1}C_0\right| =$$

$$\left|\frac{jK+1+j\sqrt{K^2-1}}{jK+1-j\sqrt{K^2-1}}\right|\left|\frac{K\tan\left(\frac{\psi(\tau_0)}{2}\right)+\left(1-j\sqrt{K^2-1}\right)}{K\tan\left(\frac{\psi(\tau_0)}{2}\right)+\left(1+j\sqrt{K^2-1}\right)}\right| = \qquad (1.3\text{-}57)$$

$$\sqrt{\frac{1+\left(K+\sqrt{K^2-1}\right)^2}{1+\left(K-\sqrt{K^2-1}\right)^2}}$$

which is less than unity for K negative. Expressions (1.3-54) and (1.3-56) thus provide convergent series representations of the solution for the phase dynamics under unlocked conditions and we note that they are actually Fourier series. As such, the coefficients are the amplitudes of the harmonics of a line spectrum representing the oscillator signal. This spectrum has a well-known classic form that is easily observed experimentally using a spectrum analyzer and is depicted schematically in Fig. 1-2.

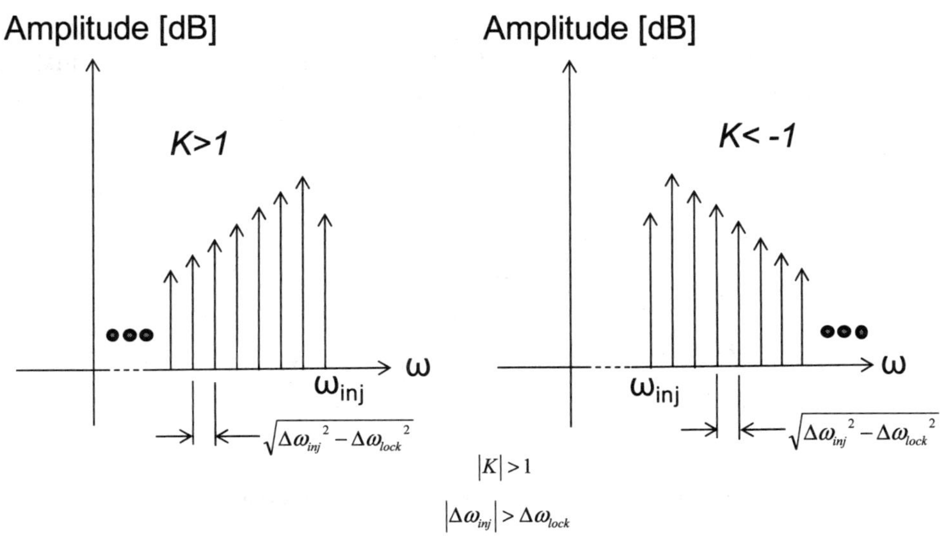

Fig. 1-2. Spectra of an unlocked injected oscillator.

(This K is Kurokawa's [15], which is the negative of Adler's [10] and Armand's [20].)

These mirror-image spectra have a number of interesting features. The most obvious feature is that they are one-sided, which may seem puzzling, but is a natural result of the analysis. Secondly, the amplitudes decrease linearly on a logarithmic scale as one progresses away from the injection frequency. This is a consequence of the geometric nature of the series representing the solution. Finally, the spacing between the spectral lines decreases with the proximity of the injection frequency to the oscillator free running frequency and, when the injection frequency differs from the free running frequency by exactly one locking range, the spacing goes to zero and the oscillator locks, reducing the spectrum to a single line at ω_{inj}.

Before we can legitimately call this analysis of injection locking complete, there remains one important issue to consider. The oscillator model shown in Fig. 1-1 exhibits a parallel resonance. It is, of course, possible to design an oscillator that exhibits a series resonance, and the question then becomes: How is this difference manifest in the formalism presented? This question has been studied in detail by Chang, Shapiro, and York [16]. They pointed out that the Taylor series for the admittance in the parallel resonant oscillator, Eq. (1.2-9), is identical in form to the Taylor expansion of the impedance in the series resonant case. We can see this by considering the series resonant oscillator shown in Fig. 1-3. In this case the resonant load, Z_L, on the active device is a series combination of an inductor, a capacitor, and a resistor.

The output signal here is the current through this resonant series combination rather than the node voltage used in the parallel case. Application of Kirchhoff's voltage law around the oscillator loop yields

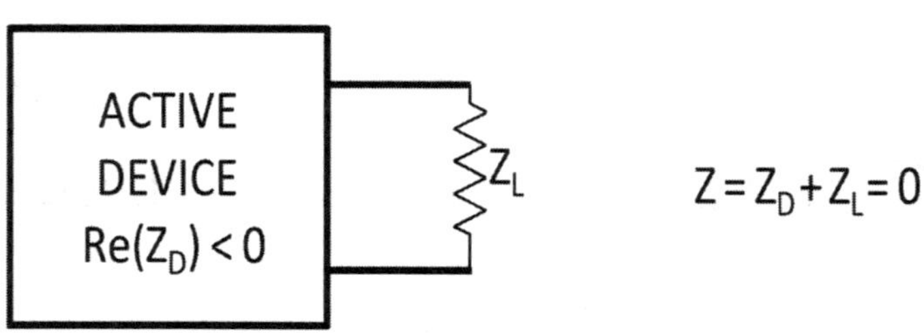

Fig. 1-3. An oscillator as a negative impedance.

$$V_D + \left(j\omega L + R + \frac{1}{j\omega C} \right) I = 0 \tag{1.3-58}$$

Using a van der Pol-type nonlinearity, the analog of Eq. (1.2-2) is

$$v_D(t) = -\varepsilon\left(r_1 i(t) - r_3 i^3(t) \right) \tag{1.3-59}$$

and the analog of Eq. (1.2-7) is

$$Z = -\varepsilon\left(r_1 - 3r_3 I^2 \right) + j\omega L + R + \frac{1}{j\omega C} = \frac{1}{Y} \tag{1.3-60}$$

Expanding Y in a Taylor series about the resonant frequency, we arrive at

$$Y \approx \frac{1}{R - \varepsilon\left(r_1 - 3r_3 I^2 \right)} - \frac{2jQ}{\omega_0 R}(\omega - \omega_0) \tag{1.3-61}$$

Comparing with Eq. (1.2-9) we see that the salient difference is the change in sign of the linear term in frequency. This in turn induces a change in the algebraic sign of the sine term in Eq. (1.3-17), resulting in

$$\frac{d\theta}{dt} = \omega_0 - \Delta\omega_{lock} \sin\left(\theta_{inj} - \theta \right) \tag{1.3-62}$$

and the remainder of the analysis proceeds as for the parallel resonant case above. We will further describe the implications of this when we consider more than one oscillator.

1.4 Mutual Injection Locking of Two Oscillators

Consider now two parallel resonant oscillators, identical except for free-running frequency, coupled together so that each injects a signal into the other. Such a system was considered by Stephan and Young [3] in which the coupling was due to free-space mutual coupling between radiating elements excited by the oscillators. We may describe this situation using Adler's Eq. (1.3-17) for each oscillator. That is,

$$\frac{d\theta_1}{dt} = \omega_{01} + \Delta\omega_{lock} \sin\left(\theta_2 - \theta_1 \right) \tag{1.4-1}$$

$$\frac{d\theta_2}{dt} = \omega_{02} + \Delta\omega_{lock} \sin(\theta_1 - \theta_2) \qquad (1.4\text{-}2)$$

where the subscripts identify the oscillators. Subtracting these equations yields

$$\frac{d(\theta_1 - \theta_2)}{dt} = (\omega_{01} - \omega_{02}) + 2\Delta\omega_{lock} \sin(\theta_2 - \theta_1) \qquad (1.4\text{-}3)$$

We now define

$$\tilde{\psi} = \theta_1 - \theta_2 \qquad (1.4\text{-}4)$$

$$\tilde{K} = \frac{\omega_{02} - \omega_{01}}{2\Delta\omega_{lock}} \qquad (1.4\text{-}5)$$

$$\tilde{\tau} = 2\Delta\omega_{lock} t \qquad (1.4\text{-}6)$$

so that Eq. (1.4-3) becomes

$$\frac{d\tilde{\psi}}{d\tilde{\tau}} = -\left(\sin\tilde{\psi} + \tilde{K}\right) \qquad (1.4\text{-}7)$$

which is identical with Eq. (1.3-20) except for the tildes, and all of the preceding results apply. Note that the locking range is replaced by twice the locking range in this equation. This happens because the injecting oscillator frequency is permitted to change under the influence of the oscillator being injected. The result is that the two oscillator frequencies can differ by nearly twice the locking range and still maintain lock. This is true because it will turn out that the steady-state oscillation frequency of the pair is the average of the two free-running frequencies, and we can show this as follows.

Recall that in steady state, if $\tilde{K} < 1$ so the oscillators are locked, $\tilde{\psi} = -\sin^{-1}\tilde{K}$, a constant, so its time derivative is zero. Further, from Eq. (1.4-4) we have

$$\theta_1 = \theta_2 + \tilde{\psi} \qquad (1.4\text{-}8)$$

so that, in steady state,

$$\frac{d\theta_1}{d\tilde{\tau}} = \frac{d\theta_2}{d\tilde{\tau}} + \frac{d\tilde{\psi}}{d\tilde{\tau}} = \frac{d\theta_2}{d\tilde{\tau}} \tag{1.4-9}$$

Therefore,

$$2\frac{d\theta_1}{d\tilde{\tau}} = \frac{d\theta_2}{d\tilde{\tau}} + \frac{d\theta_1}{d\tilde{\tau}} \tag{1.4-10}$$

or

$$\frac{d\theta_1}{d\tilde{\tau}} = \frac{d}{d\tilde{\tau}}\left(\frac{\theta_1 + \theta_2}{2}\right) = \frac{\omega_{01} + \omega_{02}}{2} \tag{1.4-11}$$

Similarly,

$$\frac{d\theta_2}{d\tilde{\tau}} = \frac{d}{d\tilde{\tau}}\left(\frac{\theta_1 + \theta_2}{2}\right) = \frac{\omega_{01} + \omega_{02}}{2} \tag{1.4-12}$$

Thus, we conclude that the steady-state frequency of the two oscillators, when mutually locked, that is, the "ensemble frequency," is the average of their free-running frequencies.

It now becomes clear how it is that the locking range for the two oscillators is twice that for one. One may visualize each oscillator differing from the ensemble frequency of the pair by one locking range so that the total difference between the free-running frequencies of the two oscillators is not one, but two, locking ranges. The term "ensemble frequency" has no relevance when one of the oscillators injection locks the other and is not influenced by the injected oscillator as discussed previously. In that case, as was demonstrated, the steady-state frequency is the injection frequency.

Now suppose that the coupling between the oscillators is accomplished via a transmission line so that there is a phase delay associated with the coupled signal. This coupling phase changes the phase relationship between the coupled signal and the oscillator that produced it and thus modifies the behavior of the oscillator pair. We can account for this in our formulation by inserting the coupling phase shift through the transmission line, Φ_{12}, into Eqs. (1.4-1) and (1.4-2), resulting in

$$\frac{d\theta_1}{dt} = \omega_{01} + \Delta\omega_{lock} \sin\left(\theta_2 - \theta_1 - \Phi_{12}\right) \tag{1.4-13}$$

$$\frac{d\theta_2}{dt} = \omega_{02} + \Delta\omega_{lock} \sin\left(\theta_1 - \theta_2 - \Phi_{12}\right) \tag{1.4-14}$$

where we have assumed that the transmission line is reciprocal so that the coupling phase is the same in both directions. Using trigonometric identities, Eqs. (1.4-13) and (1.4-14) may be rewritten in the form

$$\begin{aligned}\frac{d\theta_1}{dt} &= \left[\omega_{01} - \Delta\omega_{lock} \sin\Phi_{12} \cos\left(\theta_2 - \theta_1\right)\right] \\ &+ \left[\Delta\omega_{lock} \cos\Phi_{12}\right] \sin\left(\theta_2 - \theta_1\right)\end{aligned} \tag{1.4-15}$$

$$\begin{aligned}\frac{d\theta_2}{dt} &= \left[\omega_{02} - \Delta\omega_{lock} \sin\Phi_{12} \cos\left(\theta_1 - \theta_2\right)\right] \\ &+ \left[\Delta\omega_{lock} \cos\Phi_{12}\right] \sin\left(\theta_1 - \theta_2\right)\end{aligned} \tag{1.4-16}$$

Again by subtraction we obtain

$$\frac{d\left(\theta_1 - \theta_2\right)}{dt} = \left(\omega_{01} - \omega_{02}\right) - 2\left(\Delta\omega_{lock} \cos\Phi_{12}\right) \sin\left(\theta_1 - \theta_2\right) \tag{1.4-17}$$

Comparing with Eq. (1.4-3) we see that the locking range has been modified by the cosine of the coupling phase. We define this effective locking range to be

$$\Delta\omega_{eff} = \Delta\omega_{lock} \cos\Phi_{12} \tag{1.4-18}$$

and using this in place of the unmodified locking range, the preceding theory may be applied to the case having nonzero coupling phase. One obvious consequence of this is that, if the coupling phase is 90 degrees (deg) or an odd multiple thereof, the effective locking range becomes zero and the two oscillators cannot be made to lock.

If, instead of subtracting Eqs. (1.3-15) and (1.3-16), we add them, we obtain

$$\frac{d\left(\theta_1 + \theta_2\right)}{dt} = \left(\omega_{01} + \omega_{02}\right) - 2\left(\Delta\omega_{lock} \sin\Phi_{12}\right) \cos\left(\theta_1 - \theta_2\right) \tag{1.4-19}$$

and we note that the ensemble frequency Eq. (1.4-12) is replaced by

$$\omega_{ens} = \frac{\left(\omega_{01} + \omega_{02}\right)}{2} - \left(\Delta\omega_{lock} \sin\Phi_{12}\right) \cos(\theta_1 - \theta_2) \tag{1.4-20}$$

which varies sinusoidally with coupling phase. This variation of ensemble frequency with coupling phase has been studied in somewhat more detail by

Sancheti and Fusco in the context of an active radiator coupling with its image in a reflecting object [21, 22].

Before moving on to study arrays of oscillators we take a quick look at the stability of the behavior of two coupled oscillators. Much more detail on this subject may be found in Chapter 7. The stability of the solution can be assessed by assuming that the oscillators are evolving according to a solution of Eq. (1.4-17) and perturbing the phase difference away from that solution by a small amount, δ. This results in the following differential equation for the time dependence of the perturbation.

$$\frac{d\delta}{dt} = -\left[2\Delta\omega_{lock}\cos\Phi_{12}\cos\left(\theta_1-\theta_2\right)\right]\delta \tag{1.4-21}$$

This equation has the solution

$$\delta(t) = e^{-\left[2\Delta\omega_{lock}\cos\Phi_{12}\cos\left(\theta_1-\theta_2\right)\right]t} \tag{1.4-22}$$

The solution for the oscillator phase difference is stable against the perturbation, δ, if the exponent is negative. That is,

$$\cos\Phi_{12}\cos\left(\theta_1-\theta_2\right) > 0 \tag{1.4-23}$$

This means that, if the magnitude of the coupling phase is less than 90 deg, the oscillators will lock such that their phases differ by less than 90 deg; while if the magnitude of the coupling phase is greater than 90 deg, the oscillators will lock such that their phases differ by more than 90 deg; that is, they will tend to oscillate out of phase. This behavior was predicted and observed by Stephan and Young [3] and formulated and studied in more detail by Humphrey and Fusco [23, 24] using an earlier theoretical construct they formulated for linear chains of coupled oscillators [25].

Conversely, for series resonant oscillators, the stability condition is

$$\cos\Phi_{12}\cos\left(\theta_1-\theta_2\right) < 0 \tag{1.4-24}$$

and the behavior of the oscillators will be opposite that described above. These properties have been exploited by Lee and Dalman in switching pairs of coupled oscillators from symmetric to antisymmetric phase by changing the coupling phase [26]. All of these effects have been observed experimentally as reported by Chang, Shapiro, and York [16]. Thus, the optimum coupling phase for parallel resonant oscillators is an even multiple of 180 deg, while that for series resonant oscillators is an odd multiple of 180 deg.

Very recently, it was pointed out that a given oscillator can present either series or parallel resonance, depending upon where in the oscillator circuit the coupling is implemented [27].

1.5 Conclusion

In this chapter we have developed a theory of oscillator behavior that admits the possibility of coupling the oscillators together such that they can mutually injection lock and thus oscillate as a coherent ensemble. This behavior is central to the remainder of the book as it forms the basis of the applications to be discussed. In Chapter 2 this theoretical framework will be applied in describing the behavior of arrays containing many oscillators coupled together in linear and planar configurations. The coupling for the most part is with nearest neighbors only. More elaborate coupling schemes have been studied in mathematical biology but remain as a potentially fruitful but largely untapped resource in the arena of phased-array antennas.

Chapter 2
Coupled-Oscillator Arrays—Basic Analytical Description and Operating Principles

In this chaper we will show how to use the theory developed in Chapter 1 to mathematically describe a linear array of oscillators coupled to nearest neighbors. It was Karl Stephan who first showed that such arrays can be useful in providing excitation signals for a linear array of radiating elements in that if locking signals are injected into the end oscillators of the array, variation of the relative phase of the locking signals can be used to control the distribution of the phase of the signals across the array [1]. Later, Liao and York pointed out that by merely tuning the end oscillators of the array the phase distribution can be controlled without any external injection signals [28]. We will show that, while the equations and associated boundary conditions at the array ends can describe the nonlinear behavior of the array through numerical solution, if the inter-oscillator phase differences remain small, the equations may be linearized. The linearized version may be solved analytically for the dynamic behavior of the phase, and from this one may obtain the dynamic behavior of the beam radiated by the elements of this linear phased array antenna.

An important consideration in the analysis is the manner in which the oscillators are coupled. The coupling can be represented as a "coupling network" connected to the array of oscillators, and this network can be described in terms of its port characteristics—that is, in terms of its admittance matrix or its scattering matrix.

The above theoretical description will then be generalized to planar arrays of oscillators coupled to nearest neighbors and the phase dynamics obtained by solution of the resulting equations. Here again, the coupling can be described in terms of port characteristics.

2.1 Fundamental Equations

Recall that two oscillators coupled together as symbolized in Fig. 2-1 were described by Eqs. (1.4-1) and (1.4-2). We now consider the generalization to $2N+1$ oscillators shown in Fig. 2-2. The generalization of Eqs. (1.4-1) and (1.4-2) is

$$\frac{d\theta_i}{dt} = \omega_{0i} + \Delta\omega_{lock}\sin\left(\theta_{i+1} - \theta_i - \Phi_{i,i+1}\right) + \Delta\omega_{lock}\sin\left(\theta_{i-1} - \theta_i - \Phi_{i,i-1}\right) \tag{2.1-1}$$

where i is an integer index that identifies each oscillator and runs from $-N$ to N. We choose the number of oscillators to be odd so that there will be a center oscillator. This is not really necessary as the theory can be adapted to an even number of oscillators also. (A simple artifice for accomplishing this is to generalize N to half of an odd integer value so that $2N+1$ becomes an even number and let the index, i, take on only half integer values from $-N/2$ to $N/2$ with unit increments.)

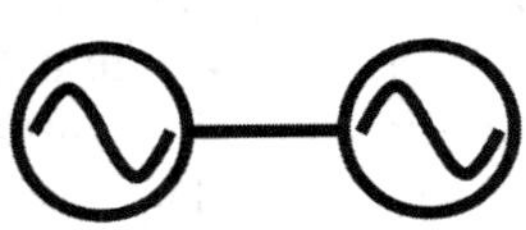

Fig. 2-1. Two coupled oscillators.

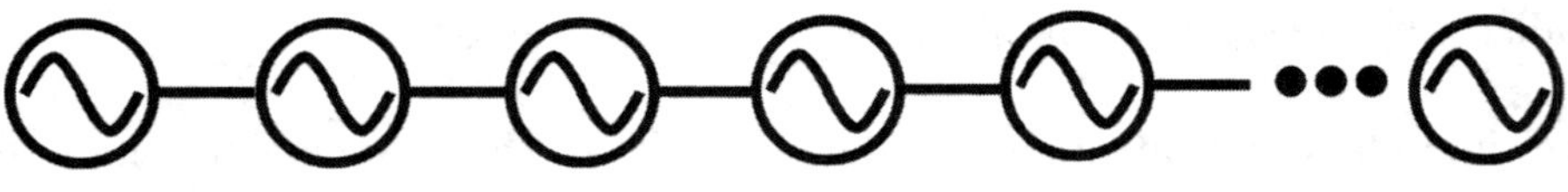

Fig. 2-2. 2*N* + 1 coupled oscillators.

However, from a practical point of view, it is convenient to have a center point at which to inject an external signal from a stable oscillator for the purpose of stabilizing the array oscillation. We therefore select the number to be odd. Note that, because the end oscillators are coupled to only one other oscillator, they are described by differential equations with only one sine term on the right side; that is,

$$\frac{d\theta_{-N}}{dt} = \omega_{0N} + \Delta\omega_{lock}\sin\left(\theta_{-N+1} - \theta_{-N} - \Phi_{-N,-N+1}\right) \tag{2.1-2}$$

$$\frac{d\theta_{N}}{dt} = \omega_{0N} + \Delta\omega_{lock}\sin\left(\theta_{N-1} - \theta_{N} - \Phi_{N,N-1}\right) \tag{2.1-3}$$

Note further that, because the maximum magnitude of the sine function is unity, the end oscillators of the array can be detuned from their nearest neighbors by a maximum of one locking range without losing lock whereas the center oscillator can be detuned up to two locking ranges. The maximum permitted detuning of the other oscillators will lie between one and two locking ranges. (See Section 3.1, Eq. (3.1-35).)

This system of simultaneous nonlinear first-order differential equations, (2.1-1)–(2.1-3), can be solved numerically beginning with an initial phase distribution and oscillator tuning, thus providing the phase distribution at all subsequent times. However, numerical solution does not provide an intuitive grasp of the behavior and how the parameters affect it. This intuitive understanding may be more easily gleaned from an approximate analytic solution. Then, later, if a more exact result is needed, the numerical approach can be applied.

Before proceeding to solve Eqs. (2.1-1) to (2.1-3) by linearization, we remark that the oscillator tuning required to produce a desired steady-state phase distribution may be easily obtained from these equations. That is, in steady state the time derivatives are zero, and from (2.1-1) to (2.1-3) the oscillator tuning is merely

$$\begin{aligned}\omega_{0i} - \omega_{ref} &= -\Delta\omega_{lock}\sin\left(\varphi_{i+1} - \varphi_{i} - \Phi_{i,i+1}\right)\\ &\quad -\Delta\omega_{lock}\sin\left(\varphi_{i-1} - \varphi_{i} - \Phi_{i,i-1}\right)\end{aligned} \tag{2.1-4}$$

$$\omega_{0,-N} - \omega_{ref} = -\Delta\omega_{lock}\sin\left(\varphi_{-N+1} - \varphi_{-N} - \Phi_{-N,-N+1}\right) \tag{2.1-5}$$

$$\omega_{0N}-\omega_{ref}=-\Delta\omega_{lock}\sin\left(\varphi_{N-1}-\varphi_N-\Phi_{N,N-1}\right) \tag{2.1-6}$$

where we have defined a new phase variable via

$$\theta_i=\varphi_i+\omega_{ref}t \tag{2.1-7}$$

and ω_{ref} is taken to be the ensemble frequency of the array.

Let us assume for the moment that the coupling phases are a multiple of π, and sum (2.1-4)–(2.1-6) over the $2N+1$ array elements. We find that under this assumption,

$$\begin{aligned}&\sum_{i=-N}^{N}\omega_{0i}-N\omega_{ref}=\\&-\Delta\omega_{lock}\sum_{i=-N+1}^{N-1}\sin\left(\varphi_{i+1}-\varphi_i-\Phi\right)+\sin\left(\varphi_{i-1}-\varphi_i-\Phi\right)\\&-\Delta\omega_{lock}\sin\left(\varphi_{N-1}-\varphi_N-\Phi\right)-\Delta\omega_{lock}\sin\left(\varphi_{-N-1}-\varphi_{-N}-\Phi\right)=0\end{aligned} \tag{2.1-8}$$

so that

$$\omega_{ref}=\frac{1}{N}\sum_{i=-N}^{N}\omega_{0i} \tag{2.1-9}$$

the average of the free-running frequencies. Thus, we have shown that *for coupling phase equal to a multiple of π, the ensemble frequency of the array is the average of the free running frequencies of the oscillators*.

As an example, in an array with zero coupling phase, a linear phase distribution with an inter-oscillator phase difference of $\delta\varphi$ requires

$$\omega_{0i}-\omega_{ref}=0 \tag{2.1-10}$$

$$\omega_{0,-N}-\omega_{ref}=-\Delta\omega_{lock}\sin\left(\delta\varphi\right) \tag{2.1-11}$$

$$\omega_{0N} - \omega_{ref} = \Delta\omega_{lock} \sin(\delta\varphi) \tag{2.1-12}$$

Thus, beginning with all the oscillators tuned to the reference frequency, tuning the leftmost oscillator down in frequency by half of the locking range and tuning the rightmost oscillator up in frequency by half of the locking range will produce a phase distribution across the array with a positive slope of $\pi/6$ radians between oscillators, $\pi/6$ being the arcsine of 1/2. If the oscillator outputs are used to excite radiating elements spaced a half wavelength (π radians) apart, the radiated beam will be directed 9.6 deg from normal to the array, that is, the arcsine of 1/6. It is this method of beam-steering that was first described by Liao and York [28]. Of course, much more general phase distributions are possible and the required oscillator tunings to produce them are given by Eqs. (2.1-4)–(2.1-6).

2.2 Discrete Model Solution (Linearization and Laplace Transformation)

In order to render the analytic solution tractable, we assume that the arguments of the sine functions in Eqs. (2.1-1)–(2.1-2) are close to an integral multiple of 2π. Specifically, we will assume that the coupling phase is zero and that the inter-oscillator phase differences are small so that the sine functions can be approximated by their arguments. In this approximation, Eq. (2.1-1) becomes

$$\frac{d\theta_i}{dt} = \omega_{0i} + \Delta\omega_{lock}\left(\theta_{i+1} - 2\theta_i + \theta_{i-1}\right) \tag{2.2-1}$$

Similarly, Eqs. (2.1-2) and (2.1-3) become

$$\frac{d\theta_{-N}}{dt} = \omega_{0,-N} + \Delta\omega_{lock}\left(\theta_{-N+1} - \theta_{-N}\right) \tag{2.2-2}$$

$$\frac{d\theta_N}{dt} = \omega_{0N} + \Delta\omega_{lock}\left(\theta_{N-1} - \theta_N\right) \tag{2.2-3}$$

Note that these approximate linearized equations would seem to imply that the end oscillators of the array can be detuned by $\pi/2$ locking ranges and the center one can be detuned by π locking ranges and still remain locked because the phase differences between oscillators remain less than or equal to $\pi/2$. However, from the full nonlinear theory of Section 2.1, we know that this is actually not true. These linearized equations only apply when the phase differences are small so that the sine functions may be accurately replaced by their arguments and $\pi/2$ is certainly not a small value in this sense.

In terms of the new phase, Eq. (2.1-7), we find that Eqs. (2.2-1)–(2.2-3) become

$$\frac{d\varphi_i}{dt} = \omega_{0i} - \omega_{ref} + \Delta\omega_{lock}\left(\varphi_{i+1} - 2\varphi_i + \varphi_{i-1}\right) \tag{2.2-4}$$

$$\frac{d\varphi_{-N}}{dt} = \omega_{0,-N} - \omega_{ref} + \Delta\omega_{lock}\left(\varphi_{-N+1} - \varphi_{-N}\right) \tag{2.2-5}$$

$$\frac{d\varphi_N}{dt} = \omega_{0N} - \omega_{ref} + \Delta\omega_{lock}\left(\varphi_{N-1} - \varphi_N\right) \tag{2.2-6}$$

Now we have a system of first-order linear differential equations that describe the dynamic behavior of the oscillator array. Unlike the system of first-order nonlinear differential equations from which it was derived, this system can be solved analytically.

We begin by writing these linear equations, Eqs. (2.2-4)–(2.2-6), in matrix form,

$$\frac{d[\varphi]}{dt} = [\omega_0] - [\omega_{ref}] + \Delta\omega_{lock}[M][\varphi] \tag{2.2-7}$$

where $[\varphi]$ is a $2N+1$ element vector of oscillator phases, $[\omega_0]$ is a similar vector of oscillator free-running frequencies, and $[M]$ is a $(2N+1)$ by $(2N+1)$ tridiagonal matrix with –2's on the diagonal, except for the –1's in the upper left and lower right corners, and 1's on the first super- and subdiagonals. Dividing by $\Delta\omega_{lock}$ yields

$$\frac{d[\varphi]}{d\tau} = [\Delta\Omega_{tune}] + [M][\varphi] \tag{2.2-8}$$

where $\tau = \Delta\omega_{lock}t$ and $[\Delta\Omega_{tune}] = \left[\dfrac{\omega_{0i} - \omega_{ref}}{\Delta\omega_{lock}}\right]$, a vector of oscillator free running frequencies relative to the reference frequency (detuning frequencies). Laplace transformation with respect to τ gives

$$[s[I] - [M]][\tilde{\varphi}] = [\Delta\tilde{\Omega}_{tune}] \tag{2.2-9}$$

with the tildes indicating transformed quantities and with $[I]$ being the identity matrix. We now define eigenvectors, $[v]_n$, and eigenvalues, λ_n, of the matrix $[M]$ to be such that

$$[M][v]_n = \lambda_n[v]_n \tag{2.2-10}$$

Our intention is to express the solution of Eq. (2.2-9) as a sum of these eigenvectors with unknown coefficients. When this sum is substituted into Eq. (2.2-9), the orthogonality of the eigenvectors will be employed to determine the coefficients of the expansion and thus obtain the solution in series form. Since the number of eigenvectors is finite, this series will be a finite sum—that is, a closed form. Moreover, as we will see in the next section, in steady state, an approximation of this sum may be carried out to produce a simple functional form for the phase distribution.

Note that Eq. (2.2-10) is a three term recurrence relation for the elements of the eigenvectors, v_i; that is,

$$v_{i-1} - (2+\lambda_n)v_i + v_{i+1} = 0 \tag{2.2-11}$$

with the two auxiliary conditions,

$$v_{N-1} - (1+\lambda_n)v_N = 0 \tag{2.2-12}$$

$$v_{-N+1} - (1+\lambda_n)v_{-N} = 0 \tag{2.2-13}$$

Now, Eq. (2.2-11) is satisfied by the Chebyshev polynomials, $T_i(x_n)$ and $U_i(x_n)$, where

$$x_n = \frac{2+\lambda_n}{2} \tag{2.2-14}$$

so that Eqs. (2.2-12) and (2.2-13) become

$$W_{N-1}(x_n) - (2x_n - 1)W_N(x_n) = 0 \tag{2.2-15}$$

$$W_{-N+1}(x_n) - (2x_n - 1)W_{-N}(x_n) = 0 \tag{2.2-16}$$

where W_i is a linear combination of T_i and U_{i-1}. Equivalently, using (2.2-11) we have

$$W_{N+1}(x_n) - W_N(x_n) = 0 \tag{2.2-17}$$

$$W_{-N-1}(x_n) - W_{-N}(x_n) = 0 \tag{2.2-18}$$

These boundary condition equations determine the permissible values, x_n. Let

$$W_i(x_n) = \alpha_T T_i(x_n) + \alpha_U U_{i-1}(x_n) \tag{2.2-19}$$

so that

$$W_{-i}(x_n) = \alpha_T T_i(x_n) - \alpha_U U_{i-1}(x_n) \tag{2.2-20}$$

Adding and subtracting Eqs. (2.2-17) and (2.2-18) using Eqs. (2.2-19) and (2.2-20) yields

$$T_{N+1}(x_n) - T_N(x_n) = 0 \tag{2.2-21}$$

$$U_N(x_n) - U_{N-1}(x_n) = 0 \tag{2.2-22}$$

Using the trigonometric expression for T, Eq. (2.2-21) yields

$$\sin\left(\left(N + \frac{1}{2}\right)\cos^{-1}(x_n)\right)\sin\left(\frac{1}{2}\cos^{-1}(x_n)\right) = 0 \tag{2.2-23}$$

which implies that

$$x_{Tn} = \cos\left(\frac{2n\pi}{(2N+1)}\right) \tag{2.2-24}$$

so that the eigenvalues are given by

$$\lambda_{Tn} = 2\cos\left(\frac{2n\pi}{(2N+1)}\right) - 2 = -4\sin^2\left(\frac{n\pi}{(2N+1)}\right) \tag{2.2-25}$$

where the subscript T indicates that the elements of the corresponding eigenvectors are $T_i(x_n)$. Conversely, using the trigonometric expression for U, Eq. (2.2-22) yields

$$\cos\left(\left(N + \frac{1}{2}\right)\cos^{-1}(x_n)\right) = 0 \tag{2.2-26}$$

which implies that

$$x_{Un} = \cos\left(\frac{(2n+1)\pi}{(2N+1)}\right) \tag{2.2-27}$$

so that the eigenvalues are given by

$$\lambda_{Un} = 2\cos\left(\frac{(2n+1)\pi}{(2N+1)}\right) - 2 = -4\sin^2\left(\frac{(2n+1)\pi/2}{(2N+1)}\right) \tag{2.2-28}$$

where the subscript U indicates that the elements of the corresponding eigenvectors are $U_{i-1}(x_n)$. In (2.2-25) and (2.2-28) the index n runs from 0 to N, after which the eigenvalues repeat. Thus, we have arrived at two sets of eigenfunctions; one set, the T's, excited by the symmetric part of the detuning function and the other set, the U's, excited by the antisymmetric part, with respect to the array center.

We may now expand the solution of Eq. (2.2-9) in these eigenvectors as

$$[\tilde{\varphi}] = \sum_{n=0}^{N} A_n[v_T]_n + B_n[v_U]_n \tag{2.2-29}$$

Substituting this expansion into Eq. (2.2-9), we obtain

$$\begin{gathered} [s[I]-[M]]\sum_{n=0}^{N} A_n[v_T]_n + B_n[v_U]_n = \\ \sum_{n=0}^{N} A_n(s-\lambda_{Tn})[v_T]_n + B_n(s-\lambda_{Un})[v_U]_n = [\Delta\tilde{\Omega}_{tune}] \end{gathered} \tag{2.2-30}$$

Using the orthogonality of the eigenvectors, we may now solve for the coefficients A_n and B_n.

$$A_n = \frac{[\Delta\tilde{\Omega}_{tune}]\bullet[v_T]_n}{(s-\lambda_{Tn})[v_T]_n \bullet [v_T]_n} \tag{2.2-31}$$

$$B_n = \frac{[\Delta\tilde{\Omega}_{tune}]\bullet[v_U]_n}{(s-\lambda_{Un})[v_U]_n \bullet [v_U]_n} \tag{2.2-32}$$

Substituting into Eq. (2.2-29), we obtain

$$[\tilde{\varphi}]=\sum_{n=0}^{N}\frac{[\Delta\tilde{\Omega}_{tune}]\bullet[v_T]_n}{(s-\lambda_{Tn})[v_T]_n\bullet[v_T]_n}[v_T]_n$$
$$+\sum_{n=0}^{N}\frac{[\Delta\tilde{\Omega}_{tune}]\bullet[v_U]_n}{(s-\lambda_{Un})[v_U]_n\bullet[v_U]_n}[v_U]_n \qquad (2.2\text{-}33)$$

and, if the detuning function is a step function at time zero, $u(\tau)$, the inverse Laplace transform is

$$[\varphi]=\frac{\tau}{2N+1}\sum_{i=-N}^{N}\Delta\Omega_{tune,i}u(\tau)$$
$$-\sum_{n=1}^{N}\frac{[\Delta\Omega_{tune}]\bullet[v_T]_n}{\lambda_{Tn}[v_T]_n\bullet[v_T]_n}[v_T]_n\left(1-e^{\lambda_{Tn}\tau}\right)u(\tau) \qquad (2.2\text{-}34)$$
$$-\sum_{n=0}^{N}\frac{[\Delta\Omega_{tune}]\bullet[v_U]_n}{\lambda_{Un}[v_U]_n\bullet[v_U]_n}[v_U]_n\left(1-e^{\lambda_{Un}\tau}\right)u(\tau)$$

The first of the three summations, the one arising from the zero eigenvalue, indicates that *the steady-state ensemble frequency of the array is shifted by the average oscillator* detuning—that is, the sum of the elements of the $[\Delta\Omega_{tune}]$ vector divided by the number of oscillators.

Recall that we assumed at the start of this section that the coupling phase is zero. Returning for a moment to Eq. (2.1-1) and using Eq. (2.1-7), we may write

$$\frac{d\varphi_i}{dt}=\omega_{0i}-\omega_{ref}+\Delta\omega_{lock}\sin\left(\varphi_{i+1}-\varphi_i-\Phi_{i,i+1}\right)$$
$$+\Delta\omega_{lock}\sin\left(\varphi_{i-1}-\varphi_i-\Phi_{i,i-1}\right) \qquad (2.2\text{-}35)$$

If the coupling phases are taken to be equal, this can be rearranged to read

$$\frac{d\varphi_i}{dt}=\left(\omega_{0i}-\omega_{ref}\right)-\Delta\omega_{lock}\sin(\Phi)\left[\cos\left(\varphi_{i+1}-\varphi_i\right)+\cos\left(\varphi_{i-1}-\varphi_i\right)\right]$$
$$+\Delta\omega_{lock}\cos(\Phi)\left[\sin\left(\varphi_{i+1}-\varphi_i\right)+\sin\left(\varphi_{i-1}-\varphi_i\right)\right] \qquad (2.2\text{-}36)$$

or

$$\frac{d\varphi_i}{dt} = \left(\omega_{0i} - \bar{\omega}_{ref}\right) + \Delta\omega_{eff}\left[\sin\left(\varphi_{i+1} - \varphi_i\right) + \sin\left(\varphi_{i-1} - \varphi_i\right)\right] \quad (2.2\text{-}37)$$

where

$$\bar{\omega}_{ref} = \omega_{ref} + \Delta\omega_{lock}\sin(\Phi)\left[\cos\left(\varphi_{i+1} - \varphi_i\right) + \cos\left(\varphi_{i-1} - \varphi_i\right)\right] \quad (2.2\text{-}38)$$

and

$$\Delta\omega_{eff} = \Delta\omega_{lock}\cos(\Phi) \quad (2.2\text{-}39)$$

which is the same as Eq. (1.4-18). Thus we conclude that, in a $2N + 1$ oscillator array, a uniform coupling phase modifies the effective locking range according to Eq. (2.2-39) just as it did for two oscillators, and the ensemble frequency is modified according to Eq. (2.2-38). Interestingly, if the inter-oscillator phase difference is 90 deg, the ensemble frequency becomes independent of the coupling phase as pointed out by Humphrey and Fusco [25].

The speed of the array response to the application of a step tuning is determined by the smallest nonzero eigenvalue. From Eq. (2.2-28) this is

$$\lambda_{U0} = -4\sin^2\left(\frac{\pi/2}{(2N+1)}\right) \approx -\left(\frac{\pi}{(2N+1)}\right)^2 \quad (2.2\text{-}40)$$

This provides the important result that the linear array response time constant is roughly proportional to the square of the number of elements, with the approximation becoming more accurate as the number of elements is increased.

While the time constant is unaffected, the effective steering speed of such arrays, as defined by the radiated beam peak neglecting aberration, may be increased by "over-steering." That is, one may apply more detuning than necessary to achieve the desired steady-state phase gradient but reduce it to the required value during the beam-steering transient. Generalizing this concept, one may apply arbitrarily time-varying detuning as suggested by Heath et al. [29]. In particular they considered sinusoidal detuning and showed that the maximum stable inter-oscillator phase shift is thereby increased from 90 to 138 deg.

2.3 Steady-State Solution

In this section we will investigate the steady-state solution for the phase distribution in a bit more detail. From Eq. (2.2-34), the steady-state solution is

$$[\varphi] = -\sum_{n=1}^{N} \frac{[\Delta\Omega_{tune}] \bullet [v_T]_n}{\lambda_{Tn}[v_T]_n \bullet [v_T]_n}[v_T]_n \\ - \sum_{n=0}^{N} \frac{[\Delta\Omega_{tune}] \bullet [v_U]_n}{\lambda_{Un}[v_U]_n \bullet [v_U]_n}[v_U]_n \tag{2.3-1}$$

in which we have suppressed the linear term in time that merely represents a shift in the ensemble frequency due to the detuning. The denominators of the terms of the series may be written explicitly as

$$[v_T]_n \bullet [v_T]_n = \sum_{i=-N}^{N} T_i(x_n)^2 = \sum_{i=-N}^{N} \cos^2\left(\frac{2ni\pi}{(2N+1)}\right) = N + \frac{1}{2} \tag{2.3-2}$$

and

$$[v_U]_n \bullet [v_U]_n = \sum_{i=-N}^{N} U_{i-1}(x_n)^2 \\ = \frac{1}{\sin^2\left(\frac{(2n+1)\pi}{(2N+1)}\right)} \sum_{i=-N}^{N} \sin^2\left(\frac{i(2n+1)\pi}{(2N+1)}\right) \\ = \frac{N+\frac{1}{2}}{\sin^2\left(\frac{(2n+1)\pi}{(2N+1)}\right)} \tag{2.3-3}$$

Suppose that one of the oscillators, say the jth one, is step detuned at time zero from the ensemble frequency by one locking range. The solution given by Eq. (2.3-1) then becomes

$$[\varphi] = \sum_{n=1}^{N} \frac{2T_j(x_{Tn})}{(-\lambda_{Tn})(2N+1)}[T_i(x_n)] \\ + \sum_{n=0}^{N} \frac{2U_j(x_{Un})}{(-\lambda_{Un})(2N+1)}[U_{i-1}(x_n)]\sin\left(\cos^{-1}(x_{Un})\right) \tag{2.3-4}$$

Thus, the elements of the vector of oscillator phases may be written

$$\varphi_i = \frac{2}{2N+1}\sum_{n=1}^{N}\frac{\cos\left(\frac{j2n\pi}{(2N+1)}\right)\cos\left(\frac{i2n\pi}{(2N+1)}\right)}{4\sin^2\left(\frac{n\pi}{(2N+1)}\right)} + \frac{2}{2N+1}\sum_{n=0}^{N}\frac{\sin\left(\frac{j(2n+1)\pi}{(2N+1)}\right)\sin\left(\frac{i(2n+1)\pi}{(2N+1)}\right)}{4\sin^2\left(\frac{(2n+1)\pi/2}{(2N+1)}\right)} \tag{2.3-5}$$

The series given by Eq. (2.3-5) has a finite number of terms so it can be summed numerically. As an example, we evaluate this series for $N = 10$, a 21-element array, with oscillator number 5 detuned one locking range. Then we plot the phase of each oscillator in Fig. 2-3 as the dots.

Noting that the lowest-order terms in n contribute most of the sum, we approximate the eigenvalues in the denominators of Eq. (2.3-5) as follows.

$$4\sin^2\left(\frac{n\pi}{(2N+1)}\right) \approx \left(\frac{2n\pi}{(2N+1)}\right)^2 \tag{2.3-6}$$

$$4\sin^2\left(\frac{(2n+1)\pi/2}{(2N+1)}\right) \approx \left(\frac{(2n+1)\pi}{(2N+1)}\right)^2 \tag{2.3-7}$$

Substituting these approximations in Eq. (2.3-5) gives

$$\varphi_i \approx \frac{2}{2N+1}\sum_{n=1}^{N}\frac{\cos\left(\frac{j2n\pi}{(2N+1)}\right)\cos\left(\frac{i2n\pi}{(2N+1)}\right)}{\left(\frac{2n\pi}{(2N+1)}\right)^2} + \frac{2}{2N+1}\sum_{n=0}^{N}\frac{\sin\left(\frac{j(2n+1)\pi}{(2N+1)}\right)\sin\left(\frac{i(2n+1)\pi}{(2N+1)}\right)}{\left(\frac{(2n+1)\pi}{(2N+1)}\right)^2} \tag{2.3-8}$$

If the upper limit of these summations is extended to infinity (adding presumably negligible terms), the sum may be written as the simple quadratic function

$$\varphi_i = \frac{1}{2(2N+1))}\left[i^2 + j^2 - (2N+1)|j-i| + \frac{1}{6}(2N+1)^2\right] \tag{2.3-9}$$

Thus, we see that the steady-state phase distribution when one oscillator is detuned is approximately parabolic with a slope discontinuity at the detuned oscillator. To compare with the earlier example, we evaluate this function for $N = 10$, a 21-element array, with oscillator number 5 detuned one locking range and plot the phase of each oscillator in Fig. 2-3 as x's. Note that the approximation is quite accurate. In fact, in the present example, the maximum error is only about 4 milliradians (mr) of phase.

Finally, we note that, since the eigenvalues repeat, if the sums in Eq. (2.3-5) are continued to an infinite number of terms instead of stopping at N, the result would be a set of delta functions, one at each oscillator, with amplitude (area) equal to the phase of that oscillator.

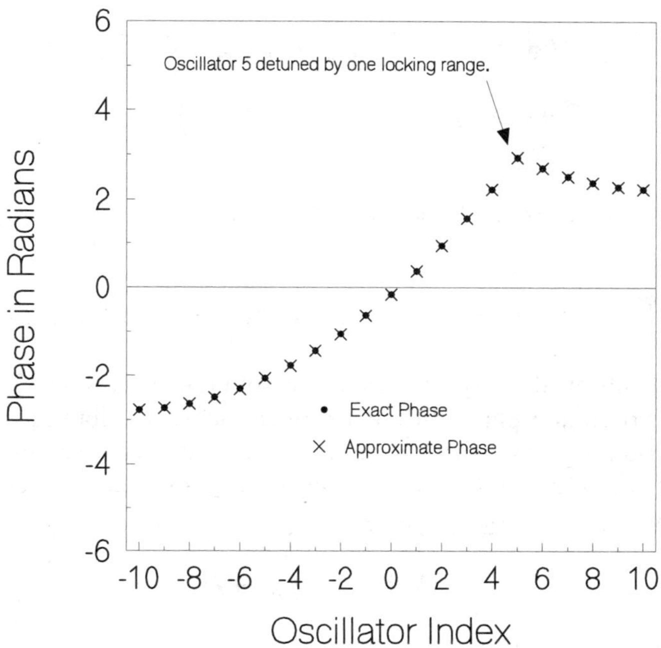

Fig. 2-3. A possible phase distribution for a 21-element array.

2.4 Stability of the Phase Solution in the Full Nonlinear Formulation

In the previous sections of this chapter, we found that linearization provided a path to analytic solution for the phase distribution across the array. It was also pointed out in passing that the full nonlinear formulation provided the oscillator tuning necessary to achieve a desired steady-state phase distribution. In this section we discuss the properties of the steady-state phase solution of the full nonlinear formulation largely as described by Heath et al. [29]. Recalling that linearization permitted solution of the problem, we expect that the effect of a small perturbation of an assumed solution of the nonlinear equations can be investigated in a similar manner. This is the approach taken by Heath et al. [29] in determining the stability of the solution in the fully nonlinear case. To place this in the framework of our previous analysis we begin with Eqs. (2.1-1)–(2.1-3) and introduce (2.1-7) to obtain

$$\frac{d\varphi_i}{dt} = \omega_{0i} - \omega_{ref} + \Delta\omega_{lock} \sin\left(\varphi_{i+1} - \varphi_i - \Phi_{i,i+1}\right) + \Delta\omega_{lock} \sin\left(\varphi_{i-1} - \varphi_i - \Phi_{i,i-1}\right) \quad (2.4\text{-}1)$$

$$\frac{d\varphi_{-N}}{dt} = \omega_{0N} - \omega_{ref} + \Delta\omega_{lock} \sin\left(\varphi_{-N+1} - \varphi_{-N} - \Phi_{-N,-N+1}\right) \quad (2.4\text{-}2)$$

$$\frac{d\varphi_N}{dt} = \omega_{0N} - \omega_{ref} + \Delta\omega_{lock} \sin\left(\varphi_{N-1} - \varphi_N - \Phi_{N,N-1}\right) \quad (2.4\text{-}3)$$

Following Heath et al. [29], we assume a solution of (2.4-1)–(2.4-3) with a uniform inter-oscillator phase difference (linear phase distribution) and uniform reciprocal coupling, as was the case in the earlier example given by Eqs. (2.1-10)–(2.1-12), and let the phase of each oscillator be changed by a small time-dependent perturbation, η_i; that is,

$$\theta_i = \varphi_i + \omega_{ref} t + \eta_i \quad (2.4\text{-}4)$$

Equations (2.4-1)–(2.4-3) then become

$$\frac{d\eta_i}{dt} = \Delta\omega_{lock} \Big[\cos(\delta\varphi - \Phi)\eta_{i+1} - 2\cos(\Phi)\cos(\delta\varphi)\eta_i + \cos(\delta\varphi + \Phi)\eta_{i-1}\Big] \quad (2.4\text{-}5)$$

$$\frac{d\eta_{-N}}{dt} = \Delta\omega_{lock} \cos(\delta\varphi - \Phi)(\eta_{-N+1} - \eta_{-N}) \quad (2.4\text{-}6)$$

$$\frac{d\eta_N}{dt} = \Delta\omega_{lock} \cos(\delta\varphi + \Phi)(\eta_{N-1} - \eta_N) \quad (2.4\text{-}7)$$

and again we note that the system coefficients matrix multiplying the vector of η's will be tridiagonal with diagonal elements $-2\cos(\Phi)\cos(\delta\varphi)$ except for the upper left and lower right corners, which are $-\cos(\delta\varphi - \Phi)$ and $-\cos(\delta\varphi + \Phi)$, respectively. The superdiagonal elements are $\cos(\delta\varphi - \Phi)$ and the subdiagonal elements are $\cos(\delta\varphi + \Phi)$.

From the analysis in Section 2.2, we recall that the stability of the system depended upon the eigenvalues of the system matrix, Eqs. (2.2-25) and

(2.2-28), being negative. So it is in the present case. Thus, we must determine the eigenvalues for this new more complicated system matrix. Heath et al. [29] provide us with a prescription for doing this. The first step is to symmetrize the matrix by defining new eigenvector elements related to the η's as follows.

First, define a new variable, γ, via

$$\gamma_i = \eta_i \left(\frac{\cos(\delta\varphi - \Phi)}{\cos(\delta\varphi + \Phi)} \right)^{\frac{i}{2}} \tag{2.4-8}$$

Now, substitution of Eq. (2.4-8) into Eqs. (2.4-5)–(2.4-7) yields

$$\begin{aligned} \frac{d\gamma_i}{dt} = \Delta\omega_{lock} \Big[& \sqrt{\cos(\delta\varphi - \Phi)\cos(\delta\varphi + \Phi)}\,\gamma_{i+1} \\ & - 2\cos(\Phi)\cos(\delta\varphi)\,\gamma_i \\ & + \sqrt{\cos(\delta\varphi - \Phi)\cos(\delta\varphi + \Phi)}\,\gamma_{i-1} \Big] \end{aligned} \tag{2.4-9}$$

$$\begin{aligned} \frac{d\gamma_{-N}}{dt} = & \Delta\omega_{lock}\sqrt{\cos(\delta\varphi - \Phi)\cos(\delta\varphi + \Phi)}\,\gamma_{-N+1} \\ & - \Delta\omega_{lock}\cos(\delta\varphi - \Phi)\,\gamma_{-N} \end{aligned} \tag{2.4-10}$$

$$\begin{aligned} \frac{d\gamma_{N}}{dt} = & \Delta\omega_{lock}\sqrt{\cos(\delta\varphi - \Phi)\cos(\delta\varphi + \Phi)}\,\gamma_{N-1} \\ & - \Delta\omega_{lock}\cos(\delta\varphi + \Phi)\,\gamma_{N} \end{aligned} \tag{2.4-11}$$

The system matrix for Eqs. (2.4-9)–(2.4-11) is symmetric. Rearranging these equations a bit results in

$$\begin{aligned} \frac{d\gamma_i}{dt} = & \Delta\omega_{lock}\sqrt{\cos(\delta\varphi - \Phi)\cos(\delta\varphi + \Phi)} \\ & \times \left[\gamma_{i+1} - \left(\sqrt{\frac{\cos(\delta\varphi - \Phi)}{\cos(\delta\varphi + \Phi)}} + \sqrt{\frac{\cos(\delta\varphi + \Phi)}{\cos(\delta\varphi - \Phi)}} \right) \gamma_i + \gamma_{i-1} \right] \end{aligned} \tag{2.4-12}$$

$$\frac{d\gamma_{-N}}{dt} = \Delta\omega_{lock}\sqrt{\cos(\delta\varphi-\Phi)\cos(\delta\varphi+\Phi)} \times\left[\gamma_{-N+1} - \sqrt{\frac{\cos(\delta\varphi-\Phi)}{\cos(\delta\varphi+\Phi)}}\gamma_{-N}\right] \quad (2.4\text{-}13)$$

$$\frac{d\gamma_{N}}{dt} = \Delta\omega_{lock}\sqrt{\cos(\delta\varphi-\Phi)\cos(\delta\varphi+\Phi)} \times\left[\gamma_{N-1} - \sqrt{\frac{\cos(\delta\varphi+\Phi)}{\cos(\delta\varphi-\Phi)}}\gamma_{N}\right] \quad (2.4\text{-}14)$$

As in Section 2.2, these equations may be written in matrix form. The stability is determined by the eigenvalues, λ_n, of the system matrix, which can be found as follows. Let

$$\alpha = \cos^{-1}\left[\frac{1}{2}\left(\sqrt{\frac{\cos(\delta\varphi-\Phi)}{\cos(\delta\varphi+\Phi)}} + \sqrt{\frac{\cos(\delta\varphi+\Phi)}{\cos(\delta\varphi-\Phi)}}\right)\right] \quad (2.4\text{-}15)$$

Now the analogs of Eqs. (2.2-11)–(2.2-13) are

$$v_{i-1} - (2\cos\alpha + \lambda_n)v_i + v_{i+1} = 0 \quad (2.4\text{-}16)$$

$$v_{N-1} - (e^{i\alpha} + \lambda_n)v_N = 0 \quad (2.4\text{-}17)$$

$$v_{-N+1} - (e^{-i\alpha} + \lambda_n)v_{-N} = 0 \quad (2.4\text{-}18)$$

and the argument of the Chebyshev polynomials is

$$x_n = \frac{2\cos\alpha + \lambda_n}{2} \quad (2.4\text{-}19)$$

Substituting Eqs. (2.2-19) and (2.2-20) into Eqs. (2.4-17) and (2.4-18) and setting the determinant of the coefficients of α_T and α_U equal to zero gives us the following transcendental equation for the eigenvalues.

$$x_n \sin\left[(2N+1)\cos^{-1}(x_n)\right] = \sin\left[(2N+1)\cos^{-1}(x_n)\right]\cos\alpha \tag{2.4-20}$$

So either

$$x_n = \cos\alpha \tag{2.4-21}$$

and, from Eq. (2.4-19) the eigenvalues are zero, or

$$\sin\left[(2N+1)\cos^{-1}(x_n)\right] = 0 \tag{2.4-22}$$

and the eigenvalues are,

$$\begin{aligned}\lambda_n &= 2\cos\left(\frac{n\pi}{2N+1}\right) - 2\cos\alpha \\ &= -4\sin^2\left(\frac{n\pi/2}{2N+1}\right) + 4\sin^2\left(\frac{\alpha}{2}\right)\end{aligned} \tag{2.4-23}$$

The characteristic time dependence of the nth perturbation mode will be

$$e^{\lambda_n \Delta\omega_{lock}\sqrt{\cos(\delta\varphi-\Phi)\cos(\delta\varphi+\Phi)}\,t} = e^{\sigma_n t} \tag{2.4-24}$$

and

$$\sigma_n = \lambda_n \Delta\omega_{lock}\sqrt{\cos(\delta\varphi-\Phi)\cos(\delta\varphi+\Phi)} \tag{2.4-25}$$

Substituting Eq. (2.4-23) into Eq. (2.4-25), we obtain

$$\begin{aligned}\sigma_n = &-4\Delta\omega_{lock}\left[\sin^2\left(\frac{n\pi/2}{2N+1}\right) - \sin^2\left(\frac{\alpha}{2}\right)\right] \\ &\times\sqrt{\cos(\delta\varphi-\Phi)\cos(\delta\varphi+\Phi)}\end{aligned} \tag{2.4-26}$$

The nth perturbation eigenmode will be stable if σ_n has a nonpositive real part. In general, the phase distribution across the array will be stable if all of the σ_n's have nonpositive real parts. Note that in the typical case where Φ is an integral multiple of π, α is zero. Then, if $|\delta\varphi|$ is less than $\pi/2$, all of the eigenvalues are real and nonpositive so the array phase distribution is stable against small perturbations.

In this section we have shown that, by linearizing the equations about an arbitrary solution for the phase distribution in a linear array in which only the end oscillator free running frequencies are controlled, we may study analytically the stability of the solution against small perturbations. Heath et al. [29] have also shown that exact stable constant phase gradient solutions of the nonlinear equations with arbitrary time dependence can be obtained if one is willing to control the free-running frequencies of all of the oscillators in the array rather than just the end ones.

2.5 External Injection Locking

It was mentioned in passing in Section 2.1 that we chose the number of oscillators in the array to be odd so as to provide a convenient center point at which to inject a stabilizing external signal. In this section we discuss the needed modifications to the mathematical formulation to accommodate an external injection signal and account for its impact on array behavior.

Beginning with Eq. (2.1-1) we envision an external signal injected into the *p*th oscillator and add a term to the equation representing this signal.

$$\begin{aligned}\frac{d\theta_i}{dt} &= \omega_{0i} + \Delta\omega_{lock}\sin\left(\theta_{i+1} - \theta_i - \Phi_{i,i+1}\right)\\ &+\Delta\omega_{lock}\sin\left(\theta_{i-1} - \theta_i - \Phi_{i,i-1}\right)\\ &+\delta_{ip}\Delta\omega_{lock,p,inj}\sin\left(\theta_{inj} - \theta_p - \Phi_{p,inj}\right)\end{aligned} \tag{2.5-1}$$

where δ_{ip} is the Kronecker delta function and $\Delta\omega_{lock,p,inj}$ is the locking range between the external oscillator and the injected oscillator in the array. Note that the phase of the injection signal must remain within $\pi/2$ radians of that of the injected array oscillator to maintain lock. For simplicity, let all of the coupling phases be zero and assume that the inter-oscillator phase differences are small to permit linearization. Then, introducing Eq. (2.1-7) we have

$$\begin{aligned}\frac{d\varphi_i}{dt} &= \omega_{0i} - \omega_{ref} + \Delta\omega_{lock}\left(\varphi_{i+1} - 2\varphi_i + \varphi_{i-1}\right)\\ &-\delta_{ip}\Delta\omega_{lock,p,inj}\left(\varphi_p - \varphi_{inj}\right)\end{aligned} \tag{2.5-2}$$

By replacing Eq. (2.2-4) with Eq. (2.5-2) while Eqs. (2.2-5) and (2.2-6) remain unchanged, Eq. (2.2-9) now becomes

$$[s[I]-[M]+[d][\tilde{\varphi}]=[\Delta\tilde{\Omega}_{tune}]+r[\tilde{\varphi}_{inj}] \tag{2.5-3}$$

where $[d]$ is a matrix with one nonzero element, r, at position pp on its diagonal, $[\varphi_{inj}]$ is a vector with one nonzero component, the pth one, and

$$r=\frac{\Delta\omega_{lock,p,inj}}{\Delta\omega_{lock}} \tag{2.5-4}$$

Here again the tilde denotes Laplace transformation with respect to the scaled time, τ. Equation (2.2-11) is thus replaced by

$$v_{i-1}-(2+r\delta_{ip}+\lambda_n)v_i+v_{i+1}=0 \tag{2.5-5}$$

And Eqs. (2.2-12) and (2.2-13) are unchanged.

We now postulate eigenvectors with two sets of elements, those to the left of and including the injection site $I = p$ labeled "L" and those to the right of and including the injection site $I = p$ labeled "R." That is,

$$W_i^{(L)}(x_n)=\alpha_T^{(L)}T_i(x_n)+\alpha_U^{(L)}U_{i-1}(x_n); \quad i\le p \tag{2.5-6}$$

$$W_i^{(R)}(x_n)=\alpha_T^{(R)}T_i(x_n)+\alpha_U^{(R)}U_{i-1}(x_n); \quad i\ge p \tag{2.5-7}$$

and we require that the pth elements match at the injection site; that is

$$W_p^{(L)}(x_n)=W_p^{(R)}(x_n) \tag{2.5-8}$$

Now Eq. (2.5-5) with $i = p$, Eq. (2.2-12), Eq. (2.2-13), and Eq. (2.5-8) are four equations in the four unknowns, $\alpha_T^{(L)}$, $\alpha_U^{(L)}$, $\alpha_T^{(R)}$, and $\alpha_U^{(R)}$. The equations are homogeneous, so the determinant of the coefficients must be zero if we are to obtain a nontrivial solution. As usual, this condition yields a transcendental equation for x_n, thus giving the eigenvalues, λ_n. The transcendental equation in this case is

$$\sin\left(\cos^{-1}x_n\right)\sin\left((2N+1)\cos^{-1}x_n\right) =$$
$$r\cos\left((N+\frac{1}{2}+p)\cos^{-1}x_n\right)\cos\left((N+\frac{1}{2}-p)\cos^{-1}x_n\right) \tag{2.5-9}$$

Note that if $r = 0$ we recover the eigenvalues for the uninjected array, Eqs. (2.2-24) and (2.2-27).

Proceeding as in the uninjected case, the solution may be expressed in terms of the eigenvectors and eigenvalues in the form

$$[\tilde{\varphi}] = \sum_{n=0}^{2N} \frac{\left([\Delta\tilde{\Omega}_{tune}] + [\tilde{\varphi}_{inj}]\right) \bullet [v]_n}{(s - \lambda_n)[v]_n \bullet [v]_n} [v]_n \tag{2.5-10}$$

and the inverse Laplace transform follows immediately. For practice, you may wish to explicitly compute the eigenvalues and eigenvectors and evaluate the solution from Eq. (2.5-10).

The beam-steering scheme proposed by Stephan [1] requires two injection points, $i = p_L$ and $i = p_R$, characterized by two locking-range ratios, r_L and r_R. The solution procedure described above can be generalized to accommodate such a situation as follows. We postulate eigenvectors in three parts, one to the left of both injection points denoted "*L*," one between the injection points denoted "*B*," and one to the right of both injection points denoted "*R*." The elements of these vectors are linear combinations of Chebyshev polynomials as before. Thus, there will be six unknown coefficients, $\alpha_T^{(L)}$, $\alpha_U^{(L)}$, $\alpha_T^{(B)}$, $\alpha_U^{(B)}$, $\alpha_T^{(R)}$, and $\alpha_U^{(R)}$. Imposing the end conditions, continuity at each injection point, and the modified three term recurrence at each injection point,

$$v_{p_L-1} - (2 + r_L + \lambda_n)v_{p_L} + v_{p_L+1} = 0 \tag{2.5-11}$$

$$v_{p_R-1} - (2 + r_R + \lambda_n)v_{p_R} + v_{p_R+1} = 0 \tag{2.5-12}$$

provides a homogeneous system of six equations for these unknown coefficients. Setting the determinant of this system equal to zero yields a transcendental equation for the eigenvalues and the solution proceeds as before. This transcendental equation is

$$\begin{aligned}&\sin\left(\cos^{-1}x_n\right)\sin\left((2N+1)\cos^{-1}x_n\right)\\&+r_L\cos\left((N+\frac{1}{2}+p_L)\cos^{-1}x_n\right)\cos\left((N+\frac{1}{2}-p_L)\cos^{-1}x_n\right)\\&+r_R\cos\left((N+\frac{1}{2}+p_R)\cos^{-1}x_n\right)\cos\left((N+\frac{1}{2}-p_R)\cos^{-1}x_n\right)\\&+\frac{r_Lr_R}{\sin\left(\cos^{-1}x_n\right)}\cos\left((2N+1+2p_L)\cos^{-1}x_n\right)\\&\times\cos\left((2N+1-2p_R)\cos^{-1}x_n\right)\sin\left((p_R-p_L)\cos^{-1}x_n\right)=0\end{aligned} \tag{2.5-13}$$

Note that if either r_L or r_R is zero, we recover Eq. (2.5-9). Here again you may want to perform the detailed calculations to obtain the explicit solution. Solutions of this type will be discussed in further detail in connection with the continuum model treated in Chapter 3.

In the extreme case where all of the oscillators are injection locked to the same external oscillator, the solution simplifies considerably. Returning to Eq. (2.5-3), we find that the elements of $[d]$ are all equal as are the elements of $[\varphi_{inj}]$. Thus Eq. (2.5-3) becomes

$$[s[I]-[\bar{M}][\tilde{\varphi}]=[\bar{\Delta}\tilde{\Omega}_{tune}] \tag{2.5-14}$$

where

$$[\bar{M}]=[M]-[d] \tag{2.5-15}$$

and

$$[\bar{\Delta}\tilde{\Omega}_{tune}]=[\Delta\tilde{\Omega}_{tune}]+r[\tilde{\varphi}_{inj}] \tag{2.5-16}$$

Equation (2.5-14) is now identical in form to Eq. (2.2-9), and the solution in the form of Eq. (2.2-34) follows immediately. However, if the injection signals differ sufficiently in phase, the elements of the right side of Eq. (2.5-16) can exceed unity and the oscillators therefore lose lock. This phenomenon has been exploited in discriminating between signals arriving at disparate angles in illuminating a phased array. For a given illumination angle the signals at each element differ from those of nearest-neighboring elements by a constant phase difference and, if used to inject the corresponding oscillators of a coupled oscillator array, represent the second term on the right side of Eq. (2.5-16).

Thus, as the incidence angle increases, the phase differences increase and eventually the array loses lock. This permits identification of signals arriving outside a given range of incidence angles [18].

Before proceeding to planar arrays, we remark at this point that one may also produce beam-steering via a variant of the Stephan approach in which the external injection signals are provided by the end oscillators of the array and their phase is controlled by adjusting the coupling phase between the end and next to end oscillators [30].

Finally, we add that, as shown by Heath, control of the coupling phase also affords the possibility of creating a so-called "difference pattern" in which a null is formed instead of a beam [31]. This is done by switching the phase of one interior coupling by π radians. Of course, such a null can also be steered via either detuning or injection of the end oscillators.

2.6 Generalization to Planar Arrays

Nearly all of the formalism presented in connection with linear arrays of oscillators can be generalized to planar arrays. The simplest of planar arrays consists of a linear array of linear arrays placed side by side as shown in Fig. 2-4. Assuming nearest-neighbor coupling, this implies that each oscillator is coupled to four others and can be described mathematically by analogy with Eq. (2.1-1). That is,

$$\begin{aligned}\frac{d\theta_{ij}}{dt} = \omega_{0ij} &+ \Delta\omega_{lock}\sin\left(\theta_{i+1,j} - \theta_{ij} - \Phi\right)\\ &+ \Delta\omega_{lock}\sin\left(\theta_{i-1,j} - \theta_{ij} - \Phi\right)\\ &+ \Delta\omega_{lock}\sin\left(\theta_{i,j-1} - \theta_{ij} - \Phi\right)\\ &+ \Delta\omega_{lock}\sin\left(\theta_{i,j+1} - \theta_{ij} - \Phi\right)\end{aligned} \tag{2.6-1}$$

where, for simplicity, we have assumed that all of the coupling phases are equal. The oscillators are indexed separately in the two orthogonal directions x and y in the plane of the array by indices i and j, respectively. The four sine terms correspond to coupling to the four nearest-neighboring oscillators, implying that for a zero-coupling phase the center oscillator may be detuned by as much as four locking ranges, and the array will still remain locked. Similarly, the corner oscillators may be detuned by two locking ranges. The largest permitted detuning of the other oscillators will lie between two and four locking ranges. The effects of a uniform coupling phase can be determined in

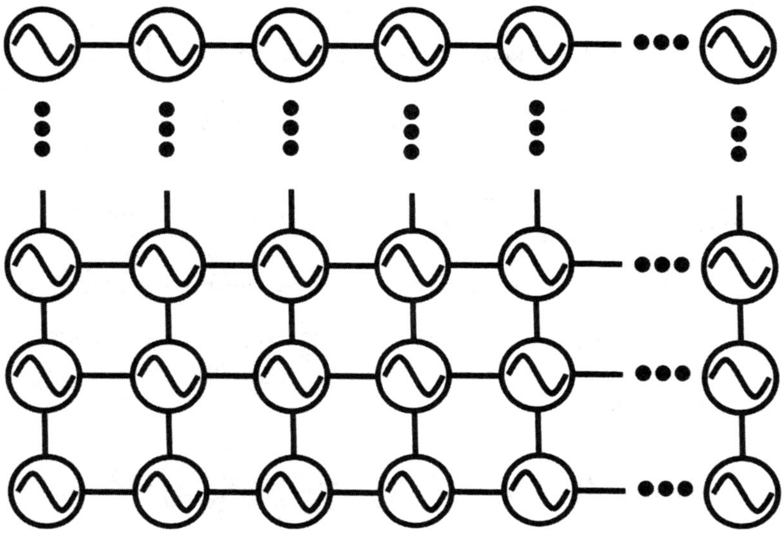

Fig. 2-4. Planar array with nearest-neighbor coupling.

the same manner as in the one-dimensional case. That is, an effective locking range and ensemble frequency shift can be determined as a function of coupling phase. Letting the coupling phase be zero and linearizing as we have done in the case of linear arrays, we obtain

$$\begin{aligned}\frac{d\theta_{ij}}{dt} &= \omega_{0ij} + \Delta\omega_{lock}\left(\theta_{i+1,j} - \theta_{ij}\right) + \Delta\omega_{lock}\left(\theta_{i-1,j} - \theta_{ij}\right) \\ &\quad + \Delta\omega_{lock}\left(\theta_{i,j-1} - \theta_{ij}\right) + \Delta\omega_{lock}\left(\theta_{i,j+1} - \theta_{ij}\right) \\ &= \omega_{0ij} + \Delta\omega_{lock}\left(\theta_{i+1,j} - 2\theta_{ij} + \theta_{i-1,j}\right) \\ &\quad + \Delta\omega_{lock}\left(\theta_{i,j-1} - 2\theta_{ij} + \theta_{i,j+1}\right)\end{aligned} \qquad (2.6\text{-}2)$$

or, using Eq. (2.1-7) and $\tau = \Delta\omega_{lock} t$, we find that

$$\begin{aligned}\frac{d\varphi_{ij}}{d\tau} &= \left(\frac{\omega_{0ij} - \omega_{ref}}{\Delta\omega_{lock}}\right) + \left(\varphi_{i+1,j} - 2\varphi_{ij} + \varphi_{i-1,j}\right) \\ &\quad + \left(\varphi_{i,j-1} - 2\varphi_{ij} + \varphi_{i,j+1}\right)\end{aligned} \qquad (2.6\text{-}3)$$

Laplace transformation leads to

$$s\tilde{\varphi}_{ij} = \Delta\tilde{\Omega}_{tune} + \left(\tilde{\varphi}_{i+1,j} - 2\tilde{\varphi}_{ij} + \tilde{\varphi}_{i-1,j}\right) + \left(\tilde{\varphi}_{i,j-1} - 2\tilde{\varphi}_{ij} + \tilde{\varphi}_{i,j+1}\right) \tag{2.6-4}$$

where $[\Delta\Omega_{tune}] = \left[\dfrac{\omega_{0ij} - \omega_{ref}}{\Delta\omega_{lock}}\right]$, elements of a rectangular matrix. This equation may now be written in matrix form by defining a matrix $[\tilde{\varphi}]$ with elements $\tilde{\varphi}_{ij}$. We now define eigenmatrices, $[v]$, with elements, v_{ij}, and eigenvalues, λ_{mn}, satisfying

$$\left(v_{i+1,j} - 2v_{ij} + v_{i-1,j}\right) + \left(v_{i,j-1} - 2v_{ij} + v_{i,j+1}\right) = \lambda_{mn} v_{ij} \tag{2.6-5}$$

The key concept enabling analytical treatment of planar arrays is separability of the behavior in the two orthogonal directions. That is, if we define the two-dimensional phase distribution to be the product of two one-dimensional distributions, we obtain

$$v_{ij} = v_i v_j \tag{2.6-6}$$

Then Eq. (2.6-5) becomes

$$v_j\left(v_{i+1} - 2v_i + v_{i-1}\right) + v_i\left(v_{j-1} - 2v_i + v_{j+1}\right) = \lambda_{mn} v_i v_j \tag{2.6-7}$$

Dividing by the product, $v_i v_j$, results in

$$\frac{\left(v_{i+1} - 2v_i + v_{i-1}\right)}{v_i} + \frac{\left(v_{j-1} - 2v_j + v_{j+1}\right)}{v_j} = \lambda_{mn} \tag{2.6-8}$$

The first term on the left is dependent only on i and is independent of j. Similarly, the second term on the left is dependent only on j and is independent of i. The right side of the equation is independent of both i and j. Thus, we have a sum of a function of i and a function of j equal to a constant which implies that each of these functions must itself be a constant. That is

$$\frac{\left(v_{i+1} - 2v_i + v_{i-1}\right)}{v_i} = \lambda_m \tag{2.6-9}$$

$$\frac{\left(v_{j-1} - 2v_j + v_{j+1}\right)}{v_j} = \lambda_n \tag{2.6-10}$$

and

$$\lambda_m + \lambda_n = \lambda_{mn} \tag{2.6-11}$$

Using the definition of [*M*] from Eq. (2.2-7), we may write Eqs. (2.6-9) and (2.6-10) in the forms

$$[M][v]_m = \lambda_m [v]_m \tag{2.6-12}$$

and

$$[M][v]_n = \lambda_n [v]_n \tag{2.6-13}$$

which are identical to Eq. (2.2-10). Thus, the eigenmatrices have been separated into the outer product of eigenvectors, one for the i dependence and one for the j dependence, and each of these eigenvectors is identical with those of the linear array of Section 2.2. That is,

$$[v]_{mn} = [v]_m \otimes [v]_n \tag{2.6-14}$$

The eigenvectors and eigenvalues of Eqs. (2.6-12) and (2.6-13) were found in section 2.2, and we will use them here to express the solution of Eq. (2.6-4) in the form

$$[\tilde{\varphi}] = \sum_m \sum_n C_{mn} [v]_{mn} \tag{2.6-15}$$

Substitution of this form into Eq. (2.6-4) gives

$$\begin{aligned} s \sum_m \sum_n C_{mn} [v]_{mn} = \Delta\tilde{\Omega}_{tune} + \lambda_m \sum_m \sum_n C_{mn} [v]_{mn} \\ + \lambda_n \sum_m \sum_n C_{mn} [v]_{mn} \end{aligned} \tag{2.6-16}$$

We now make use of the orthogonality of the eigenvectors. Premultiplying by $[v]_p$ and postmultiplying by $[v]_q$, we have

$$sC_{pq}\left([v]_p \bullet [v]_p\right)\left([v]_q \bullet [v]_q\right) = [v]_p \bullet \Delta\tilde{\Omega}_{tune} \bullet [v]_q + \left(\lambda_p + \lambda_q\right)\left([v]_p \bullet [v]_p\right)\left([v]_q \bullet [v]_q\right) \tag{2.6-17}$$

so that

$$C_{pq} = \frac{[v]_p \bullet \Delta\tilde{\Omega}_{tune} \bullet [v]_q}{\left(s - \lambda_p - \lambda_q\right)\left([v]_p \bullet [v]_p\right)\left([v]_q \bullet [v]_q\right)} \tag{2.6-18}$$

And Eq. (2.6-15) becomes

$$[\tilde{\varphi}] = \sum_m \sum_n \frac{[v]_m \bullet \Delta\tilde{\Omega}_{tune} \bullet [v]_n}{\left(s - \lambda_m - \lambda_n\right)\left([v]_m \bullet [v]_m\right)\left([v]_n \bullet [v]_n\right)}[v]_{mn} \tag{2.6-19}$$

the planar analog of Eq. (2.2-33). The stability analysis of Section 2.4 also carries over to the planar case as discussed by Heath et al. [29]. One may similarly derive a planar analog of (2.5-10) should there be external injection [2]. The dynamic behavior of the phase distribution for these cases will be discussed in greater detail in connection with the continuum model presented in Chapter 3. However, we remark here that Karl Stephan and his student, William Morgan, reported application of his external injection beam-steering technique to a four-by-four planar array of mutually injection-locked oscillators [2]. They also developed a theory for such arrays in which the coupling is accomplished via a general multiport coupling network described by an admittance matrix as will be further described in Section 2.7 [2–3]. They considered theoretically the use of such an oscillator array to excite an array of tapered slot radiators reasoning that the higher gain of these elements would mitigate grating lobes if the array size were increased by using element spacing greater than a half wavelength.

2.7 Coupling Networks

So far we have focused primarily on the behavior of the oscillators in the array but very little on the manner in which they are coupled. We merely asserted that the coupling was present with a certain assumed strength and coupling phase. In addition, two other parameters are important in the design of coupled-oscillator arrays: the network quality factor or Q, which is related to the bandwidth, and the load presented to the oscillators by the network. Although not essential, two simplifying assumptions are quite commonly made. The coupling strength is assumed to be weak in a sense to be detailed shortly, and the coupling network Q is assumed to be small relative to the oscillator Q; thus the network can be assumed to be frequency independent over the operating bandwidth of the oscillators. Finally, it is essential that the load resistance presented to the oscillators be smaller than the maximum negative resistance the oscillator can produce so that oscillation can be sustained. In order to systematically design appropriate coupling networks, it is necessary to derive relationships between these three parameters and the values of the components used in constructing the network. In this section, such relationships will be derived.

Before proceeding, we remark that the consequences of violation of the above simplifying assumptions have in fact been studied. The case of strong coupling was treated in this context by Nogi et al. [17]. They showed that strongly coupled arrays exhibit many modes in which the oscillator amplitudes as well as the phases vary across the array and that only one mode has constant amplitude. They further suggested that all other modes can be suppressed by placing a series resistor at the center point of each coupling line. The principle underlying this approach was pointed out very early by Stephan and Young [3]. The implications of narrow-band coupling networks were studied by Lynch and York [32]. The analysis becomes more complicated than in the broadband case [33], but useful results can still be obtained. Very recently these issues were re-examined by Seetharam and Pearson [19]. They showed that strongly coupled oscillator arrays exhibit wider locking ranges and lower phase noise levels but that the broadband assumption concerning the coupling network is violated, necessitating the use of the more complicated theoretical formalism.

Generally, the oscillators may be viewed as being coupled by a multiport passive network to which an oscillator is to be connected at each port. This situation was analyzed by Pogorzelski [34]. The network is characterized by its complex admittance matrix; and for a linear array, if the desired coupling is to nearest neighbors, then the admittance matrix will be tridiagonal. For analytical simplicity, we assume an infinitely long array. We want the oscillators to operate in identical environments, so we design the network to be periodic with period unity in the oscillator index. Its admittance matrix will therefore have equal diagonal elements, Y_{11}, and equal off-diagonal elements, Y_{12}.

Focusing now on the network alone, in terms of the complex impedance matrix we may write the network equations in the form

$$[Z]\left[I_{port}\right]=\left[V_{port}\right] \tag{2.7-1}$$

where the vector components are the port currents and port voltages. Because the network is periodic, the elements of the impedance matrix, Z_{mn}, will have the form

$$Z_{mn} = Z_{n-m} = Z_{\ell} \tag{2.7-2}$$

where the port indices, *m* and *n*, extend from minus infinity to plus infinity for this infinite network. In this sense, ℓ denotes the "distance of the element from the main diagonal of the matrix." Furthermore, the network periodicity implies that the components of the eigenvectors, $[w]_{\Delta\varphi}$, of the impedance matrix will have uniform phase progression and uniform amplitude across the network ports. That is, the elements of the eigenvector with inter-port phase difference $\Delta\varphi$ are

$$w_m = e^{jm\Delta\varphi} \tag{2.7-3}$$

Defining

$$\left[I_{port}\right] = I_{\Delta\varphi}[w]_{\Delta\varphi} \tag{2.7-4}$$

and

$$\left[V_{port}\right] = V_{\Delta\varphi}[w]_{\Delta\varphi} \tag{2.7-5}$$

every equation in the system given by Eq. (2.7-1) becomes

$$\left(\sum_{n=-\infty}^{\infty} Z_{\ell} e^{j\ell\Delta\varphi}\right) I_{\Delta\varphi} = V_{\Delta\varphi} \tag{2.7-6}$$

Now, the tridiagonal admittance matrix gives us

$$I_{\Delta\varphi} = V_{\Delta\varphi}\left(Y_{11} + Y_{12}e^{j\Delta\varphi} + Y_{12}e^{-j\Delta\varphi}\right) \tag{2.7-7}$$

Combining Eqs. (2.7-6) and (2.7-7), we have

$$\sum_{n=-\infty}^{\infty} Z_{\ell} e^{j\ell\Delta\varphi} = \frac{1}{(Y_{11} + 2Y_{12}\cos\Delta\varphi)} \tag{2.7-8}$$

a Fourier series for which the coefficients, Z_{ℓ}, can be found by means of the usual integration. That is,

$$Z_\ell = \frac{1}{2\pi}\int_{-\pi}^{\pi}\frac{e^{-j\ell\alpha}d\alpha}{\left(Y_{11}+2Y_{12}\cos\alpha\right)} \tag{2.7-9}$$

The integration can be carried out analytically to yield

$$Z_\ell = \frac{1}{\sqrt{Y_{11}^2-4Y_{12}^2}}\left(\frac{\sqrt{Y_{11}^2-4Y_{12}^2}-Y_{11}}{2Y_{12}}\right)^\ell \tag{2.7-10}$$

This approach is quite similar to that used in the analysis of phased-array antennas to obtain the well-known relationship between the mutual coupling coefficients among the elements and the active reflection coefficients of the array [35]. The result is that we have expressed the elements of the impedance matrix and admittance matrix of the coupling network in terms of the two parameters, Y_{11} and Y_{12}.

We now define the complex coupling coefficient of the coupling network in the following way. Let the voltage at the nth port be V_n and the current into the nth port be I_n. We can establish a Norton equivalent circuit at the $(n + 1)$st port as follows. The open circuit voltage is

$$V_{oc} = I_n Z_1 \tag{2.7-11}$$

and the short circuit current is

$$I_{sc} = -V_n Y_{12} \tag{2.7-12}$$

Thus, the Norton admittance is

$$Y_N = \frac{V_n Y_{12}}{I_n Z_1} = \frac{-Y_{12}}{Y_{11} Z_1} = -\frac{Y_{11}^2-4Y_{12}^2-Y_{11}\sqrt{Y_{11}^2-4Y_{12}^2}}{2Y_{11}} \tag{2.7-13}$$

Now, using the Norton equivalent circuit and connecting load admittance G_L to the port, the voltage at port $n + 1$ is

$$V_{n+1} = I_{sc}\frac{1}{G_L+Y_N} = -V_n\frac{Y_{12}}{G_L+Y_N} \tag{2.7-14}$$

The complex coupling coefficient, κ, is defined to be the ratio of the voltage at port $n + 1$ to the voltage at port n. That is,

$$\kappa = \frac{-Y_{12}}{G_L + Y_N} \tag{2.7-15}$$

In the weak coupling approximation, Y_{12} is small and to first order in Y_{12} we have

$$\kappa \approx \frac{-Y_{12}}{G_L} \tag{2.7-16}$$

This coupling factor determines the locking range because the injection signal arriving at an oscillator, i, from a neighboring one, j, is the amplitude of the oscillator signal, A_j, multiplied by κ, so the locking range is given by,

$$\Delta\omega_{lock} = \frac{\omega_{0i}}{2Q}\frac{\kappa A_j}{A_i} \approx \frac{-Y_{12}\omega_{0i}}{2QG_L} \tag{2.7-17}$$

For comparison, see Eq. (1.3-16).

Before proceeding, we wish to highlight an interesting point regarding the nature of the coupling. If a current is injected into the nth port of the network with all other ports open circuited, the voltage appearing at port $n + m$ is, from the impedance matrix, just Z_m / Z_o times the voltage at the injected port. (See Eq. (2.7-10).) That is, open-circuit voltages appear at *all ports throughout the network*, not just at the adjacent ports. In this sense, each oscillator really influences all the others, and the coupling is "all to all" rather than "nearest neighbor" in nature. The fact that the admittance matrix is banded might seem to imply nearest-neighbor coupling, but the banded nature of the matrix merely implies that, when a voltage is applied to the nth port *with all the other ports shorted*, short-circuit current flows only in the adjacent ports. The limited influence results from shorting the ports, not from limited coupling. That said, we proceed to define the coupling factor of the network as the ratio of the open-circuit voltages at adjacent ports when a current is injected into the nth port of the network with all other ports open circuited. From Eq. (2.7-10), that ratio is

$$\varepsilon = \frac{\sqrt{Y_{11}^2 - 4Y_{12}^2} - Y_{11}}{2Y_{12}} \tag{2.7-18}$$

For weak coupling, this becomes

$$\varepsilon \approx \frac{-Y_{12}}{Y_{11}} \tag{2.7-19}$$

where we have effectively neglected the coupling to the non-nearest-neighbor ports by working only to first order in this ratio. This same result is obtained if we define the coupling factor by applying a voltage to the nth port, shorting all the others and taking the ratio of the $(n+1)$st port current to the nth port current. This obtains because if G_L is large compared to Y_{12}, the coupling is weak and the ports are all nearly shorted.

The coupling factor appearing in the expression for the locking range given by Eq. (2.7-17) depends on both the oscillator load and the coupling network. Based on the discussion above, we can separate these by writing Eq. (2.7-16) in the form

$$\kappa \approx \left(\frac{-Y_{12}}{Y_{11}}\right)\left(\frac{Y_{11}}{G_L}\right) = \varepsilon\eta \tag{2.7-20}$$

so that the first factor, ε, characterizes the network coupling and the second factor, η, characterizes the coupling of the oscillators to the network.

The second important parameter in network design is the network quality factor or Q. Fundamentally, Q is defined in terms of energy stored and energy lost per unit time, but equivalently, Q can also be defined in terms of the frequency dependence of the port admittance near resonance; that is,

$$Q = \frac{\omega}{2}\left|\frac{\frac{\partial Y}{\partial \omega}}{Y}\right|_{\omega=\omega_{res}} \tag{2.7-21}$$

a unitless quantity. For our coupling network we thus have the formula

$$Q_{net} = \frac{\omega}{2}\left|\frac{\frac{\partial}{\partial \omega}\left(Y_{11} + 2Y_{12}\cos\Delta\varphi\right)}{Y_{11} + 2Y_{12}\cos\Delta\varphi}\right|_{\omega=\omega_{res}} \tag{2.7-22}$$

And, for our formulation to apply, this must be much smaller than the Q of the oscillators.

Finally, the third parameter in the design of these arrays is the amount of negative resistance that must be provided by the oscillators. When operating normally, the current entering each port of the network is related to the port voltage by the port admittance; that is,

$$Y_{\Delta\varphi} = \frac{I_{\Delta\varphi}}{V_{\Delta\varphi}} = Y_{11} + 2Y_{12}\cos\Delta\varphi \qquad (2.7\text{-}23)$$

Since the inter-oscillator phase difference cannot exceed $\pi/2$ and Y_{12} has a negative real part, the maximum susceptance presented to the oscillator by the network is $\mathrm{Re}(Y_{11})$. Thus, we conclude that the oscillator must be designed to provide a minimum of this amount of negative susceptance plus an amount sufficient to compensate for the internal load susceptance, G_L, to maintain oscillation when connected to the network.

Let us now consider a concrete example of a network of the sort commonly used in experimental studies of linear coupled oscillator arrays. Each unit cell consists of a one-wavelength-long transmission line of characteristic impedance Z_C, two parallel resistors to reduce the network Q by reducing reflections at the transmission line ends, and two series resistors to control the coupling strength. Such a network is shown in Fig. 2-5 wherein the circles indicate terminals where the oscillators are connected at each end of the unit cell. Using the definitions of the elements of the admittance matrix, we may determine that

$$\begin{aligned} Y_{11} &= \frac{1}{R_S}\left(\frac{R_P + 2R_S}{R_P + R_S}\right) \\ Y_{12} &= -\frac{1}{2R_S}\left(\frac{R_P}{R_P + R_S}\right) \end{aligned} \qquad (2.7\text{-}24)$$

so that from Eq. (2.7-19) for weak coupling, we obtain

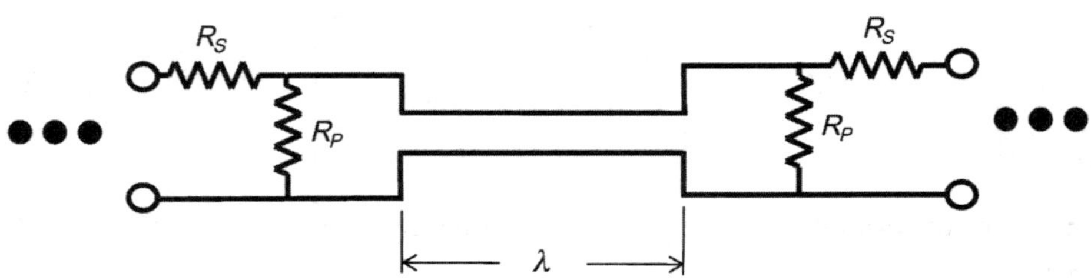

Fig. 2-5. Unit cell of an infinite one-dimensional coupling network.

$$\varepsilon \approx \frac{-Y_{12}}{Y_{11}} = \frac{1}{2}\left(\frac{R_P}{R_P + 2R_S}\right) \tag{2.7-25}$$

and the oscillators must provide negative susceptance, $-G_{osc}$, where

$$-G_{osc} > \mathrm{Re}(Y_{11}) = \frac{1}{R_S}\left(\frac{R_P + 2R_S}{R_P + R_S}\right) \tag{2.7-26}$$

In order to determine the network Q using Eq. (2.7-22), it is necessary to explicitly display the frequency dependence of the admittance parameters induced by the transmission lines. That is, from transmission line theory, we have

$$Y_{11} = 2\frac{\frac{1}{R_p}(Z_R \cosh\gamma\ell + Z_C \sinh\gamma\ell) + \frac{1}{Z_C}(Z_R \sinh\gamma\ell + Z_C \cosh\gamma\ell)}{\frac{R_s}{Z_R}(Z_R \cosh\gamma\ell + Z_C \sinh\gamma\ell) + \frac{R_s}{Z_C}(Z_R \sinh\gamma\ell + Z_C \cosh\gamma\ell)} \tag{2.7-27}$$

$$Y_{12} = \frac{-\frac{Z_R}{R_s}}{\frac{R_s}{Z_R}(Z_R \cosh\gamma\ell + Z_C \sinh\gamma\ell) + \frac{R_s}{Z_C}(Z_R \sinh\gamma\ell + Z_C \cosh\gamma\ell)} \tag{2.7-28}$$

where Z_R is the impedance of the parallel combination of R_p and R_s, and γ is the propagation constant of the transmission line. The frequency dependence arises because γ is linear in ω. Using these expressions in Eq. (2.7-22), we find that

$$Q_{net} = \frac{\pi}{2}\left|\frac{\left(\frac{Z_R}{Z_C} - \frac{Z_C}{Z_R}\right) + \frac{1}{2}\left(\frac{Z_R}{Z_C} + \frac{Z_C}{Z_R}\right)\cos\Delta\varphi}{\left(\frac{R_s}{Z_R} + \frac{R_s}{R_p}\right) - \cos\Delta\varphi}\right| \tag{2.7-29}$$

(Note that this does not agree with equation (18) of Ref. 34 due to an algebraic error in the derivation of that equation.) Typical oscillators used with this network have Q's on the order of 100. To minimize the Q of the network, we chose parameter values to minimize reflections at the ends of the transmission lines. That is, we chose $Z_R \approx Z_C$ so that for small inter-oscillator phase differences we have

$$Q_{net} \approx \frac{\pi}{4}\frac{R_p}{R_s} \tag{2.7-30}$$

and for resistors of comparable value, the network Q is of order unity, clearly much smaller than the oscillator Q as assumed in the mathematical model.

Another particularly simple coupling network was proposed by Humphrey and Fusco; that of a single capacitor between adjacent ports of the network [36]. The corresponding parameters can be derived from the previous example by setting R_p to infinity and replacing R_s by $1/(2j\omega C)$. Thus,

$$\begin{aligned} Y_{11} &= 2j\omega C \\ Y_{12} &= -j\omega C \end{aligned} \tag{2.7-31}$$

Note that the approximation of Eq. (2.7-19) is not valid for this network so Eq. (2.7-18) must be used and we obtain

$$\begin{aligned} \varepsilon &\approx \frac{\sqrt{Y_{11}^2 - 4Y_{12}^2} - Y_{11}}{2Y_{12}} = 1 \\ Q_{net} &= \frac{1}{2} \\ -G_{osc} &> 0 \end{aligned} \tag{2.7-32}$$

Note further that the coupling phase is zero.

The approach outlined above for coupling networks designed for linear arrays of oscillators can be generalized to the case of planar arrays in which, for example, each oscillator is coupled to its four nearest neighbors via the network unit cell shown in Fig. 2-5. The admittance matrix is then block tridiagonal, and each diagonal block is tridiagonal while the off-diagonal blocks are diagonal. The matrix is symmetric, and the elements along any diagonal are equal. Thus, the admittance matrix has only three independent elements. The diagonal elements are denoted by Y_d, the off-diagonal elements of the diagonal blocks by Y_x, and the diagonal elements of the off-diagonal blocks by Y_y. As shown in Ref. 34, the integral in Eq. (2.7-9) then becomes the two-dimensional integral

$$Z_{mn} = \frac{1}{4\pi^2}\int_{-\pi}^{\pi}\int_{-\pi}^{\pi}\frac{e^{-jmu}e^{-jnv}\,du\,dv}{\left(Y_d + 2Y_x\cos u + 2Y_y\cos v\right)} \tag{2.7-33}$$

The impedance matrix, though full, also has a block structure. The elements along any diagonal within a block are equal. Similarly, the blocks along any block diagonal are equal. Using a generalization of the notation used previously for the linear case, the first subscript indicates the "distance" from the diagonal within each block and the second subscript indicates the "distance" of the block from the block diagonal.

The integrals given by Eq. (2.7-33) for nearest neighbors, Z_{01} and Z_{10}, as well as the integral for the diagonal elements Z_{00}, can be expressed in terms of elliptic integrals. That is,

$$Z_{00} = \frac{2}{\pi\sqrt{Y_d^2 - 4\left(Y_x - Y_y\right)^2}} K\left[\frac{16Y_xY_y}{Y_d^2 - 4\left(Y_x - Y_y\right)^2}\right] \tag{2.7-34}$$

$$\begin{aligned} Z_{01} = & \frac{-2}{\pi\sqrt{Y_d^2 - 4\left(Y_x - Y_y\right)^2}} \left\{\left(\frac{Y_d - 2Y_x}{2Y_y}\right) K\left[\frac{16Y_xY_y}{Y_d^2 - 4\left(Y_x - Y_y\right)^2}\right]\right. \\ & \left. -\left(\frac{Y_d - 2Y_x - 2Y_y}{2Y_y}\right)\Pi\left[\frac{\pi}{2}, \frac{4Y_y}{Y_d - 2Y_x - 2Y_y}, \sqrt{\frac{16Y_xY_y}{Y_d^2 - 4\left(Y_x - Y_y\right)^2}}\right]\right\} \end{aligned} \tag{2.7-35}$$

$$\begin{aligned} Z_{10} = & \frac{-2}{\pi\sqrt{Y_d^2 - 4\left(Y_x - Y_y\right)^2}} \left\{\left(\frac{Y_d - 2Y_y}{2Y_x}\right) K\left[\frac{16Y_xY_y}{Y_d^2 - 4\left(Y_x - Y_y\right)^2}\right]\right. \\ & \left. -\left(\frac{Y_d - 2Y_x - 2Y_y}{2Y_x}\right)\Pi\left[\frac{\pi}{2}, \frac{4Y_x}{Y_d - 2Y_x - 2Y_y}, \sqrt{\frac{16Y_xY_y}{Y_d^2 - 4\left(Y_x - Y_y\right)^2}}\right]\right\} \end{aligned} \tag{2.7-36}$$

where K is the complete elliptic integral of the first kind and Π is the elliptic integral of the third kind [37]. The analogs of the voltage ratio of Eq. (2.7-18) are

$$\varepsilon_x = \frac{Z_{10}}{Z_{00}} \tag{2.7-37}$$

and

$$\varepsilon_y = \frac{Z_{01}}{Z_{00}} \tag{2.7-38}$$

These expressions are quite cumbersome. However, as shown in Ref. 34, one may obtain more manageable expressions by evaluating the integrals asymptotically for large subscript via the method of stationary phase when the other subscript is zero. While technically only valid for large index, the form of these expressions exhibits a common ratio between the ports which may be taken to be a measure of the coupling. That is, one obtains

$$\varepsilon_x \approx -\left(\frac{Y_d + 2Y_y}{2Y_x}\right) + \sqrt{\left(\frac{Y_d + 2Y_y}{2Y_x}\right)^2 - 1} \tag{2.7-39}$$

and

$$\varepsilon_y \approx -\left(\frac{Y_d + 2Y_x}{2Y_y}\right) + \sqrt{\left(\frac{Y_d + 2Y_x}{2Y_y}\right)^2 - 1} \tag{2.7-40}$$

as the analogs of Eq. (2.7-18). For weak coupling, Y_x and Y_y are small compared with Y_d and we obtain

$$\varepsilon_x \approx -\frac{Y_x}{Y_d + 2Y_y} \tag{2.7-41}$$

and

$$\varepsilon_y \approx -\frac{Y_y}{Y_d + 2Y_x} \tag{2.7-42}$$

Similarly, the analogs of Eq. (2.7-20) are

$$\kappa_x \approx \left(\frac{-Y_x}{Y_d + 2Y_y}\right)\left(\frac{Y_d + 2Y_y}{G_L}\right) \tag{2.7-43}$$

$$\kappa_y \approx \left(\frac{-Y_y}{Y_d + 2Y_x}\right)\left(\frac{Y_d + 2Y_x}{G_L}\right) \tag{2.7-44}$$

The network Q can again be expressed as the logarithmic derivative of the port admittance with respect to frequency. That is,

$$Q_{net} = \frac{\omega}{2}\left|\frac{\frac{\partial}{\partial\omega}\left(Y_d + 2Y_x \cos\Delta\varphi_x + 2Y_y \cos\Delta\varphi_y\right)}{Y_d + 2Yx\cos\Delta\varphi_x + 2Y_y \cos\Delta\varphi_y}\right|_{\omega=\omega_{res}} \tag{2.7-45}$$

Finally, the oscillators must provide negative resistance such that

$$-G_{osc} > \mathrm{Re}\left(Y_d\right) \tag{2.7-46}$$

For a network using the coupling configuration shown in Fig. 2-5, we have

$$\begin{aligned} Y_d &= \frac{2}{R_S}\left(\frac{R_P + 2R_S}{R_P + R_S}\right) \\ Y_x = Y_y &= -\frac{1}{2R_S}\left(\frac{R_P}{R_P + R_S}\right) \end{aligned} \tag{2.7-47}$$

so that

$$\varepsilon_x = \varepsilon_y \approx \frac{1}{2}\left(\frac{R_p}{R_p + 4R_s}\right) \tag{2.7-48}$$

$$-G_{osc} > \frac{2}{R_S}\left(\frac{R_P + 2R_S}{R_P + R_S}\right) \tag{2.7-49}$$

and

$$Q_{net} = \frac{\pi}{2}\left|\frac{\left(\frac{Z_R}{Z_C} - \frac{Z_C}{Z_R}\right) + \frac{1}{2}\left(\frac{Z_R}{Z_C} + \frac{Z_C}{Z_R}\right)\left(\frac{\cos\Delta\varphi_x + \cos\Delta\varphi_y}{2}\right)}{\left(\frac{R_s}{Z_R} + \frac{R_s}{R_p}\right) - \left(\frac{\cos\Delta\varphi_x + \cos\Delta\varphi_y}{2}\right)}\right| \tag{2.7-50}$$

For $Z_R \approx Z_C$ and small inter-oscillator phase differences we again have

$$Q_{net} \approx \frac{\pi}{4}\frac{R_p}{R_s} \tag{2.7-51}$$

which is typically much smaller than the oscillator Q.

In this section we have discussed the analysis of coupling networks for infinite arrays both linear and planar. Although, in practice, the arrays are of course finite, the analysis of infinite arrays is more tractable and provides insight into how the circuit parameters affect the array behavior. Thus, relatively simple approximate formulas obtain for the coupling strength, network Q, and needed oscillator negative resistance in terms of circuit element values facilitating the design of such networks.

2.8 Conclusion

In this chapter, the oscillators of the arrays were treated as individual circuits capable of oscillation in themselves. These were coupled to form a mutually injection-locked system of oscillators. In that sense, the modeling was discrete, and the phase distributions studied were distributions of the phases of the individual oscillator outputs, which (of course) have meaning only in terms of the individual oscillator output signals. In the next chapter, however, we will introduce the concept of the continuum model in which the phase distributions are continuous functions. It is emphasized that the values of these continuous functions still only have physical meaning when the functions are evaluated for arguments corresponding to individual oscillators. Arguments between these are for mathematical convenience and the corresponding function values have no physical significance.

Chapter 3
The Continuum Model for Linear Arrays

All of the analysis presented so far has treated each oscillator as a discrete device with an injection port and an output port from which a signal emanates having a discrete phase value relative to a phase reference. For this reason, the mathematical model represented has been termed the discrete model. We emphasize that the discrete model encompasses the dynamic behavior of the oscillator array both nonlinear and, if desired, linearized. No new phenomena are added to this range of capability by means of the formulation to be discussed in the present chapter. However, it will be shown that, provided one is willing to linearize, the so-called "continuum model" offers considerable advantage in terms of insight and applicability of familiar mathematical techniques. Although the continuum model is fundamentally approximate primarily because of the linearization, it nevertheless provides intuitive understanding of the behavior of coupled oscillator arrays with small inter-oscillator phase differences, an important special case in terms of practical application. Moreover, it provides a basis for understanding the impact of nonlinearity when the inter-oscillator phase differences increase beyond the limits of accurate linear approximation.

The continuum model in this context was suggested by Pogorzelski et al. [38]. In essence we replace the index identifying the oscillators with a continuous variable such that, when the continuous variable takes on the value of the index for a given oscillator, a continuous function of that variable takes on the value of the phase of that oscillator. Thus, only the values of the function at integer values of its argument have physical meaning. The values between integer values of the argument serve only to facilitate the formulation in terms of a differential equation.

3.1 The Linear Array without External Injection

To derive the continuum model of a simple linear array of oscillators coupled to nearest neighbors, we begin with Eq. (2.2-4) for the linearized discrete model *with zero coupling phase* and replace the discrete index i with a continuous variable, x.

$$\frac{d\varphi(x,t)}{dt} = \omega_0(x,t) - \omega_{ref} + \Delta\omega_{lock}\left[\varphi(x+\Delta x,t) - 2\varphi(x,t) + \varphi(x-\Delta x,t)\right] \tag{3.1-1}$$

where Δx=1. Now treating $\varphi(x,t)$ as a continuous function of x, expanding each term in a Taylor series about x, and retaining terms up to second order in Δx, we obtain

$$\frac{\partial\varphi(x,t)}{\partial t} = \omega_0(x,t) - \omega_{ref} + \Delta\omega_{lock}\frac{\partial^2\varphi(x,t)}{\partial x^2} \tag{3.1-2}$$

Finally, dividing by the locking range and using the normalized time variable, $\tau = \Delta\omega_{lock}t$, we have

$$\frac{\partial\varphi(x,\tau)}{\partial\tau} = \Delta\Omega_{tune}(x,\tau) + \frac{\partial^2\varphi(x,\tau)}{\partial x^2} \tag{3.1-3}$$

This is the fundamental equation for the continuum model of a simple linear array of oscillators with nearest-neighbor coupling and no external injection. It is the well-known diffusion equation. Laplace transformation with respect to time results in

$$\frac{d^2\tilde{\varphi}(x,s)}{dx^2} - s\tilde{\varphi}(x,s) = -\Delta\tilde{\Omega}_{tune}(x,s) \tag{3.1-4}$$

a simple second-order linear differential equation for the transform of the phase distribution.

Suppose that the array is infinitely long and that one oscillator is step detuned at time zero by C locking ranges where C is less than two. Without loss of generality, we may select the detuned oscillator to be the one at $x = 0$. For this situation, Eq. (3.1-4) becomes

$$\frac{d^2\tilde{\varphi}(x,s)}{dx^2} - s\tilde{\varphi}(x,s) = -\frac{C}{s}\delta(x) \tag{3.1-5}$$

As discussed in Ref. 38, it might be considered more correct to use, in place of the delta function, a square pulse one unit wide to represent the detuning. However, it is shown in Ref. 38 that the difference in the results is very small, and (in the spirit of the continuum model) the use of the delta function affords considerable convenience with minor impact on the results.

The differential equation given by Eq. (3.1-5) has an exact solution in closed form. It is

$$\tilde{\varphi}(x,s) = \frac{C}{2s\sqrt{s}} e^{-|x|\sqrt{s}} \tag{3.1-6}$$

and the inverse Laplace transform is

$$\varphi(x,\tau) = C\sqrt{\frac{\tau}{\pi}} e^{-x^2/(4\tau)} u(\tau) - \frac{C}{2}|x| \, erfc\left(\frac{|x|}{2\sqrt{\tau}}\right) u(\tau) \tag{3.1-7}$$

Figure 3-1 shows a plot of this function over the range $-10 \le x \le 10$ from time zero to time equal to 250 inverse locking ranges for $C = 1$. Note that as time goes to infinity, the phase diverges as the square root of the time, never reaching a steady state. This may be viewed as a manifestation of the branch cut of Eq. (3.1-6) in the complex s plane. However, differentiating the phase with respect to time gives the simple expression for the frequency,

$$\frac{\omega(x,\tau) - \omega_{ref}}{\Delta\omega_{lock}} = \frac{C}{2}\sqrt{\frac{1}{\pi\tau}} e^{-x^2/(4\tau)} u(\tau) \tag{3.1-8}$$

and thus the frequency converges to the reference frequency at infinite time as one over the square root of the time. This function is plotted in Fig. 3-2 for C equal to unity.

Next, let us consider a finite-length array over the range $-a - \frac{1}{2} \le x \le a + \frac{1}{2}$. For example, if $a = 10$, there will be 21 oscillators in the array and the overall length will be $2a + 1$ or 21 unit cells. Now, in addition to using Eq.(3.1-4), we

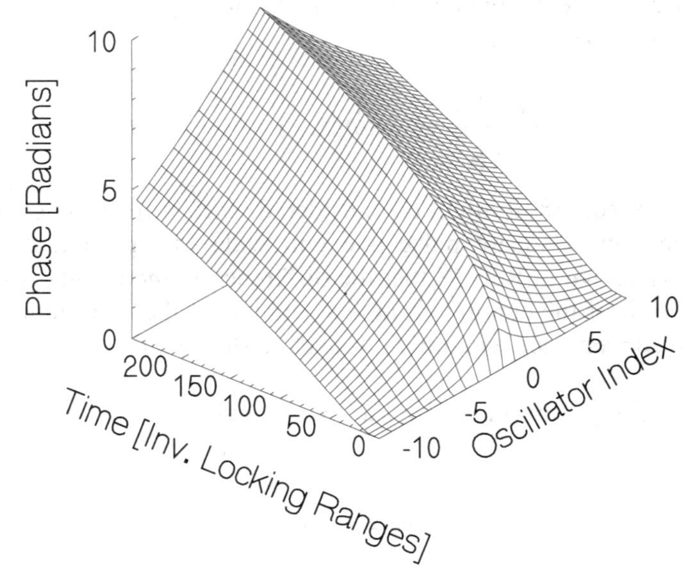

Fig. 3-1. Dynamic phase behavior of an infinite linear array.

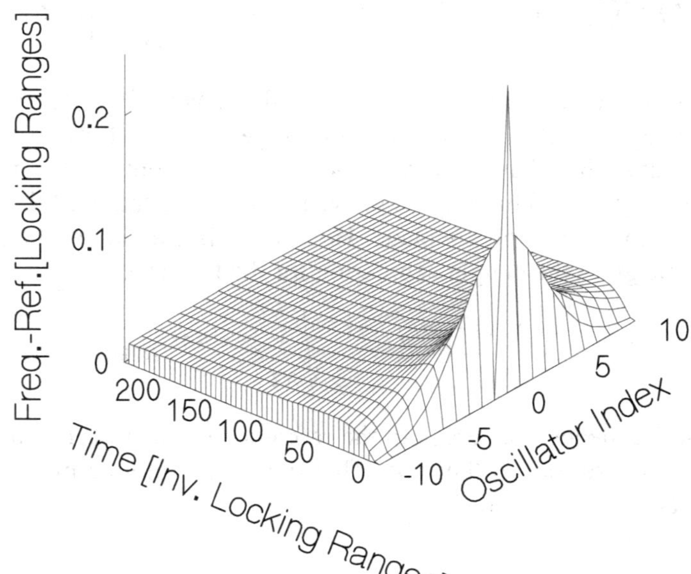

Fig. 3-2. Dynamic frequency behavior of an infinite linear array.

must determine the boundary conditions at the ends of the array in order to obtain the solution. These conditions can be easily obtained via an artifice outlined in Ref. 38. That is, we imagine two additional fictitious oscillators added to the array, one at each end and coupled to the corresponding end oscillator. These oscillators are dynamically tuned so that at all times their phase is maintained equal to the phase of the corresponding end oscillator of the true array. Under these conditions, as may be seen from Eqs. (1.4-1) and (1.4-2), there will be no mutual injection between the end oscillators and the fictitious ones. Thus, the fictitious ones may be removed without effect. However, since the phase of the end oscillator and the corresponding fictitious oscillator are always equal so that the phase difference is zero, and since in the continuum model this difference is represented by the derivative with respect to x, one may conclude that the appropriate boundary condition is that the derivative of the phase with respect to x must be zero—that is, a Neumann boundary condition. At this point, having both the differential equation Eq. (3.1-4) and the boundary conditions, we are in a position to treat the case of a finite length linear array via the continuum model. This will be accomplished using two alternative approaches described below; both of which, of course, yield the same result.

Before proceeding on this course however, we note an interesting result obtainable directly from the differential equation and the boundary conditions. Suppose we integrate Eq. (3.1-3) over the length of the array.

$$\int_{-a-\frac{1}{2}}^{a+\frac{1}{2}} \frac{\partial^2 \varphi(x,\tau)}{\partial x^2} dx - \frac{\partial}{\partial \tau} \int_{-a-\frac{1}{2}}^{a+\frac{1}{2}} \varphi(x,\tau) dx = - \int_{-a-\frac{1}{2}}^{a+\frac{1}{2}} \Delta\Omega_{tune}(x,\tau) dx \qquad (3.1\text{-}9)$$

The first term is zero by virtue of the Neumann boundary conditions at the array ends. Thus, we may write

$$\frac{1}{2a+1} \frac{\partial}{\partial \tau} \int_{-a-\frac{1}{2}}^{a+\frac{1}{2}} \varphi(x,\tau) \, dx = \frac{1}{2a+1} \int_{-a-\frac{1}{2}}^{a+\frac{1}{2}} \Delta\Omega_{tune}(x,\tau) \, dx \qquad (3.1\text{-}10)$$

or

$$\frac{1}{2a+1}\int_{-a-\frac{1}{2}}^{a+\frac{1}{2}}\frac{\partial\varphi(x,\tau)}{\partial t}dx=\frac{1}{2a+1}\int_{-a-\frac{1}{2}}^{a+\frac{1}{2}}\omega_{tune}(x,\tau)dx-\omega_{ref} \qquad (3.1\text{-}11)$$

Now from Eq. (1.3-6), neglecting amplitude variation, we have that the instantaneous frequencies of the oscillators are given by

$$\omega_{inst}=\omega_{ref}+\frac{\partial\varphi}{\partial t} \qquad (3.1\text{-}12)$$

Substituting this into (3.1-11), we obtain

$$\frac{1}{2a+1}\int_{-a-\frac{1}{2}}^{a+\frac{1}{2}}\omega_{inst}dx=\frac{1}{2a+1}\int_{-a-\frac{1}{2}}^{a+\frac{1}{2}}\omega_{tune}(x,\tau)dx \qquad (3.1\text{-}13)$$

That is, the average over the array of the instantaneous oscillator frequencies is equal to the average over the array of the oscillator tuning (or free running) frequencies. In steady state the instantaneous frequency is equal to the ensemble frequency. So, we can conclude that *the steady-state ensemble frequency of the array is the average of the oscillator tuning frequencies.* (Recall the assumption of zero coupling phase.)

We now set ourselves the problem of determining the phase dynamics of a finite linear array when one oscillator in the array is step detuned at time zero. The solution of this problem will be a Green's function permitting solution for an arbitrary distribution of detuning, including the antisymmetrical detuning of the end oscillators for beam-steering as suggested by Liao and York [28]. The first approach will be to construct a solution as a superposition of a particular integral and two homogeneous solutions of the differential equation. The particular integral is known from the solution of the infinite array problem. It is essentially Eq. (3.1-6) generalized to accommodate detuning an arbitrary oscillator at $x = b$ instead of the one at $x = 0$. That is,

$$\tilde{\varphi}_p(x,s)=\frac{C}{2s\sqrt{s}}e^{-|x-b|\sqrt{s}} \qquad (3.1\text{-}14)$$

Adding to this two independent homogeneous solutions with unknown coefficients, C_R and C_L, we postulate the desired solution in the form

$$\tilde{\varphi}(x,s) = \frac{C}{2s\sqrt{s}} e^{-|x-b|\sqrt{s}} + C_R e^{-x\sqrt{s}} + C_L e^{x\sqrt{s}} \qquad (3.1\text{-}15)$$

The two unknown coefficients are now determined by applying the boundary conditions at the two ends of the array, $x = a + \frac{1}{2}$ and $x = a + \frac{1}{2}$, resulting in the two simultaneous linear equations,

$$-C_R \sqrt{s}u + C_L \sqrt{s}\frac{1}{u} = -\frac{C}{2su} e^{-b\sqrt{s}} \qquad (3.1\text{-}16)$$

$$-C_R \sqrt{s}\frac{1}{u} + C_L \sqrt{s}u = \frac{C}{2su} e^{b\sqrt{s}} \qquad (3.1\text{-}17)$$

where

$$u = e^{(a+\frac{1}{2})\sqrt{s}} \qquad (3.1\text{-}18)$$

Solving Eqs. (3.1-16) and (3.1-17) simultaneously for C_R and C_L, we obtain

$$C_R = \frac{-\dfrac{C}{2s\sqrt{s}}\left(\dfrac{1}{u^2} e^{b\sqrt{s}} + e^{-b\sqrt{s}}\right)}{\left(\dfrac{1}{u^2} - u^2\right)} \qquad (3.1\text{-}19)$$

and

$$C_L = \frac{-\dfrac{C}{2s\sqrt{s}}\left(\dfrac{1}{u^2} e^{-b\sqrt{s}} + e^{b\sqrt{s}}\right)}{\left(\dfrac{1}{u^2} - u^2\right)} \qquad (3.1\text{-}20)$$

The solution given by Eq. (3.1-15) is then

$$\tilde{\varphi}(x,s) = \frac{C}{2s\sqrt{s}} e^{|x-b|\sqrt{s}} - \frac{\frac{C}{2s\sqrt{s}}\left(\frac{1}{u^2}e^{b\sqrt{s}} + e^{-b\sqrt{s}}\right)}{\left(\frac{1}{u^2} - u^2\right)} e^{-x\sqrt{s}} - \frac{\frac{C}{2s\sqrt{s}}\left(\frac{1}{u^2}e^{-b\sqrt{s}} + e^{b\sqrt{s}}\right)}{\left(\frac{1}{u^2} - u^2\right)} e^{x\sqrt{s}} \tag{3.1-21}$$

which simplifies to

$$\tilde{\varphi}(x,s) = \frac{C\cosh\left[\left(2a+1-|x-b|\right)\sqrt{s}\right] + C\cosh\left[\left(x+b\right)\sqrt{s}\right]}{s\sqrt{s}\sinh\left[\left(2a+1\right)\sqrt{s}\right]} \tag{3.1-22}$$

Note that, despite the presence of square roots of s, there are no branch cuts in the s plane because this function is even in the square root of s. Thus, the inverse Laplace transform can be computed purely via residue calculus. The poles, s_n, are located by

$$s_n\sqrt{s_n}\sinh\left[\left(2a+1\right)\sqrt{s_n}\right] = 0 \tag{3.1-23}$$

Thus,

$$s_n = -\left(\frac{n\pi}{2a+1}\right)^2 = -\sigma_n \tag{3.1-24}$$

Except for the double pole at $s = 0$, the residues at these poles are

$$residue_n = (-1)^n C \frac{\cosh\left[\left(2a+1-|x-b|\right)\sqrt{s_n}\right] + \cosh\left[\left(x+b\right)\sqrt{s_n}\right]}{s_n(2a+1)} \tag{3.1-25}$$

and the residue at the double pole is,

$$residue_0 = \frac{C}{2a+1} \tag{3.1-26}$$

The inverse Laplace transform is thus

$$\varphi(x,\tau) = \frac{C\tau}{2a+1}u(\tau) + \frac{C}{2a+1}\sum_{n=1}^{\infty}\frac{\cos\left[(x-b)\left(\frac{n\pi}{2a+1}\right)\right]+(-1)^n\cos\left[(x+b)\left(\frac{n\pi}{2a+1}\right)\right]}{\left(\frac{n\pi}{2a+1}\right)^2} \times\left(1-e^{-\left(\frac{n\pi}{2a+1}\right)^2\tau}\right)u(\tau) \tag{3.1-27}$$

This may be rewritten in the form

$$\begin{aligned}\varphi(x,\tau) &= \frac{C\tau}{2a+1}u(\tau) \\ &+ \frac{C}{2a+1}\sum_{m=1}^{\infty}\frac{2\cos\left[b\left(\frac{2m\pi}{2a+1}\right)\right]\cos\left[x\left(\frac{2m\pi}{2a+1}\right)\right]}{\left(\frac{2m\pi}{2a+1}\right)^2}\left(1-e^{-\left(\frac{2m\pi}{2a+1}\right)^2\tau}\right)u(\tau) \\ &+ \frac{C}{2a+1}\sum_{n=0}^{\infty}\frac{2\sin\left[b\left(\frac{(2n+1)\pi}{2a+1}\right)\right]\sin\left[x\left(\frac{(2n+1)\pi}{2a+1}\right)\right]}{\left(\frac{(2n+1)\pi}{2a+1}\right)^2}\left(1-e^{-\left(\frac{(2n+1)\pi}{2a+1}\right)^2\tau}\right)u(\tau)\end{aligned} \tag{3.1-28}$$

The overall time constant of the array dynamics is determined by the smallest eigenvalue. In general, this is given by the $n = 0$ term in Eq. (3.1-28); that is,

$$\sigma_0 = \left(\frac{\pi}{2a+1}\right)^2 \tag{3.1-29}$$

However, if the detuned oscillator happens to be the center one, the residues of the n series are zero and the smallest eigenvalue is the one for $m = 1$; that is,

$$\sigma_0 = \left(\frac{2\pi}{2a+1} \right)^2 \tag{3.1-30}$$

Thus, when the center oscillator is detuned, the array responds four times faster than if any other oscillator is detuned. (There is an error in Ref. 38 where this response is claimed to be only twice as fast.)

Recall now that from Eq. (3.1-13) the ensemble frequency of the array is the average of the tuning frequencies. When one oscillator out of the $2a+1$ oscillator array is detuned by C locking ranges, the ensemble frequency of the array measured in locking ranges will thus change by $C/(2a+1)$ locking ranges. This is manifest in the solution Eq. (3.1-28) as the linear time dependence of slope $C/(2a+1)$ as a function of the scaled time, τ. Aside from this linear term, from Eq. (3.1-28) we see that the steady-state phase distribution across the array is given by

$$\varphi_{ss}(x) = \frac{C}{2a+1} \sum_{n=1}^{\infty} \frac{2\cos\left[b\left(\frac{2n\pi}{2a+1} \right) \right] \cos\left[x\left(\frac{2n\pi}{2a+1} \right) \right]}{\left(\frac{2n\pi}{2a+1} \right)^2} + \frac{C}{2a+1} \sum_{n=0}^{\infty} \frac{2\sin\left[b\left(\frac{(2n+1)\pi}{2a+1} \right) \right] \sin\left[x\left(\frac{(2n+1)\pi}{2a+1} \right) \right]}{\left(\frac{(2n+1)\pi}{2a+1} \right)^2} \tag{3.1-31}$$

a Fourier series which can be summed in closed form to yield the simple expression

$$\varphi_{ss}(x) = \frac{C}{2(2a+1)} \left[x^2 + b^2 - (2a+1)|b-x| + \frac{1}{6}(2a+1)^2 \right] \tag{3.1-32}$$

This may be compared with the result from the discrete model where we approximated the eigenvalues and extended the sums to an infinite number of terms to arrive at the simple approximate result Eq. (2.3-9). Recall that in the linearized discrete model the eigenvalues repeat; so, if the sums are continued to an infinite number of terms, a set of delta functions results. Here, in contrast, the sums are in fact infinite and result in a smooth function passing through the correct value of oscillator phase as x passes through the corresponding index of that oscillator. Thus, the two results, discrete and continuum, are only equal at the oscillators and not in between.

As indicated in Ref. 38, because the inter-oscillator phase difference cannot exceed, $\pi/2$, this steady-state result indicates that the detuning C is limited by

$$C < \frac{\pi(2a+1)}{2\left(a+|b|\right)} \tag{3.1-33}$$

However, when operating near the limits of lock, this is not a very good approximation so it is suggested in Ref. 38 that the sine terms be approximated by defining an effective locking range, $\Delta\tilde{\omega}_{lock}$, as follows.

$$\Delta\omega_{lock}\sin(\Delta\varphi) \approx \Delta\omega_{lock}\frac{\sin(\Delta\varphi)}{\Delta\varphi}\Delta\varphi = \Delta\tilde{\omega}_{lock}\Delta\varphi \tag{3.1-34}$$

For small phase differences the effective locking range will be nearly equal to the true locking range; but near the limits of lock, it will be $2/\pi$ times the true locking range. Thus, as pointed out in Ref. 38, though still approximate, the maximum detuning is more accurately given by

$$\Delta\omega_{\max} \approx \frac{(2a+1)}{\left(a+|b|\right)}\Delta\omega_{lock} \tag{3.1-35}$$

Let us now return to the problem of determining the phase dynamics of a finite linear array when one oscillator in the array is step detuned at time zero and solve it via an alternative approach. We wish to solve Eq. (3.1-5) subject to Neumann boundary conditions at the array ends. Following Pogorzelski et al. [38] in this alternate approach, we first determine the eigenfunctions and eigenvalues defined by

$$\frac{d^2 w_\ell}{dx^2} = \lambda_\ell w_\ell \tag{3.1-36}$$

such that

$$\left.\frac{dw_n}{dx}\right|_{x=a+\frac{1}{2}} = 0 \tag{3.1-37}$$

and

$$\left.\frac{dw_n}{dx}\right|_{x=-a-\frac{1}{2}}=0 \tag{3.1-38}$$

Clearly, the appropriately normalized eigenfunctions are

$$u_m=\frac{\sqrt{2}\cosh\left(x\sqrt{\lambda_m}\right)}{\sqrt{2a+1}} \tag{3.1-39}$$

and

$$v_n=\frac{\sqrt{2}\sinh\left(x\sqrt{\lambda_n}\right)}{i\sqrt{2a+1}} \tag{3.1-40}$$

and the eigenvalues are given by

$$\sinh\left[\sqrt{\lambda_m}\left(a+\frac{1}{2}\right)\right]=0 \tag{3.1-41}$$

and

$$\cosh\left[\sqrt{\lambda_n}\left(a+\frac{1}{2}\right)\right]=0 \tag{3.1-42}$$

Thus the explicit eigenvalues are

$$\lambda_m=-\left(\frac{2m\pi}{2a+1}\right)^2 \tag{3.1-43}$$

and

$$\lambda_n=-\left(\frac{(2n+1)\pi}{2a+1}\right)^2 \tag{3.1-44}$$

We now express the solution of Eq. (3.1-5) as a sum of these eigenfunctions. That is,

$$\tilde{\varphi}=\sum_{m=0}^{\infty}A_m u_m+\sum_{n=0}^{\infty}B_n v_n \tag{3.1-45}$$

Substituting this into Eq. (3.1-5) generalized to an arbitrary detuned oscillator at $x = b$ gives

$$A_m \lambda_m u_m + B_n \lambda_n v_n - sA_m u_m + B_n v_n = -\frac{C}{s}\delta(x-b) \tag{3.1-46}$$

Now using the orthogonality of the eigenfunctions over the length of the array, we obtain

$$A_m = -\frac{Cu_m(b)}{s(\lambda_m - s)} \tag{3.1-47}$$

and

$$B_n = -\frac{Cv_n(b)}{s(\lambda_n - s)} \tag{3.1-48}$$

The solution is then immediately written as

$$\tilde{\varphi} = -C\sum_{m=0}^{\infty}\frac{u_m(b)u_m(x)}{s(\lambda_m - s)} + \sum_{n=0}^{\infty}\frac{v_n(b)v_n(x)}{s(\lambda_n - s)} \tag{3.1-49}$$

or, inserting the explicit expressions for the eigenfunctions, we obtain

$$\begin{aligned}\tilde{\phi} = \frac{C}{(2a+1)s^2} - \frac{C}{s}\sum_{m=1}^{\infty}\frac{2\cosh\left(b\sqrt{\lambda_m}\right)\cosh\left(x\sqrt{\lambda_m}\right)}{(2a+1)(\lambda_m - s)} \\ -\frac{C}{s}\sum_{n=0}^{\infty}\frac{2\sinh\left(b\sqrt{\lambda_m}\right)\sinh\left(x\sqrt{\lambda_m}\right)}{(2a+1)(\lambda_n - s)}\end{aligned} \tag{3.1-50}$$

Except for the zero eigenvalue term, $m = 0$, each term of these series has one simple pole at s equal to the corresponding eigenvalue. Thus, the inverse Laplace transform follows immediately as the sum of the residues at the pole in each term of the series,

$$\varphi(x,\tau) = \frac{C\tau}{2a+1}u(\tau)$$
$$+\frac{C}{2a+1}\sum_{m=1}^{\infty}\frac{2\cos\left[b\left(\frac{2m\pi}{2a+1}\right)\right]\cos\left[x\left(\frac{2m\pi}{2a+1}\right)\right]}{\left(\frac{2m\pi}{2a+1}\right)^2}\left(1-e^{-\left(\frac{2m\pi}{2a+1}\right)^2\tau}\right)u(\tau)$$
$$+\frac{C}{2a+1}\sum_{n=0}^{\infty}\frac{2\sin\left[b\left(\frac{(2n+1)\pi}{2a+1}\right)\right]\sin\left[x\left(\frac{(2n+1)\pi}{2a+1}\right)\right]}{\left(\frac{(2n+1)\pi}{2a+1}\right)^2}\left(1-e^{-\left(\frac{(2n+1)\pi}{2a+1}\right)^2\tau}\right)u(\tau) \tag{3.1-51}$$

which is, of course, identical to Eq. (3.1-28). For the case where $a = 10$ and $b = 5$, this solution is plotted as a function of time in Fig. 3-3. Note that the shape of the distribution at late times is very much like the corresponding steady-state solution shown in Fig. 2-3. Being the solution for a delta function source on the right side of the differential equation, this is the Green's function for the problem and as such it can be used to obtain solutions for arbitrary detuning distributions.

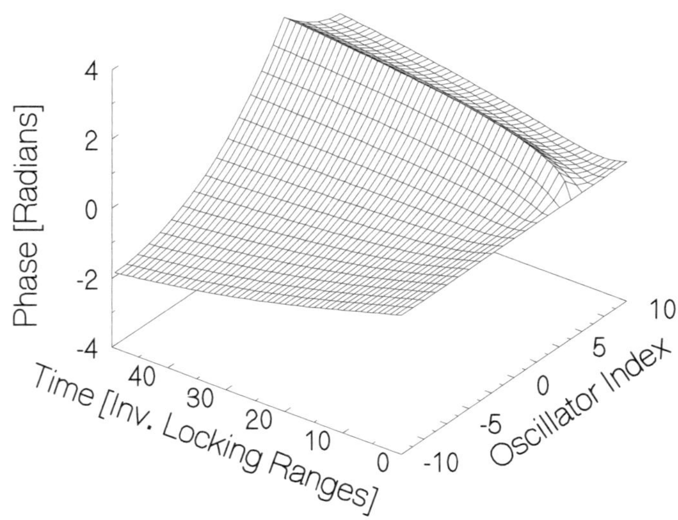

Fig. 3-3. Linear-array phase distribution under step detuning of the oscillator at *x* = 5.

To summarize, we have described two methods of solving the continuum-model partial-differential equation for the dynamic behavior of the phase across a linear array of mutually injection locked oscillators. Both methods entailed Laplace transformation with respect to the scaled time. The first method was a direct solution of the resulting second-order ordinary differential equation by postulating a solution as a superposition of a particular integral and two homogeneous solutions with unknown amplitude coefficients. The coefficients were determined by the Neumann boundary conditions at the array ends. The inverse Laplace transform was obtained as a sum of the residues of at the poles of the transform. In the second method, the Laplace transformed equation was solved by postulating a solution as a sum of eigenfunctions of the second order differential operator each satisfying the Neumann boundary conditions at the array ends. Recognizing this to be a self-adjoint boundary value problem of Sturm–Liouville type, it should not be surprising that the solution for the desired Green's function can be written as a sum of these eigenfunctions. Conveniently, each term of the sum, except the one corresponding to the zero eigenvalue, has one simple pole so that the inverse Laplace transform is immediately obtainable as a sum of the corresponding residues, one for each term of the eigenfunction series.

3.2 The Linear Array with External Injection

Thus far, the continuum model has been applied to arrays in which the phase control is accomplished by detuning one of the oscillators. The beam-steering method proposed by Stephan [1] requires that two or more array oscillators be injected with an externally derived signal. Thus, to accommodate this, it is necessary to generalize the continuum model along the lines followed in Section 2.5. Following Pogorzelski, et al. [39], we begin with Eq. (2.5-2) rewritten in terms of the continuous variable, x, and the scaled time, τ, as

$$\begin{aligned}\frac{d\varphi(x,\tau)}{d\tau} &= \frac{\omega_{0i}-\omega_{ref}}{\Delta\omega_{lock}} + \left(\varphi(x+\Delta x,\tau) - 2\varphi(x,\tau) + \varphi(x-\Delta x,\tau)\right) \\ &\quad -\delta(x-p)\frac{\Delta\omega_{lock,p,inj}}{\Delta\omega_{lock}}\left(\varphi(x,\tau) - \varphi_{inj}(\tau)\right)\end{aligned} \tag{3.2-1}$$

Now we define

$$V(x) = \delta(x-p)\frac{\Delta\omega_{lock,p,inj}}{\Delta\omega_{lock}} \tag{3.2-2}$$

and expand in a Taylor series about x keeping terms up to second order in Δx so that Eq. (3.2-1) becomes

$$\frac{\partial^2 \varphi}{\partial x^2} - V(x)\varphi - \frac{d\varphi}{d\tau} = -\Delta\Omega_{tune} - V(x)\varphi_{inj}(\tau) \tag{3.2-3}$$

Here the spatial distribution of the external injection signals is given by *V(x)* while the temporal dependence is given by $\varphi_{inj}(\tau)$, so we have implicitly assumed that these dependences are separable; that is, all of the injection signal phases have the same time dependence. While this is a convenient simplification, it is not essential in that one could include more than one such injection term in the equation and obtain a solution albeit somewhat more complicated than the one presented here. Equation (3.2-3) is the generalization of Eq. (3.1-3) required to accommodate external injection for our purposes and we will use it to study the phase dynamics of such an externally injected array.

Suppose we consider an infinitely long linear array wherein all of the oscillators are tuned to the ensemble or reference frequency and the oscillator at $x = b$ is externally injection locked to an oscillator of strength *C* with C_0 radian step time dependence of its phase. Our generalized differential equation then becomes

$$\frac{\partial^2 \varphi}{\partial x^2} - C\delta(x-b)\varphi - \frac{d\varphi}{d\tau} = -\Delta\Omega_{tune} - CC_0\delta(x-b)u(\tau) \tag{3.2-4}$$

where

$$C = \frac{\Delta\omega_{lock,p,inj}}{\Delta\omega_{lock}} \tag{3.2-5}$$

Laplace transformation with respect to the scaled time results in

$$\frac{\partial^2 \tilde{\varphi}}{\partial x^2} - C\delta(x-b)\tilde{\varphi} - s\tilde{\varphi} = -\frac{C_0}{s}C\delta(x-b) \tag{3.2-6}$$

We now define

$$\tilde{\varphi}_1 = \tilde{\varphi} - \frac{C_0}{s} \tag{3.2-7}$$

so that Eq. (3.2-6) becomes

$$\frac{\partial^2 \tilde{\varphi}_1}{\partial x^2} - C\delta(x-b)\tilde{\varphi}_1 - s\tilde{\varphi}_1 = C_0 \tag{3.2-8}$$

The particular integral of this equation is

$$\tilde{\varphi}_{1p} = -\frac{C_0}{s} \tag{3.2-9}$$

We postulate a homogeneous solution of the form

$$\tilde{\varphi}_{1h} = C_1 e^{-\sqrt{s}|x-b|} \tag{3.2-10}$$

so that our proposed solution is

$$\tilde{\varphi}_1 = -\frac{C_0}{s} + C_1 e^{-\sqrt{s}|x-b|} \tag{3.2-11}$$

Now integrating Eq. (3.2-8) across the delta function at $x = b$, we find that

$$\left.\frac{d\tilde{\varphi}_1}{dx}\right|_{x=b^-}^{x=b^+} = C\tilde{\varphi}_1(b) \tag{3.2-12}$$

Imposing this condition on the solution given by Eq. (3.2-11), we obtain

$$C_1 = \frac{C_0 C}{s\left(2\sqrt{s}+C\right)} \tag{3.2-13}$$

Substituting this into the solution given by Eq. (3.2-11) gives

$$\tilde{\varphi}_1 = \frac{C_0}{s}\left[\frac{C}{2\sqrt{s}+C}e^{-\sqrt{s}|x-b|} - 1\right] \tag{3.2-14}$$

and from Eq. (3.2-7) we obtain

$$\tilde{\varphi}(x,s) = \frac{C_0 C}{s\left(2\sqrt{s}+C\right)}e^{-\sqrt{s}|x-b|} \tag{3.2-15}$$

Finally, the inverse Laplace transform of Eq. (3.2-15) is,

$$\begin{aligned}\varphi(x,\tau) = C_0\Bigg[& erfc\left(\frac{|x-b|}{2\sqrt{\tau}}\right) \\ & - e^{C|x-b|/2}e^{C^2\tau/4}erfc\left(C\frac{\sqrt{\tau}}{2}+\frac{|x-b|}{2\sqrt{\tau}}\right)\Bigg]u(\tau)\end{aligned} \tag{3.2-16}$$

(See Ref. 37, equation (29.3.89).) This is the phase distribution across the infinite array as a function of time. It is zero at time zero and smoothly evolves to a final value of C_0 at infinite time as shown in Fig. 3-4 for $C_0 = 1$ radian and $C = 1$. Note that the injection frequency as well as the initial and final ensemble frequencies are all the same. Because it is the solution for injection at a single point in the array, you might think that it is a Green's function that can be used to construct solutions for arrays injected at multiple points. However, as we shall see in Section 3.4 when we discuss Stephan's beam-steering scheme [1] involving two injection points, this is not the case because the form of differential equation itself differs from Eq. (3.2-6) when there are multiple injection points.

The corresponding problem where the injected *frequency* is step shifted by C_0 locking ranges at time zero was treated by Pogorzelski et al. [39]. In that case the array oscillator frequencies evolve from the ensemble frequency at time zero to the injection frequency at infinite time.

Next, we consider an array of finite length, $2a + 1$, in which all of the oscillators are tuned to the same frequency, taken to be the reference frequency and one of the oscillators, the one at $x = b$, is injected with an externally generated signal of strength C defined by Eq. (3.2-5) that is step phase shifted at time zero by C_0 radians. Equation (3.2-6) applies, but this time we wish to solve it subject to Neumann boundary conditions at the array ends. Here again we have a choice of two methods of solution. Let us begin by postulating the solution in the form of a particular integral plus two complementary functions that are solutions of the homogeneous equation. That is, using Eqs. (3.2-7), (3.2-9), and (3.2-10), we have

$$\tilde{\varphi}_1 = C_b e^{-\sqrt{s}|x-b|} + C_R e^{-x\sqrt{s}} + C_L e^{x\sqrt{s}} - \frac{C_0}{s} \qquad (3.2\text{-}17)$$

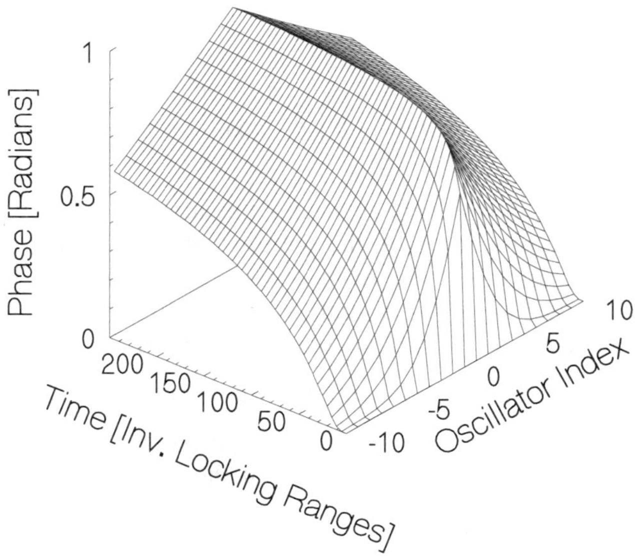

Fig. 3-4. Phase distribution versus time for an infinite linear array with one oscillator externally injected.

with the three conditions,

$$\left.\frac{d\tilde{\varphi}_1}{dx}\right|_{x=b^-}^{x=b^+} = C\tilde{\varphi}_1(b) \tag{3.2-18}$$

$$\left.\frac{d\tilde{\varphi}_1}{dx}\right|_{x=a} = 0 \tag{3.2-19}$$

$$\left.\frac{d\tilde{\varphi}_1}{dx}\right|_{x=-a} = 0 \tag{3.2-20}$$

Now, Eqs. (3.2-18), (3.2-19), and (3.2-20) can be used to determine the three constants, C_b, C_R, and C_L. Then, using Eq. (3.2-7), we get

$$\tilde{\varphi}(x,s) = \frac{C_0}{2sD(s)}\left\{ C\cosh\left[\left(2a+1-|x-b|\right)\sqrt{s}\right] + C\cosh\left[\left(x+b\right)\sqrt{s}\right]\right\} \tag{3.2-21}$$

where

$$D(s) = \sqrt{s}\sinh\left[(2a+1)\sqrt{s}\right] + C\cosh\left[\left(a+\frac{1}{2}+b\right)\sqrt{s}\right]\cosh\left[\left(a+\frac{1}{2}-b\right)\sqrt{s}\right] \tag{3.2-22}$$

Here again there are no branch cuts, and the inverse Laplace transform is expressible as a sum of residues at the poles—that is, the zeros of $D(s)$, all of which lie on the negative real axis of the s plane. Note that Eq. (3.2-22) is very reminiscent of Eq. (2.5-9) of the discrete model of this array. Comparing these two equations, we may ascertain that the continuum approximation is particularly accurate for small values of s when $\sqrt{s} \approx \sinh\left(\sqrt{s}\right)$, which, of course, corresponds to late time. In fact, the pole closest to the origin of the s plane provides us with the time constant of the array which determines the late time behavior. Let us examine Eq. (3.2-22) to see if we can estimate the location of this pole.

In anticipation of the fact that the pole lies on the negative real axis, we define ξ so that

$$\sqrt{s} = \sqrt{-\sigma} = i\sqrt{\sigma} = i\xi \tag{3.2-23}$$

Then,

$$D = -\xi\sin\left[(2a+1)\xi\right] + C\cos\left[\left(a+\frac{1}{2}+b\right)\xi\right]\cos\left[\left(a+\frac{1}{2}-b\right)\xi\right] \tag{3.2-24}$$

Setting D equal to zero, yields the transcendental equation

$$2\xi\sin\left[\left(a+\frac{1}{2}\right)\xi\right]\cos\left[\left(a+\frac{1}{2}\right)\xi\right] = C\cos\left[\left(a+\frac{1}{2}+b\right)\xi\right]\cos\left[\left(a+\frac{1}{2}-b\right)\xi\right] \tag{3.2-25}$$

For small ξ, the solution occurs where the cosine functions are near zero and the sine function is near unity. Thus, we define a new variable,

$$\eta = \xi - \frac{\pi}{2a+1} \tag{3.2-26}$$

and write Eq. (3.2-25) in the form

$$
\begin{aligned}
&2\left(\eta-\frac{\pi}{2a+1}\right)\sin\left[\left(a+\frac{1}{2}\right)\eta+\frac{\pi}{2}\right]\cos\left[\left(a+\frac{1}{2}\right)\eta+\frac{\pi}{2}\right]= \\
&\quad C\cos\left[\left(a+\frac{1}{2}+b\right)\eta+\frac{\pi}{2}+\frac{b\pi}{2a+1}\right] \\
&\quad \times\cos\left[\left(a+\frac{1}{2}-b\right)\eta+\frac{\pi}{2}-\frac{b\pi}{2a+1}\right]
\end{aligned}
\tag{3.2-27}
$$

or

$$
\begin{aligned}
&2\left(\eta-\frac{\pi}{2a+1}\right)\cos\left[\left(a+\frac{1}{2}\right)\eta\right]\sin\left[\left(a+\frac{1}{2}\right)\eta\right]= \\
&\quad -C\sin\left[\left(a+\frac{1}{2}+b\right)\eta+\frac{b\pi}{2a+1}\right] \\
&\quad \times\sin\left[\left(a+\frac{1}{2}-b\right)\eta-\frac{b\pi}{2a+1}\right]
\end{aligned}
\tag{3.2-28}
$$

Using the identity for the sine of a sum, we arrive at

$$
\begin{aligned}
&2\left(\eta-\frac{\pi}{2a+1}\right)\cos\left[\left(a+\frac{1}{2}\right)\eta\right]\sin\left[\left(a+\frac{1}{2}\right)\eta\right]= \\
&\quad -C\left\{\sin\left[\left(a+\frac{1}{2}+b\right)\eta\right]\cos\left(\frac{b\pi}{2a+1}\right)\right. \\
&\qquad \left.+\cos\left[\left(a+\frac{1}{2}+b\right)\eta\right]\sin\left(\frac{b\pi}{2a+1}\right)\right\} \\
&\quad \times\left\{\sin\left[\left(a+\frac{1}{2}-b\right)\eta\right]\cos\left(\frac{b\pi}{2a+1}\right)\right. \\
&\qquad \left.-\cos\left[\left(a+\frac{1}{2}-b\right)\eta\right]\sin\left(\frac{b\pi}{2a+1}\right)\right\}
\end{aligned}
\tag{3.2-29}
$$

Near $\eta = 0$, we have

$$\begin{aligned}\pi\eta =\ & C\left\{\left[\left(a+\frac{1}{2}+b\right)\eta\right]\cos\left(\frac{b\pi}{2a+1}\right)\right.\\ & \left.+\sin\left(\frac{b\pi}{2a+1}\right)\right\}\\ & \times\left\{\left[\left(a+\frac{1}{2}-b\right)\eta\right]\cos\left(\frac{b\pi}{2a+1}\right)\right.\\ & \left.-\sin\left(\frac{b\pi}{2a+1}\right)\right\}\end{aligned} \tag{3.2-30}$$

which is a quadratic equation for η. That is,

$$\begin{aligned}&\left[\left(a+\frac{1}{2}\right)^2-b^2\right]\eta^2\\ &-\left[\frac{\pi}{C}\sec^2\left(\frac{b\pi}{2a+1}\right)+2b\tan\left(\frac{b\pi}{2a+1}\right)\right]\eta-\tan^2\left(\frac{b\pi}{2a+1}\right)=0\end{aligned} \tag{3.2-31}$$

We can now look at two limiting cases. First, if C is small, the solution becomes that of the uninjected array, namely, $\eta = 0$. If, on the other hand, C is large, we obtain

$$\eta=\frac{\left[b\pm\left(a+\frac{1}{2}\right)\right]}{\left[\left(a+\frac{1}{2}\right)^2-b^2\right]}\tan\left(\frac{b\pi}{2a+1}\right) \tag{3.2-32}$$

and

$$\xi=\frac{\left[b\pm\left(a+\frac{1}{2}\right)\right]}{\left[\left(a+\frac{1}{2}\right)^2-b^2\right]}\tan\left(\frac{b\pi}{2a+1}\right)+\frac{\pi}{2a+1} \tag{3.2-33}$$

If b is small; that is, if the injection point is near the center of the array,

$$\xi \approx \frac{\left[b \pm \left(a+\frac{1}{2}\right)\right]}{\left[\left(a+\frac{1}{2}\right)^2 - b^2\right]}\left(\frac{b\pi}{2a+1}\right) + \frac{\pi}{2a+1} \tag{3.2-34}$$

Choosing the sign in the numerator to obtain the solution nearest the origin of the s plane, we have

$$\xi \approx \frac{\pi}{2a+1+2|b|} \tag{3.2-35}$$

Thus,

$$s_{\min} \approx -\left(\frac{\pi}{2(a+|b|+1)}\right)^2 \tag{3.2-36}$$

and the late time behavior of the array goes as

$$e^{-\left(\frac{\pi}{2(a+|b|+1)}\right)^2 \tau} \tag{3.2-37}$$

The formula given by Eq. (3.2-33) fails if b is at either end of the array because we have effectively divided by zero in the derivation. We can no longer assume that C is infinite. Retaining a finite value of C and rewriting the transcendental equation results in

$$\xi \tan\left[(2a+1)\xi\right] = C \tag{3.2-38}$$

If C is small, the solution is approximately

$$\xi \approx \sqrt{\frac{C}{2a+1}} \tag{3.2-39}$$

but if C is large, we obtain

$$\xi \approx \frac{\frac{\pi}{2}}{(2a+1)+\frac{1}{C}} \tag{3.2-40}$$

Interestingly, for large a, Eq. (3.2-40) is consistent with Eq. (3.2-35) if b is at the either end of the array so, for large C and large a, these formulas agree.

Returning now to Eq. (3.2-21), the poles are easily found by iterative bisection because they are all on the negative real axis. The residues are easily computed once the poles are known and the residue series gives the inverse Laplace transform. As an example, this inverse transform is plotted in Fig. 3-5 for the case where $a = 10$, $b = 5$, $C_0 = 1$, and $C = 10$. The time constant of this array is 96.12 inverse locking ranges, whereas the approximate formula Eq. (3.2-36) gives 103.75 inverse locking ranges. Note that for the injected oscillator $x = 5$, the response is much faster than that of the entire array. This is because for this oscillator, the residues of the poles close to the origin of the s plane are small and the more distant poles hold sway.

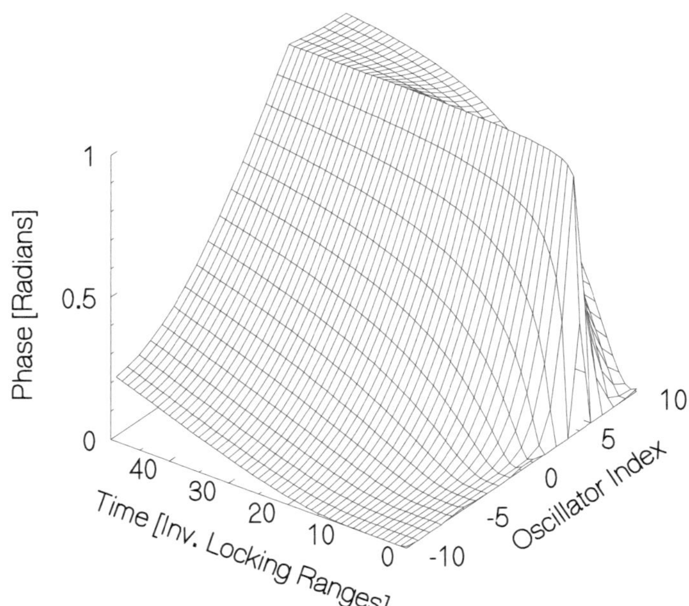

Fig. 3-5. Oscillator phases for oscillator 5 externally injected.

As discussed in connection with the detuned linear array, the above analysis can also be performed by expanding the solution in eigenfunctions of the differential operator. The relevant operator in this case is

$$\frac{\partial^2}{\partial x^2} - C\delta(x-b) \tag{3.2-41}$$

The presence of the delta function produces a slope discontinuity in the eigenfunctions which must satisfy

$$\frac{\partial^2 w_n}{\partial x^2} - C\delta(x-b)w_n = \lambda_n w_n \tag{3.2-42}$$

and the boundary conditions

$$\left.\frac{dw_n}{dx}\right|_{x=a} = 0 \tag{3.2-43}$$

$$\left.\frac{dw_n}{dx}\right|_{x=-a} = 0 \tag{3.2-44}$$

The solution is postulated in the form

$$w_n = C_b e^{-\sqrt{\lambda_n}|x-b|} + C_R e^{-x\sqrt{\lambda_n}} + C_L e^{x\sqrt{\lambda_n}} \tag{3.2-45}$$

Now we note something interesting about Eq. (3.2-42); that is, it is essentially Eq. (3.2-6) with C_0 set to zero and s set to λ_n. Therefore, we can obtain the eigenfunctions by means of a limiting process applied to Eq. (3.2-21) instead of solving for the three constants using Eqs. (3.2-42), (3.2-43), and (3.2-44). Suppose we set

$$C_0 = \alpha \tag{3.2-46}$$

and

$$s = \lambda_n + \alpha \tag{3.2-47}$$

in Eq. (3.2-21) and take the limit as α approaches zero where λ_n is the nth value of s for which $D(s)$ equals zero. In this limit both the numerator and denominator of Eq. (3.2-21) approach zero, but the ratio is finite and approaches w_n. That is,

$$w_n(x) = \left. \frac{\frac{\partial}{\partial\alpha}\left\{ \alpha\, C\cosh\left[\left(2a+1-|x-b|\right)\sqrt{\lambda_n+\alpha}\right] + \alpha\, C\cosh\left[\left(x+b\right)\sqrt{\lambda_n+\alpha}\right]\right\}}{2\frac{\partial}{\partial\alpha}\left[(\lambda_n+\alpha)D(\lambda_n+\alpha)\right]} \right|_{\alpha=0} \quad (3.2\text{-}48)$$

But, except for a factor of C_0, this is nothing but the residue of Eq. (3.2-21) at the nth pole. Not only have we found the eigenfunctions, but they are already multiplied by the coefficients needed to form the solution by summation except for an overall multiplicative constant of C_0. In effect, in Eq. (3.2-48) we are computing

$$w_n(x) = \frac{Cf_n(b)}{\lambda_n < f_n, f_n >} f_n(x) \quad (3.2\text{-}49)$$

where the bracketed expression in the denominator is the normalization integral; that is, the integral of the square of the arbitrarily normalized eigenfunction, f_n, over the array length and

$$f_n(x) = C\cosh\left[\left(2a+1-|x-b|\right)\sqrt{\lambda_n}\right] + C\cosh\left[\left(x+b\right)\sqrt{\lambda_n}\right] \quad (3.2\text{-}50)$$

The desired solution is therefore

$$\tilde{\varphi}(x,s) = C_0\sum_n w_n(x) = C_0 C\sum_n \frac{f_n(b)f_n(x)}{\lambda_n < f_n, f_n >} \quad (3.2\text{-}51)$$

the well-known form of the solution as a sum of eigenfunctions.

Thus, we see that the inverse Laplace transform of the eigenfunction sum representing the solution, $\varphi(x,\tau)$, is just the sum of the residues of Eq. (3.2-21) multiplied by the Laplace transform kernel, $e^{s\tau}$. This same property was evident in the treatment of the linear array with one oscillator detuned. It is the reason why Eqs. (3.1-28) and (3.1-51) are identical. Thus, in the present case, we can rest assured that, had we pursued the eigenfunction expansion approach to completion, the result would have been exactly that plotted in Fig. 3-5. The two approaches—the residue series based on the eigenfunction sum and the residue series based on the particular integral and complementary function—are not just equivalent, they are in fact identical.

3.3 Beam-Steering via End Detuning

The beam-steering concept suggested by Liao and York [28] involves antisymmetric detuning of the end oscillators of the linear array. The phase dynamics produced in this situation can be analyzed by means of the continuum model presented in Section 3.1. Beginning with Eq. (3.1-51), we may superpose two such solutions, one with b equal to minus a and the other with b equal to plus a and with C's of opposite sign. Let, $C=\Delta\omega_T/\Delta\omega_{lock}$ and

$$\omega_{tune}(x)=\omega_{ref}+\Delta\omega_T\delta(x-a)-\Delta\omega_T\delta(x+a) \tag{3.3-1}$$

Then we obtain

$$\varphi(x,\tau)=\frac{\Delta\omega_T}{\Delta\omega_{lock}}\sum_{n=0}^{\infty}\frac{2\sin\left[b\left(\frac{(2n+1)\pi}{2a+1}\right)\right]\sin\left[x\left(\frac{(2n+1)\pi}{2a+1}\right)\right]}{(2a+1)\left(\frac{(2n+1)\pi}{2a+1}\right)^2}\left(1-e^{-\left(\frac{(2n+1)\pi}{2a+1}\right)^2\tau}\right)u(\tau) \tag{3.3-2}$$

The steady-state phase distribution is then given by

$$\varphi(x,\infty)=\frac{\Delta\omega_T}{\Delta\omega_{lock}}\sum_{n=0}^{\infty}\frac{2\sin\left[b\left(\frac{(2n+1)\pi}{2a+1}\right)\right]\sin\left[x\left(\frac{(2n+1)\pi}{2a+1}\right)\right]}{(2a+1)\left(\frac{(2n+1)\pi}{2a+1}\right)^2} \tag{3.3-3}$$

which can be summed in closed form to yield

$$\varphi(x,\infty)=\frac{\Delta\omega_T}{\Delta\omega_{lock}}x \tag{3.3-4}$$

a linear phase distribution as indicated in Ref. 28.

The function given by Eq. (3.3-2) is plotted in Fig. 3-6 for end oscillators of a 21-oscillator array step detuned at time zero by one-half locking range. Figure 3-7 shows the corresponding far-zone radiated field if the oscillator outputs are used to excite the elements of a half-wavelength-spaced array of isotropically radiating elements. It shows that the beam is steered from normal to the array initially, to 9.16 deg from normal corresponding to the steady-state inter-element phase difference of a half radian or 28.65 deg given by Eq. (3.3-4) when $\Delta\omega_T = \frac{1}{2}\Delta\omega_{lock}$. The linearization of the sine functions in the full nonlinear theory introduces some error, but the qualitative behavior is well represented. In fact, the actual steady-state inter-element phase difference is 30 deg, resulting in beam-steering to 9.59 deg rather than the 9.16 deg given by the linearized theory.

These plots depict the dynamic behavior for an interval just a little longer than one array time constant.

We have shown that the beam-steering scheme suggested by Liao and York [28] is indeed treatable using the continuum model of coupled oscillators and that the phase transient ensuing from antisymmetric step detuning of the end oscillators produces a smoothly scanning beam in the far zone. The maximum

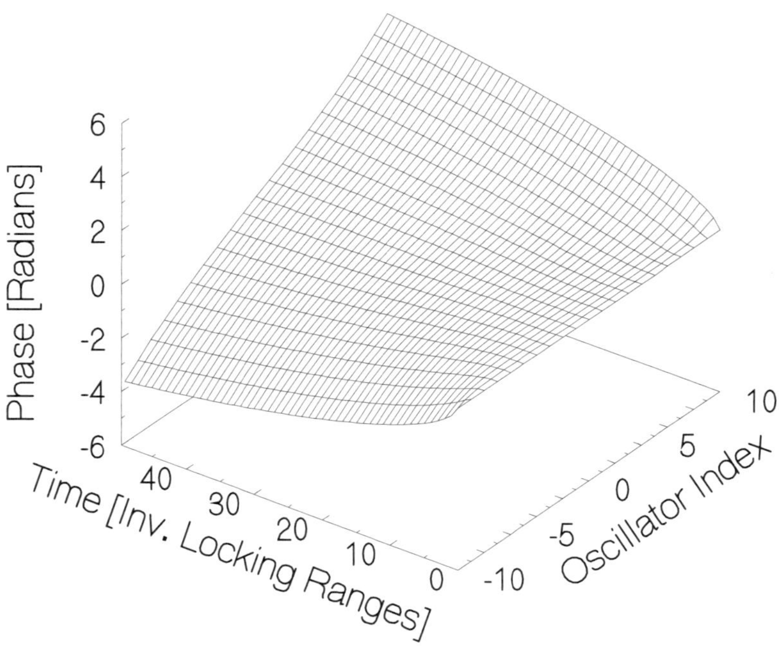

Fig. 3-6. Oscillator phases for a 21-oscillator linear array with end elements antisymmetrically detuned by half the locking range.

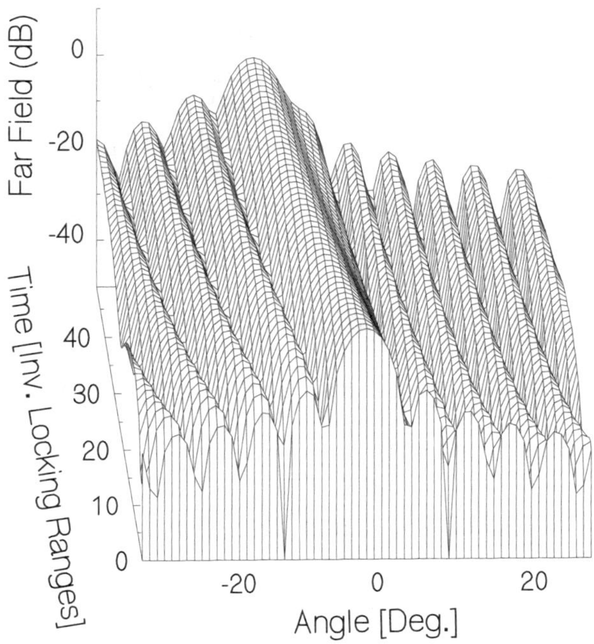

Fig. 3-7. Far-zone radiated field of a 21-element half-wavelength-spaced phased array excited by the oscillators of Fig. 3-6.

scan angle is limited by the maximum permissible inter-oscillator phase difference. However, this can be mitigated by frequency multiplication of the oscillator outputs, which similarly multiplies the phase excursion [40].

3.4 Beam-Steering via End Injection

The beam-steering scheme proposed by Stephan [1] requires that each of the end oscillators be externally injected. The phase distribution across the array is then controlled by adjusting the relative phase of these injection signals by means of a phase shifter which thus controls the beam direction. The dynamic behavior in this situation can be analyzed using the continuum model, but the analysis presented in Section 3.2 for a single injection point cannot be directly applied. If, for example, we represent the solution as a sum of eigenfunctions, the eigenfunctions for two injection points differ from those for one. Similarly, if we approach the analysis using a particular integral and complementary function, both of these will differ from those for one injection point. Thus, it will be necessary to reformulate the problem for two injection points from the beginning.

To be definite, we assume that the oscillators of the array are all initially tuned to the reference frequency and are thus in-phase with each other and that two arbitrary oscillators in the array at $x = b_1$ and $x = b_2$ are injection locked to external signals which are initially in-phase with the oscillators of the array and that at time zero the phase of each of these signals is stepped to a finite constant value. The strengths of the two injection signals are denoted by B_1 and B_2, and the amplitude of the corresponding temporal step functions are denoted by p_1 and p_2, respectively. Then, Eq. (3.2-3) becomes

$$\begin{gathered}\frac{\partial^2\varphi}{\partial x^2}-\left[B_1\delta(x-b_1)+B_2\delta(x-b_2)\right]\varphi-\frac{d\varphi}{d\tau}\\ =-B_1\delta(x-b_1)p_1u(\tau)-B_2\delta(x-b_2)p_2u(\tau)\end{gathered} \tag{3.4-1}$$

Laplace transformation results in

$$\begin{gathered}\frac{\partial^2\tilde{\varphi}}{\partial x^2}-\left[B_1\delta(x-b_1)+B_2\delta(x-b_2)\right]\varphi-s\tilde{\varphi}\\ =-B_1\delta(x-b_1)\frac{p_1}{s}-B_2\delta(x-b_2)\frac{p_2}{s}\end{gathered} \tag{3.4-2}$$

Now, as shown previously, we may solve this equation either by means of an eigenfunction expansion or by means of superposition of a particular integral and a complementary function. In the former approach, the complexity arises in the normalization of the eigenfunctions, which involves integration of the square of the eigenfunctions of the array. In the latter, this is automatically taken care of by the residues. Thus, we elect to proceed with the latter approach as was done in Ref. 39.

The solution of (3.4-2) is postulated in the form

$$\tilde{\varphi}=C_1e^{-\sqrt{s}|x-b_1|}+C_2e^{-\sqrt{s}|x-b_2|}+C_Re^{-x\sqrt{s}}+C_Le^{x\sqrt{s}} \tag{3.4-3}$$

The four unknown constants are determined by the boundary conditions at the array ends, Eqs. (3.2-19) and (3.2-20), and the conditions on the derivatives at the injection points, Eq. (3.2-18). These four constraints yield four equations for the four unknowns in Eq. (3.4-3). The solution is

$$
\begin{aligned}
\tilde{\varphi}(x,s) &= \frac{1}{sD_2(s)} \\
&\times\left\{ B_2 p_2 \cosh\left[\frac{\sqrt{s}}{2}\left((2a+1)+(b_2+x)-|b_2-x|\right)\right]\right. \\
&\quad\times\cosh\left[\frac{\sqrt{s}}{2}\left((2a+1)-(b_2+x)-|b_2-x|\right)\right] \\
&\quad+B_1 p_1 \cosh\left[\frac{\sqrt{s}}{2}\left((2a+1)+(b_1+x)-|b_1-x|\right)\right] \\
&\quad\times\cosh\left[\frac{\sqrt{s}}{2}\left((2a+1)-(b_1+x)-|b_1-x|\right)\right] \\
&\quad+\frac{B_1 B_2 p_2}{\sqrt{s}}\cosh\left[\frac{\sqrt{s}}{2}\left(2a+1+2b_1\right)\right] \\
&\quad\times\cosh\left[\frac{\sqrt{s}}{2}\left(2a+1-(b_2+b_1)-|b_2-x|-|b_1-x|\right)\right] \\
&\quad\times\sinh\left[\frac{\sqrt{s}}{2}\left((b_2-b_1)-|b_2-x|+|b_1-x|\right)\right] \\
&\quad+\frac{B_1 B_2 p_1}{\sqrt{s}}\cosh\left[\frac{\sqrt{s}}{2}\left(2a+1-2b_2\right)\right] \\
&\quad\times\cosh\left[\frac{\sqrt{s}}{2}\left(2a+1+(b_2+b_1)-|b_2-x|-|b_1-x|\right)\right] \\
&\quad\left.\times\sinh\left[\frac{\sqrt{s}}{2}\left((b_2-b_1)+|b_2-x|-|b_1-x|\right)\right]\right\}
\end{aligned}
\tag{3.4-4}
$$

where

$$\begin{aligned}D_2(s) &= \sqrt{s}\sinh\left[(2a+1)\sqrt{s}\right] \\ &+ B_1\cosh\left[\sqrt{s}\left(a+\frac{1}{2}+b_1\right)\right]\cosh\left[\sqrt{s}\left(a+\frac{1}{2}-b_1\right)\right] \\ &+ B_2\cosh\left[\sqrt{s}\left(a+\frac{1}{2}+b_2\right)\right]\cosh\left[\sqrt{s}\left(a+\frac{1}{2}-b_2\right)\right] \\ &+ \frac{B_1B_2}{\sqrt{s}}\sinh\left[\sqrt{s}\left(b_2-b_1\right)\right] \\ &\times\cosh\left[\sqrt{s}\left(a+\frac{1}{2}+b_1\right)\right]\cosh\left[\sqrt{s}\left(a+\frac{1}{2}-b_2\right)\right]\end{aligned} \tag{3.4-5}$$

Note that if either of the B's is zero, we recover Eqs. (3.2-21) and (3.2-22) for a single injection point. The form of the solution presented in Ref. 39 is slightly different but fully equivalent except for a typographical error in the $\sinh\left[\sqrt{s}\left(2b_1-|b_2-x|\right)\right]$ term, which should have been $\sinh\left[\sqrt{s}\left(2b_1+|b_2-x|\right)\right]$. The pole locations on the negative real axis of the s plane are easily found by iterative bisection, and the inverse Laplace transform is then obtainable as a residue series.

As a first example, we compute the solutions when unit strength injection signals are applied to the end oscillators of a 21-oscillator linear array, and at time zero their phase is step shifted antisymmetrically by one radian producing a phase difference of two radians. The dynamic behavior of the resulting phase distribution is shown in Fig. 3-8.

An analytic expression for the steady-state solution for the phase can be obtained by application of the final value theorem to the transform (3.4-4) and (3.4-5). The result is

$$\begin{aligned}\varphi(x,\infty) &= \frac{1}{2B_1+2B_2+2B_1B_2\left(b_2-b_1\right)}\{2B_1p_1+2B_2p_2 \\ &+B_1B_2\left[\left(p_2+p_1\right)\left(b_2-b_1\right)-\left(p_2-p_1\right)\left(|b_2-x|-|b_1-x|\right)\right]\}\end{aligned} \tag{3.4-6}$$

For the case shown in Fig. 3-8, this expression reduces to

$$\varphi(x,\infty) = \frac{x}{11} \tag{3.4-7}$$

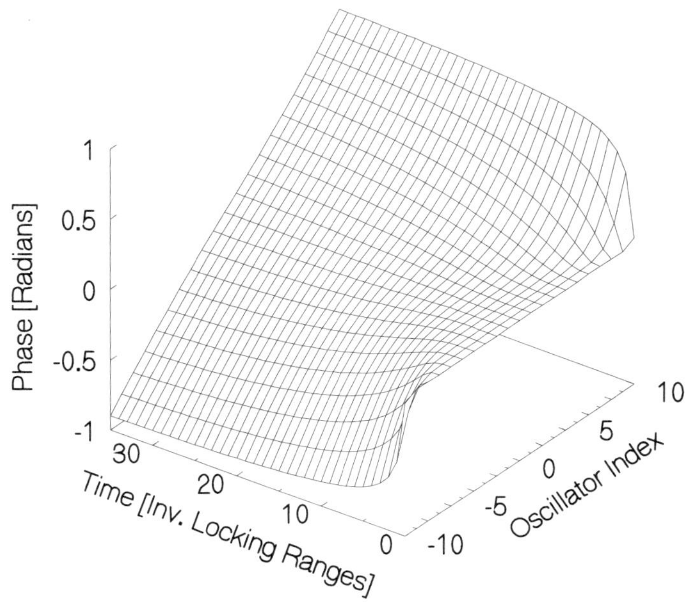

Fig. 3-8. Phase dynamics for injected end elements.

Notice that the steady-state phases of the injected oscillators at $x = 10$ and at $x = -10$ are not equal to the phases of the corresponding injection signals, plus and minus one radian. This is because the end oscillators are also injected by virtue of their coupling to their nearest neighbor in the array, and the phase of that neighbor differs from the phase of the external injection signal. Thus, the total injection of the end oscillator is not in phase with the external injection signal. However, as the strength of the injection signals is increased (large values of the B's are used), the steady-state phase of the end oscillators will approach the phase of the corresponding injection signals because the signal from the corresponding neighboring oscillators becomes negligible.

We again remark, as in Section 2.5, that the injection signals may be derived from the end oscillators of the array and used to inject the next to end oscillators to achieve beam-steering. The continuum model has been used to study this approach also [41].

Recall that the phase of the injection signals can differ from the initial phase of the injected oscillators by no more than $\pi/2$ radians for a maximum total phase difference of π radians across the array. Thus, for strong injection, the beam-steering angle is limited to a maximum of

$$\theta_{\max} = \sin^{-1}\left(\frac{\pi}{(2a+1)S}\right) \tag{3.4-8}$$

where S is the electrical radiating element spacing in radians. In our present example, if the element spacing is a half-wavelength so $S = \pi$, then the maximum steering angle is 2.73 deg, a disappointingly small angle. Fortunately, this problem is easily eliminated by gradually increasing the injection phase instead of stepping it [1]. That way, the phase difference between the injected oscillator and the injection signal can be maintained less than $\pi/2$ radians while the phase difference between the two injection signals is increased to a large value. The new limit on steering angle is now imposed by the requirement that the inter-oscillator phase difference be less than $\pi/2$ radians to maintain overall lock, a limitation also present in the detuning case. In the present example, this limits the steering angle to 30 degrees, a certainly more acceptable limit.

As an example of this enhanced beam-steering scheme, we compute the response of the array of the previous example, but this time we gradually increase the injection-signal phase difference by convolving the step function with a temporal Gaussian. By virtue of the linearity of the p dependence of the equation, we may obtain the corresponding phase response by convolving the step response with the same Gaussian. Since the solution is a residue series, each term has simple exponential time dependence, so the convolution can be carried out analytically term by term as described in detail in Ref. 39.

Let the Gaussian be

$$g(\tau) = e^{-(\tau-6)^2/100} \tag{3.4-9}$$

Then, setting p_2 equal to 2π radians and p_1 equal to -2π radians for a total phase difference of 4π radians, the expected steady-state beam-steering angle of a half-wavelength-spaced array will be 10.48 deg. The steady-state inter-oscillator phase difference is 0.628 radians, for which the sine functions are approximated by their argument with about 7 percent accuracy. However, there are times during the transient at which this difference becomes as large as 0.878 radians near the array ends. At these times, the sine functions are approximated with only 14 percent accuracy. Thus, the actual inter-oscillator phase difference will be somewhat larger. The phase behavior for these parameters and unit amplitude injection as predicted by the continuum model is shown in Fig. 3-9, and the corresponding far zone beam is shown in Fig. 3-10.

We have shown the utility of the continuum model in analyzing the transient behavior of linear arrays of mutually injection locked oscillators with external injection. Beam-steering of linear phased arrays of radiating elements can be achieved by externally injecting the end oscillators of the array and varying the relative phase if the injection signals as suggested by Stephan [1]. In order to achieve significant beam-steering angles via this approach, it is necessary to apply the phase shift to the injection signals gradually so as to avoid excessive inter-oscillator phase differences resulting in loss of lock. Here, as in the detuning approach, the steering angle range may be extended via frequency multiplication.

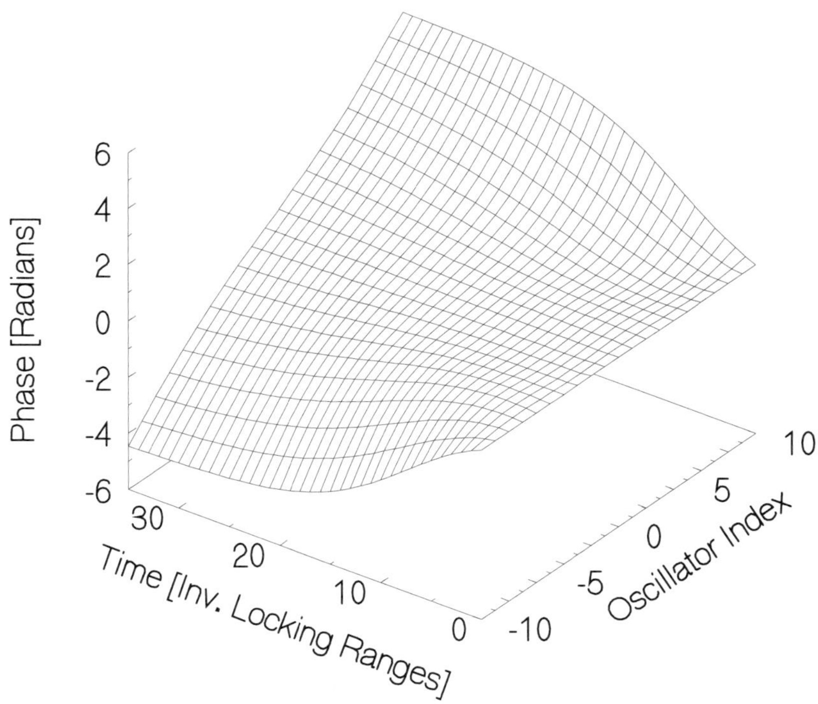

Fig. 3-9. Phase dynamics for gradually changing injection phase.

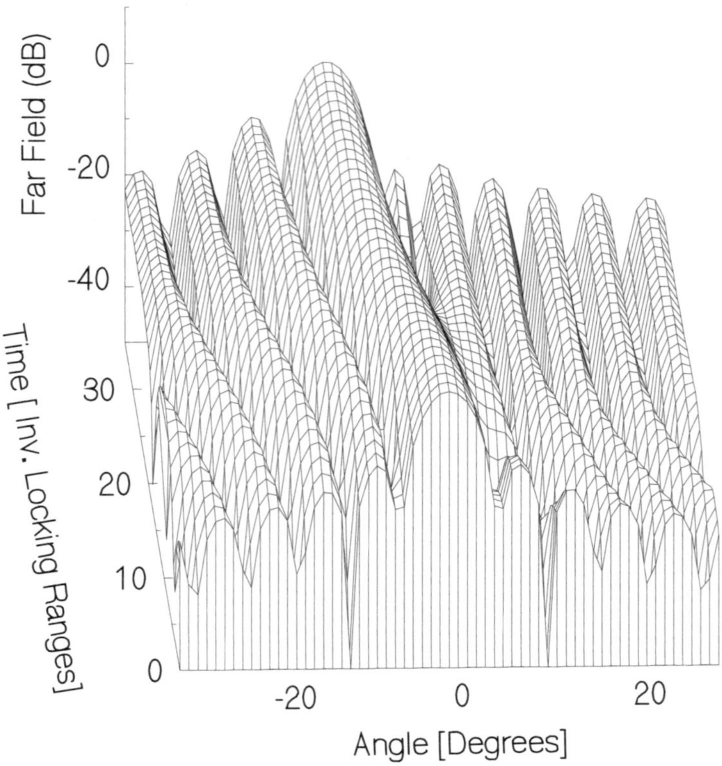

Fig. 3-10. Far-zone radiated field of a 21-element half-wavelength-spaced phased array excited by the oscillators of Fig. 3-9.

3.5 Conclusion

In this chapter, the continuum model was shown to provide considerable physical insight into the general behavior of one-dimensional coupled oscillator arrays. It highlights the fact that the phase behavior is governed by the diffusion equation; and as a consequence, the transient response time is proportional to the square of the array length. In the next chapter we extend the continuum model to planar arrays. This broadens the nearest-neighbor coupling concept to a wider range of topologies. That is, in the planar case we can envision not only the Cartesian scheme discussed in Chapter 2, in which each oscillator is coupled to its four nearest neighbors, but also hexagonal and triangular schemes in which each oscillator is coupled to three or six nearest neighbors, respectively. By means of the continuum model, we will see that these coupling topologies produce similar phase behavior but result in differing response times for the arrays.

Chapter 4
The Continuum Model for Planar Arrays

As described in Section 2.6, coupled oscillator arrays can be constructed in a planar geometry in which each oscillator is coupled to more than the two nearest neighbors of the linear array case. In that section a Cartesian coupling topology is described in which each oscillator is coupled to four nearest neighbors, and the array boundary is rectangular. In such an arrangement, the phase distributions suitable for beam-steering are obtainable either by detuning the edge oscillators [42] or by injecting them with external signals with adjustable phase [43]. Both of these approaches are treatable via the continuum model. Further generalizing the planar arrangement, one may use alternative coupling topologies such as the triangular lattice in which each oscillator is coupled to six nearest neighbors and the array boundary is triangular or the hexagonal lattice in which each oscillator is coupled to three nearest neighbors and the array boundary is again triangular [44, 45]. As will be shown in this chapter, these coupling topologies are also treatable using the continuum model.

4.1 Cartesian Coupling in the Continuum Model without External Injection

We begin with Eq. (2.6-3) for a $2M + 1$ by $2N + 1$ rectangular array with zero coupling phase replacing the discrete indices i and j with the continuous variables x and y, respectively; and we expand the phase function in a two-dimensional Taylor series retaining terms to second order. By this process, we obtain the two-dimensional analog of Eq. (3.1-3); that is,

$$\frac{d\varphi}{d\tau} = \left(\frac{\omega_0(x,y,\tau) - \omega_{ref}}{\Delta\omega_{lock}} \right) + \frac{\partial^2 \varphi}{\partial x^2} + \frac{\partial^2 \varphi}{\partial y^2} \tag{4.1-1}$$

or

$$\frac{\partial^2 \varphi}{\partial x^2} + \frac{\partial^2 \varphi}{\partial y^2} - \frac{d\varphi}{d\tau} = -\Delta\Omega_{tune} \tag{4.1-2}$$

subject to Neumann boundary conditions at the array edges. (These boundary conditions may be ascertained via the fictitious additional oscillator artifice described in Section 3.1.) Averaging Eq. (4.1-2) over the two-dimensional array and using the boundary conditions as in Eqs. (3.1-9) to (3.1-13), it can be shown that the ensemble frequency of the array is the average of the tuning (free running) frequencies of the oscillators.

Laplace transformation of Eq. (4.1-2) with respect to the scaled time, τ, results in

$$\frac{\partial^2 \tilde{\varphi}}{\partial x^2} + \frac{\partial^2 \tilde{\varphi}}{\partial y^2} - s\tilde{\varphi} = -\Delta\tilde{\Omega}_{tune} \tag{4.1-3}$$

where the tilde denotes the transformed function. As in the one-dimensional case, this equation can be solved by postulating a solution as a sum of eigenfunctions of the two-dimensional differential operator, the Laplacian operator, and solving for the coefficients of this expansion. As indicated in [42], the eigenfunctions are

$$\begin{aligned}
f_{ee,mn} &= \frac{1}{N_{ee,mn}} \cosh\left(\sqrt{s_m}\, x\right) \cosh\left(\sqrt{s_n}\, y\right) \\
f_{oo,k\ell} &= \frac{1}{N_{oo,k\ell}} \sinh\left(\sqrt{s_k}\, x\right) \sinh\left(\sqrt{s_\ell}\, y\right) \\
f_{eo,m\ell} &= \frac{1}{N_{ee,m\ell}} \cosh\left(\sqrt{s_m}\, x\right) \sinh\left(\sqrt{s_\ell}\, y\right) \\
f_{oe,kn} &= \frac{1}{N_{oo,kn}} \sinh\left(\sqrt{s_k}\, x\right) \cosh\left(\sqrt{s_n}\, y\right)
\end{aligned} \tag{4.1-4}$$

where the eigenvalues are

$$s_k = -\left(\frac{(2k+1)\pi}{2a+1}\right)^2 \qquad s_m = -\left(\frac{2m\pi}{2a+1}\right)^2$$
$$s_\ell = -\left(\frac{(2\ell+1)\pi}{2b+1}\right)^2 \qquad s_n = -\left(\frac{2n\pi}{2a+1}\right)^2 \tag{4.1-5}$$

and the normalization constants are

$$\begin{aligned}
N_{ee,mn} &= \frac{1}{2}\sqrt{(2a+1)(2b+1)\eta_m\eta_n} \\
N_{oo,k\ell} &= \frac{1}{2}\sqrt{(2a+1)(2b+1)} \\
N_{ee,m\ell} &= \frac{1}{2}\sqrt{(2a+1)(2b+1)\eta_m} \\
N_{oo,kn} &= \frac{1}{2}\sqrt{(2a+1)(2b+1)\eta_n}
\end{aligned} \tag{4.1-6}$$

where

$$\eta_m = \begin{matrix} 2; & m=0 \\ 1; & m\neq 0 \end{matrix} \tag{4.1-7}$$

The general solution procedure follows that used in the case of the linear array. That is, we postulate a two-dimensional delta function source to obtain the Green's function as an expansion in the two-dimensional eigenfunctions. Then, we integrate the product of this Green's function and the actual source function over the array to obtain the phase distribution as an expansion in the eigenfunctions. This solution is presented in Ref. 42.

The Green's function, $\tilde{g}(x, y, x', y', s)$, is a solution of

$$\frac{\partial^2 \tilde{g}}{\partial x^2} + \frac{\partial^2 \tilde{g}}{\partial y^2} - s\tilde{\varphi} = -\delta(x-x')\delta(y-y') \tag{4.1-8}$$

The solution of this equation expressed as a sum of eigenfunctions is

$$\begin{aligned}\tilde{g} &= \sum_{n=0}^{\infty}\sum_{m=0}^{\infty}\frac{f_{ee,mn}(x',y')f_{ee,mn}(x,y)}{s_m+s_n-s}\\ &+\sum_{\ell=0}^{\infty}\sum_{k=0}^{\infty}\frac{f_{oo,k\ell}(x',y')f_{oo,k\ell}(x,y)}{s_k+s_\ell-s}\\ &+\sum_{\ell=0}^{\infty}\sum_{m=0}^{\infty}\frac{f_{eo,m\ell}(x',y')f_{eo,m\ell}(x,y)}{s_m+s_\ell-s}\\ &+\sum_{n=0}^{\infty}\sum_{k=0}^{\infty}\frac{f_{oe,kn}(x',y')f_{ee,mn}(x,y)}{s_k+s_n-s}\end{aligned} \tag{4.1-9}$$

This Green's function can then be multiplied by the detuning function, $\Delta\tilde{\Omega}_{tune}(x',y',s)$, and integrated over the array. The inverse Laplace transform is easily obtained as the sum of the residues at the poles as in the one-dimensional case. Recall that the detuning of the oscillators required to produce a desired phase distribution across the array can be determined by merely substituting the desired phase distribution into Eq. (4.1-2). From the theory of uniformly spaced phased-array antennas, the steady-state phase distribution necessary to produce a beam steered to spherical coordinate angles, θ_0, φ_0, with the polar axis normal to the plane of the array, is

$$\varphi_{ss}(x,y)=\frac{\Omega_x}{2}\left(|x+a|-|x-a|\right)+\frac{\Omega_y}{2}\left(|y+b|-|y-b|\right) \tag{4.1-10}$$

where

$$\begin{aligned}\Omega_x &= -2\pi\frac{h}{\lambda}\sin\theta_0\cos\varphi_0\\ \Omega_y &= -2\pi\frac{h}{\lambda}\sin\theta_0\sin\varphi_0\end{aligned} \tag{4.1-11}$$

and h is the element spacing while λ is the wavelength. Substituting into Eq. (4.1-2) with the time derivative set to zero to obtain the steady-state result, we find that

$$\Delta\Omega_{tune}=-\Omega_x\left[\delta(x+a)-\delta(x-a)\right]-\Omega_y\left[\delta(y+b)-\delta(y-b)\right] \tag{4.1-12}$$

Thus, we discover that beam-steering requires detuning of only the edge oscillators and that the needed detuning is constant along each edge. This leads us to seek dynamic solutions of Eq. (4.1-2) that result from a temporal step detuning of the edge oscillators that is constant along each edge. That is, we limit ourselves to detuning functions of the general form,

$$\begin{aligned}\Delta\Omega_{tune} &= \Omega_{x1}\delta(x+a)u(\tau)+\Omega_{x2}\delta(x-a)u(\tau)\\ &+\Omega_{y1}\delta(y+b)u(\tau)+\Omega_{y2}\delta(y-b)u(\tau)\end{aligned} \tag{4.1-13}$$

For step temporal dependence, the Laplace transform of the detuning is,

$$\begin{aligned}\Delta\tilde{\Omega}_{tune} &= \frac{\Omega_{x1}}{s}\delta(x+a)+\frac{\Omega_{x2}}{s}\delta(x-a)\\ &+\frac{\Omega_{y1}}{s}\delta(y+b)+\frac{\Omega_{y2}}{s}\delta(y-b)\end{aligned} \tag{4.1-14}$$

The presence of the delta functions facilitates integration of the product of the Green's function and the tuning function, leading to the solution

$$\begin{aligned}\varphi(x,y,\tau) &= \left(\frac{\Omega_{x1}+\Omega_{x2}}{2a+1}+\frac{\Omega_{y1}+\Omega_{y2}}{2b+1}\right)\tau\, u(\tau)\\
&+\frac{1}{2a+1}\sum_{p=1}^{\infty}\frac{\left(\Omega_{x1}+(-1)^p\Omega_{x2}\right)\cos\left(\frac{p\pi}{2a+1}(x+a)\right)}{\left(\frac{p\pi}{2a+1}\right)^2}\left(1-e^{-\left(\frac{p\pi}{2a+1}\right)^2\tau}\right)u(\tau)\\
&+\frac{1}{2a+1}\sum_{p=1}^{\infty}\frac{\left((-1)^p\Omega_{x1}+\Omega_{x2}\right)\cos\left(\frac{p\pi}{2a+1}(x-a)\right)}{\left(\frac{p\pi}{2a+1}\right)^2}\left(1-e^{-\left(\frac{p\pi}{2a+1}\right)^2\tau}\right)u(\tau)\\
&+\frac{1}{2b+1}\sum_{p=1}^{\infty}\frac{\left(\Omega_{y1}+(-1)^p\Omega_{y2}\right)\cos\left(\frac{p\pi}{2b+1}(x+b)\right)}{\left(\frac{p\pi}{2b+1}\right)^2}\left(1-e^{-\left(\frac{p\pi}{2b+1}\right)^2\tau}\right)u(\tau)\\
&+\frac{1}{2b+1}\sum_{p=1}^{\infty}\frac{\left((-1)^p\Omega_{y1}+\Omega_{y2}\right)\cos\left(\frac{p\pi}{2b+1}(x-b)\right)}{\left(\frac{p\pi}{2b+1}\right)^2}\left(1-e^{-\left(\frac{p\pi}{2b+1}\right)^2\tau}\right)u(\tau)\end{aligned} \tag{4.1-15}$$

which is the solution for the special case of constant detuning along each edge of the array. In steady state, this reduces to Fourier series that can be summed in closed form, resulting in

$$
\begin{aligned}
\varphi(x,y,\tau) &= \left(\frac{\Omega_{x1}+\Omega_{x2}}{2a+1} + \frac{\Omega_{y1}+\Omega_{y2}}{2b+1} \right) \tau \\
&+ \left[\frac{\Omega_{x1}+\Omega_{x2}}{2} \right] \left[\frac{2a+1}{6} + \frac{a^2+x^2}{2a+1} - \left(|x+a| + |x-a| \right) \right] \\
&- \left[\frac{\Omega_{x1}-\Omega_{x2}}{2} \right] \left[\left(|x+a| - |x-a| \right) \right] \\
&+ \left[\frac{\Omega_{y1}+\Omega_{y2}}{2} \right] \left[\frac{2b+1}{6} + \frac{b^2+y^2}{2b+1} - \left(|y+b| + |y-b| \right) \right] \\
&- \left[\frac{\Omega_{y1}-\Omega_{y2}}{2} \right] \left[\left(|y+b| - |y-b| \right) \right]
\end{aligned}
\qquad (4.1\text{-}16)
$$

which clearly shows that symmetric detuning gives rise to parabolic steady-state phase distributions whereas antisymmetric detuning results in linear steady-state phase distributions. In the antisymmetric case where $\Omega_{x1} = -\Omega_{x2} = -\Omega_x$ and $\Omega_{y1} = -\Omega_{y2} = -\Omega_y$, Eq. (4.1-16) reduces to Eq. (4.1-10).

The phase distribution as function of time for beam-steering to 10 deg of polar angle at –110 deg of azimuth as given by Eq. (4.1-15) is shown in Fig. 4-1. Note that during the transient, the phase distribution is not planar, but in steady state at infinite time it becomes planar. Figure 4-2 shows the beam peak and 3-dB contour of the far-zone radiated field of a half-wavelength-spaced array of isotropic radiating elements excited by this two-dimensional oscillator array during the transient period at intervals of 10 inverse locking ranges. Because the phase during the transient is not planar, the directivity of the antenna is decreased. Assuming no losses, the gain is equal to this directivity. The gain is plotted as a function of time in Fig. 4-3 and compared with the ideal gain were the phase planar. The gain reduction observed in steady state relative to the initial gain is characteristic of phased-array antennas and is commonly referred to as "projected aperture loss" due to scanning. This term derives from the fact that for large arrays this loss is quite accurately approximated by the cosine of the beam-steering angle from normal as if the effective aperture of the array is reduced by projection in the direction of the beam peak.

Finally, Fig. 4-4 shows the behavior of the far-zone beam as a sequence of step beam-steering detunings is applied.

4.2 Cartesian Coupling in the Continuum Model with External Injection

If beam-steering of a planar array is to be accomplished through external injection as proposed by Stephan [1], the continuum model is based on the two dimensional generalization of equation Eq. (3.2-3); that is,

$$\frac{\partial^2\varphi}{\partial x^2}+\frac{\partial^2\varphi}{\partial y^2}-V(x,y)\varphi-\frac{d\varphi}{d\tau}= -\Delta\Omega_{tune}(x,y,\tau)-V(x,y)\varphi_{inj}(x,y,\tau) \qquad (4.2\text{-}1)$$

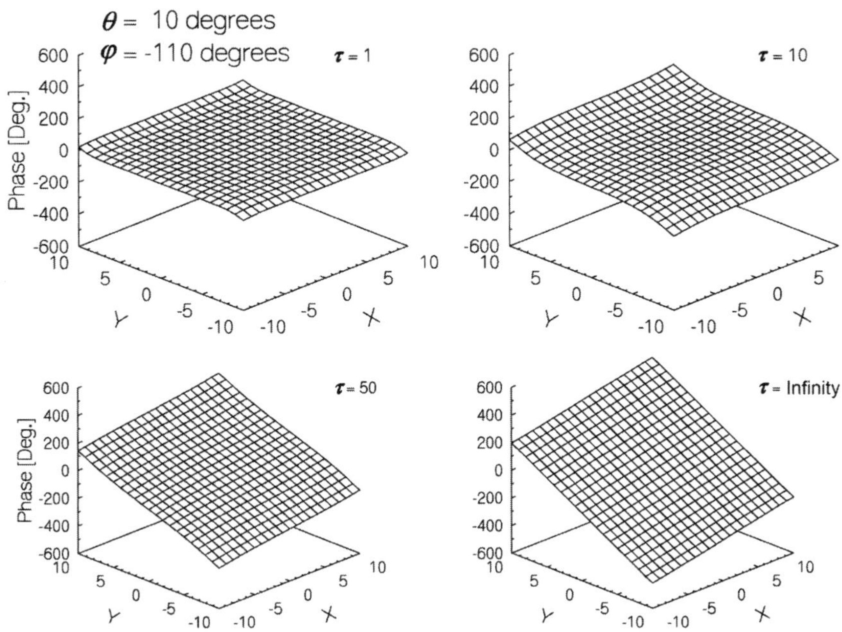

Fig. 4-1. Aperture phase distributions versus time in a two-dimensional array (edge oscillators detuned for beam-steering). (Reprinted from Ref. 42 with permission, ©2001 IEEE.)

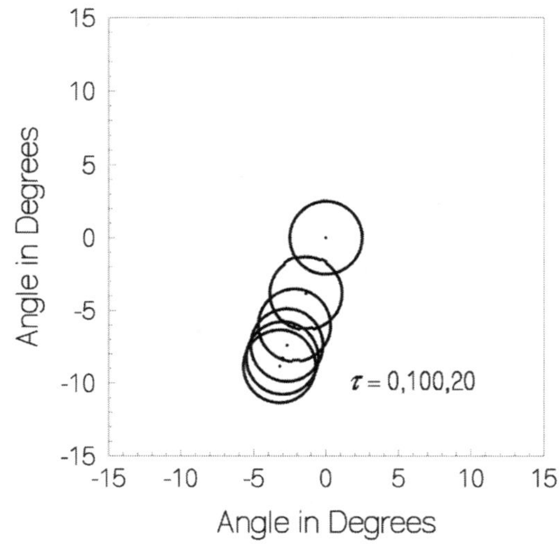

Fig. 4-2. Beam trajectory detuning during the beam-steering transient. (Reprinted from Ref. 42 with permission, ©2001 IEEE.)

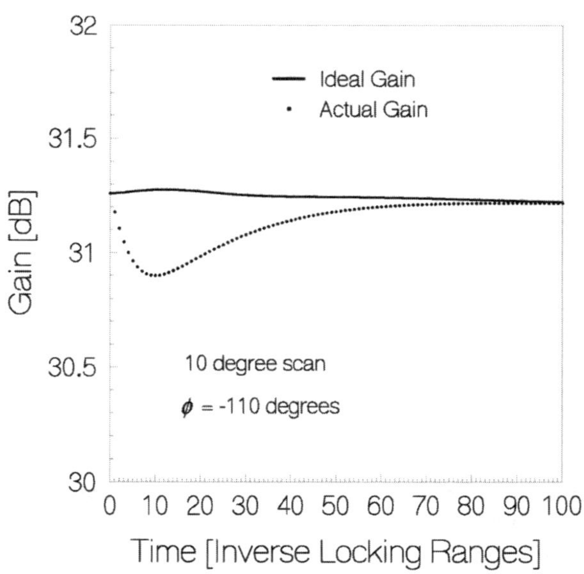

Fig. 4-3. Peak gain dynamics during the beam-steering transient. (Reprinted from Ref. 42 with permission, ©2001 IEEE.)

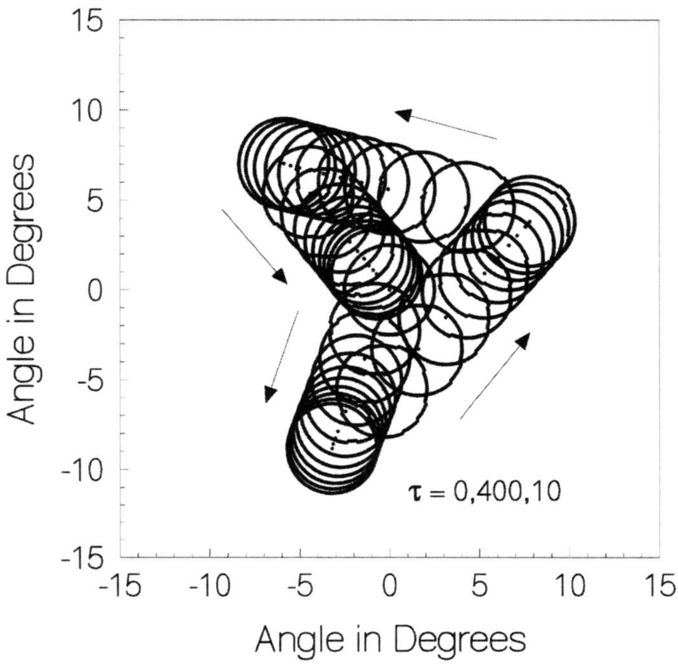

Fig. 4-4. Sequential bream-steering. (Reprinted from Ref. 42 with permission, ©2001 IEEE.)

Now, because none of the oscillators are to be detuned, $\Delta\Omega_{tune} = 0$. Following Stephan [1], the perimeter oscillators are to be injection locked to external signals that are phase-shifted with respect to one another to steer the beam. Thus, we choose the form

$$V(x,y) = CP(x) + CQ(y) \tag{4.2-2}$$

where

$$C = \frac{\Delta\omega_{inj}}{\Delta\omega_{lock}} \tag{4.2-3}$$

and

$$\begin{aligned} CP(x) &= C_{x1}\delta(x - x_1') + C_{x2}\delta(x - x_2') \\ CQ(y) &= C_{x1}\delta(y - y_1') + C_{y2}\delta(y - y_2') \end{aligned} \tag{4.2-4}$$

so that Eq. (4.2-1) becomes

$$\frac{\partial^2 \varphi}{\partial x^2}+\frac{\partial^2 \varphi}{\partial y^2}-CP(x)\varphi-CQ(y)\varphi-\frac{d\varphi}{d\tau}= \\ -CP(x)\varphi_{inj,x}(x,y,\tau)-CQ(y)\varphi_{inj,y}(x,y,\tau) \tag{4.2-5}$$

where for notational convenience we have separated the injection phases associated with the *P(x)* and *Q(x)* distributions of injection signals into the two functions, $\varphi_{inj,x}$ and $\varphi_{inj,y}$. Laplace transformation gives

$$\frac{\partial^2 \tilde{\varphi}}{\partial x^2}+\frac{\partial^2 \tilde{\varphi}}{\partial y^2}-CP(x)\tilde{\varphi}-CQ(y)\tilde{\varphi}-s\tilde{\varphi}= \\ -CP(x)\tilde{\varphi}_{inj,x}(x,y,s)-CQ(y)\tilde{\varphi}_{inj,y}(x,y,s) \tag{4.2-6}$$

Following Ref. 43, we now determine the eigenfunctions of the differential operator,

$$\frac{\partial^2}{\partial x^2}+\frac{\partial^2}{\partial y^2}-CP(x)-CQ(y) \tag{4.2-7}$$

subject to the Neumann boundary conditions at the array edges. Let the eigenfunctions be products of an x dependence and a y dependence; that is, $X(x,s_x)Y(y,s_y)$ so that by separation of variables we have,

$$X''-CPX-s_xX=0 \\ Y''-CQY-s_yY=0 \tag{4.2-8}$$

where the double primes indicate the second spatial derivative. Using Eq. (4.2-4), we obtain

$$X''-\Omega_{x1}\delta(x-x_1')X-\Omega_{x2}\delta(x-x_2')X-s_xX=0 \\ Y''-\Omega_{y1}\delta(y-y_1')Y-\Omega_{y2}\delta(y-y_2')Y-s_yY=0 \tag{4.2-9}$$

First, consider the x-dependent part. As in Ref. 43 the x region is divided into three parts, and a solution is postulated in each of these ranges of x. That is,

$$X = A_1 \cosh\left[\sqrt{s_x}\left(a+\frac{1}{2}+x\right)\right]\cosh\left[\sqrt{s_x}\left(a+\frac{1}{2}-x_1'\right)\right]$$
$$+ A_2 \cosh\left[\sqrt{s_x}\left(a+\frac{1}{2}+x\right)\right]\cosh\left[\sqrt{s_x}\left(a+\frac{1}{2}-x_2'\right)\right] \quad (4.2\text{-}10)$$

for $-a-\frac{1}{2}\leq x\leq x_1'$

$$X = A_1 \cosh\left[\sqrt{s_x}\left(a+\frac{1}{2}-x\right)\right]\cosh\left[\sqrt{s_x}\left(a+\frac{1}{2}+x_1'\right)\right]$$
$$+ A_2 \cosh\left[\sqrt{s_x}\left(a+\frac{1}{2}+x\right)\right]\cosh\left[\sqrt{s_x}\left(a+\frac{1}{2}-x_2'\right)\right] \quad (4.2\text{-}11)$$

for $x_2'\leq x\leq x_1'$

$$X = A_1 \cosh\left[\sqrt{s_x}\left(a+\frac{1}{2}-x\right)\right]\cosh\left[\sqrt{s_x}\left(a+\frac{1}{2}+x_1'\right)\right]$$
$$+ A_2 \cosh\left[\sqrt{s_x}\left(a+\frac{1}{2}-x\right)\right]\cosh\left[\sqrt{s_x}\left(a+\frac{1}{2}+x_2'\right)\right] \quad (4.2\text{-}12)$$

for $x_2'\leq x\leq a+\frac{1}{2}$

This postulated solution satisfies the Neumann conditions at the array edges. The constants, A_1 and A_2 and the eigenvalues S_x, are determined by imposing the slope discontinuities across the injection points, x_1' and x_2'. That is,

$$X'\Big|_{x_1'^-}^{x_1'^+} = C_{x1}X(x_1')$$
$$X'\Big|_{x_2'^-}^{x_2'^+} = C_{x2}X(x_2') \quad (4.2\text{-}13)$$

These conditions lead to two homogeneous linear equations for A_1 and A_2, which may be written in the form

$$\begin{bmatrix} M_{11} & M_{12} \\ M_{21} & M_{22} \end{bmatrix} \begin{bmatrix} A_1 \\ A_2 \end{bmatrix} = \begin{bmatrix} 0 \\ 0 \end{bmatrix} \tag{4.2-14}$$

in which

$$\begin{aligned} M_{11} &= \sqrt{s_x} \sinh\left[\sqrt{s_x}(2a+1)\right] \\ &\quad + C_{x1} \cosh\left[\sqrt{s_x}\left(a+\frac{1}{2}+x_1'\right)\right] \cosh\left[\sqrt{s_x}\left(a+\frac{1}{2}-x_1'\right)\right] \end{aligned} \tag{4.2-15}$$

$$M_{12} = C_{x1} \cosh\left[\sqrt{s_x}\left(a+\frac{1}{2}+x_1'\right)\right] \cosh\left[\sqrt{s_x}\left(a+\frac{1}{2}-x_2'\right)\right] \tag{4.2-16}$$

$$M_{21} = C_{x2} \cosh\left[\sqrt{s_x}\left(a+\frac{1}{2}+x_1'\right)\right] \cosh\left[\sqrt{s_x}\left(a+\frac{1}{2}-x_2'\right)\right] \tag{4.2-17}$$

$$\begin{aligned} M_{22} &= \sqrt{s_x} \sinh\left[\sqrt{s_x}(2a+1)\right] \\ &\quad + C_{x2} \cosh\left[\sqrt{s_x}\left(a+\frac{1}{2}+x_2'\right)\right] \cosh\left[\sqrt{s_x}\left(a+\frac{1}{2}-x_2'\right)\right] \end{aligned} \tag{4.2-18}$$

Setting the determinant of the two-by-two matrix in Eq. (4.2-14) equal to zero to permit a nontrivial solution for the A's provides a transcendental equation for the eigenvalues, s_x. The eigenvalues all lie on the negative real axis of the s plane and can thus be easily computed numerically by any one-dimensional root finding method, such as the Newton–Raphson method [46]. For each value of s_x for which the determinant is zero, s_m, we have either that

$$\begin{aligned} A_1 &= M_{12} \\ A_2 &= -M_{22} \end{aligned} \tag{4.2-19}$$

or that

$$\begin{aligned} A_1 &= -M_{11} \\ A_2 &= M_{21} \end{aligned} \tag{4.2-20}$$

These two possible solutions are, in fact, the same to within a multiplicative constant, but this constant has no effect once the eigenfunctions are normalized. Thus, either Eq. (4.2-19) or (4.2-20) may be used and the ultimate result will be the same. Normalization of the eigenfunctions is, of course, accomplished by integrating their square over the range of x—that is, from $-(2a + 1)/2$ to $(2a + 1)/2$. This integration can be carried out giving a rather complicated but nevertheless closed-form result for the eigenfunction, $X(x,s_m)$.

Proceeding in the same manner one may obtain a corresponding closed form expression for $Y(y,s_n)$ and the Green's function, $\tilde{g}$, which satisfies

$$\frac{\partial^2\tilde{g}}{\partial x^2}+\frac{\partial^2\tilde{g}}{\partial y^2}-[C_{x1}\delta(x-x_1')+C_{x2}\delta(x-x_2')]\tilde{g} \\ -\left[C_{y1}\delta(y-y_1')+C_{y2}\delta(y-y_2')\right]\tilde{g} \\ -s\tilde{g}=-\frac{1}{s}\delta(x-x')\delta(y-y') \tag{4.2-21}$$

may then be expressed in the form

$$\tilde{g}=\sum_{n=0}^{\infty}\sum_{m=0}^{\infty}\frac{X(x',s_m)Y(y',s_n)X(x,s_m)Y(y,s_n)}{s(s-s_m+s_n)} \tag{4.2-22}$$

Now the solution to Eq. (4.2-6) is

$$\tilde{\varphi}(x,y,s)=-\int_{-b-\frac{1}{2}}^{b+\frac{1}{2}}\int_{-a-\frac{1}{2}}^{a+\frac{1}{2}}\tilde{g}(x',y',x,y,s) \\ \left[P(x')\tilde{\varphi}_{inj,x}(x',y',s)+Q(y')\tilde{\varphi}_{inj,y}(x',y',s)\right]dx'dy' \tag{4.2-23}$$

where P and Q are given by Eq. (4.2-4). The presence of the Dirac delta functions in Eq. (4.2-4) facilitates the integration. Let

$$\tilde{\varphi}_{inj}(x,y,s)=\frac{1}{s} \tag{4.2-24}$$

so that the injection phase time dependence is a step function. The inverse Laplace transform is then computable as a sum of the residues, $R_{mn}(x,y)$, at the poles in Eq. (4.2-22) where $s = s_m + s_n$. Thus, the solution takes the form

$$\varphi(x,y,\tau)=\sum_{n=0}^{\infty}\sum_{m=0}^{\infty}R_{mn}(x,y)\left[1-e^{(s_m+s_n)\tau}\right]u(\tau) \tag{4.2-25}$$

Note that while there is a pole at $s = 0$, its residue is zero so Eq. (4.2-22) does not have a double pole at $s = 0$ and the inverse Laplace transform does not have a term linear in time. Therefore, there is no frequency shift as there was in the case of perimeter detuning. Because there is no detuning, the ensemble frequency of the array does not change. The injection frequencies are all equal to this ensemble frequency, so all oscillation remains at this same frequency. Were the injection frequency different from the ensemble frequency, the steady-state oscillation frequency would be equal to the injection frequency and a term linear in time would appear in the solution.

The desired steady-state solution (for infinite time) is a planar phase distribution. We can determine the injection phases needed to produce that steady state directly from Eq. (4.2-5). Let us use uniform-strength injection signals so that

$$C_{x2} = C_{x1} = C_{y2} = C_{y1} = C_{inj} \tag{4.2-26}$$

At infinite time, Eq. (4.2-5) becomes

$$\begin{aligned}\frac{\partial^2 \varphi_{ss}}{\partial x^2} + \frac{\partial^2 \varphi_{ss}}{\partial y^2} &= \left[C_{x1}\delta(x - x_1') + C_{x2}\delta(x - x_2')\right]\left(\varphi_{inj} - \varphi_{ss}\right) \\ &\quad + \left[C_{y1}\delta(y - y_1') + C_{y2}\delta(y - y_2')\right]\left(\varphi_{inj} - \varphi_{ss}\right)\end{aligned} \tag{4.2-27}$$

The right side of this equation is zero except on the extended perimeter of a rectangle defined by

$$\left(x - x_1'\right)\left(x - x_2'\right)\left(y - y_1'\right)\left(y - y_2'\right) = 0 \tag{4.2-28}$$

Thus, the left side must also be zero except on these four lines. The solution we seek is linear in x and y, and it will have slope discontinuities on the rectangle defined by Eq. (4.2-28). The slope of the phase surface will be set by the desired beam direction as in Eq. (4.1-11). Thus,

$$\varphi_{ss}(x, y) = \frac{\Omega_x}{2}\left(\left|x - x_1'\right| - \left|x - x_2'\right|\right) + \frac{\Omega_y}{2}\left(\left|y - y_1'\right| - \left|y - y_2'\right|\right) \tag{4.2-29}$$

Substituting Eq. (4.2-29) into Eq. (4.2-27), we obtain

$$
\begin{aligned}
&\varphi_{inj,x}(x,y)\left[C_{x1}\delta(x-x_1')+C_{x2}\delta(x-x_2')\right] \\
&\quad+\varphi_{inj,y}(x,y)\left[C_{y1}\delta(y-y_1')+C_{y2}\delta(y-y_2')\right]= \\
&\left[-\Omega_x-C_{x1}\frac{\Omega_x}{2}\left|x_2'-x_1'\right|+C_{x1}\frac{\Omega_y}{2}\left(\left|y-y_1'\right|-\left|y-y_2'\right|\right)\right]\delta(x-x_1')+ \\
&\left[\Omega_x+C_{x2}\frac{\Omega_x}{2}\left|x_2'-x_1'\right|+C_{x2}\frac{\Omega_y}{2}\left(\left|y-y_1'\right|-\left|y-y_2'\right|\right)\right]\delta(x-x_2')+ \\
&\left[-\Omega_y-C_{y1}\frac{\Omega_y}{2}\left|y_2'-y_1'\right|+C_{y1}\frac{\Omega_x}{2}\left(\left|x-x_1'\right|-\left|x-x_2'\right|\right)\right]\delta(y-y_1')+ \\
&\left[\Omega_y+C_{y2}\frac{\Omega_y}{2}\left|y_2'-y_1'\right|+C_{y2}\frac{\Omega_x}{2}\left(\left|x-x_1'\right|-\left|x-x_2'\right|\right)\right]\delta(y-y_2')
\end{aligned}
\tag{4.2-30}
$$

so that the required injection phases may be written

$$
\begin{aligned}
\varphi_{inj,x}(x,y)=\frac{\Omega_x}{2}\Bigg[&\left(\left|x-x_1'\right|-\left|x-x_2'\right|\right) \\
&+\frac{2}{C_{x1}}\frac{(x-x_2')}{\left|x_1'-x_2'\right|}+\frac{2}{C_{x2}}\frac{(x-x_1')}{\left|x_1'-x_2'\right|}\Bigg]
\end{aligned}
\tag{4.2-31}
$$

$$
\begin{aligned}
\varphi_{inj,y}(x,y)=\frac{\Omega_y}{2}\Bigg[&\left(\left|y-y_1'\right|-\left|y-y_2'\right|\right) \\
&+\frac{2}{C_{y1}}\frac{(y-y_2')}{\left|y_1'-y_2'\right|}+\frac{2}{C_{y2}}\frac{(y-y_1')}{\left|y_1'-y_2'\right|}\Bigg]
\end{aligned}
\tag{4.2-32}
$$

For these injection phases, at late times, Eq. (4.2-25) is very slowly converging. However, we may remedy this as follows. If the Fourier series for the steady-state solution Eq. (4.2-29) is subtracted from the solution Eq. (4.2-25), the resulting series converges rapidly for late times. Then, to obtain the complete late time solution, one merely adds the steady-state solution Eq. (4.2-29) to this rapidly converging series. This solution conveniently complements the form given by Eq. (4.2-25) that converges rapidly for early times. (Convergence acceleration of this sort may also be applied in steering via detuning and in the one-dimensional cases treated in Chapter 3 if desired.)

If C_{inj} is large (strong injection), the last two terms in each of Eqs. (4.2-31) and (4.2-32) are negligible, and the injection phase equals the desired steady-state phase at the injection points as in the one-dimensional case. This strong injection assumption was implicit in the analyses presented in Refs. 39 and 43 because the small terms were neglected in accelerating the series, but this fact was not explicitly noted.

As an example, the solution given by Eq. (4.2-25) was computed for a 21-by-21 element array with injection signals of strength parameter C equal to 0.7 on the perimeter phased to steer the beam 10 deg from normal at –110 deg of azimuth. This requires that we apply the phase shift gradually as discussed in Section 3.4. We choose to do this linearly over an interval of 50 inverse locking ranges, after which the injection phases become constant. The solution for linear-injection phase can be obtained from that for constant-injection phase by integration with respect to time or division by s in the Laplace domain. The phase distributions across the array at four instants of time are shown in Fig. 4-5. Figure. 4-6 shows the corresponding trajectory of the beam peak and 3-dB contour during the beam-steering transient at intervals of 10 inverse locking ranges. Here again, as shown in Fig. 4-7, because of the phase aberration (deviation from planarity) across the aperture during the transient, the directivity of the antenna decreases, but this loss is recovered in steady state when the phase distribution again becomes planar. The so-called "projected aperture loss" discussed in Section 4.1 is also clearly visible. Finally, Fig. 4-8 shows the result of applying a sequence of injection phases resulting in sequential beam-steering to several angles.

4.3 Non-Cartesian Coupling Topologies

The planar arrays presented so far have made use of a Cartesian coupling topology in which oscillators on a Cartesian lattice were coupled to four nearest neighbors. However, this is by no means the only coupling topology leading to planar arrays that admit beam-steering. In this section we treat, via the continuum formulation, two other possible topologies, triangular (Fig. 4-9) and hexagonal (Fig. 4-10).

In the triangular case, shown in Fig. 4-9, the unit cells are hexagons and each interior oscillator is coupled to six nearest neighbors [44]. The oscillators are identified with pairs of integer values of the coordinates p and q ranging from 1 to N.

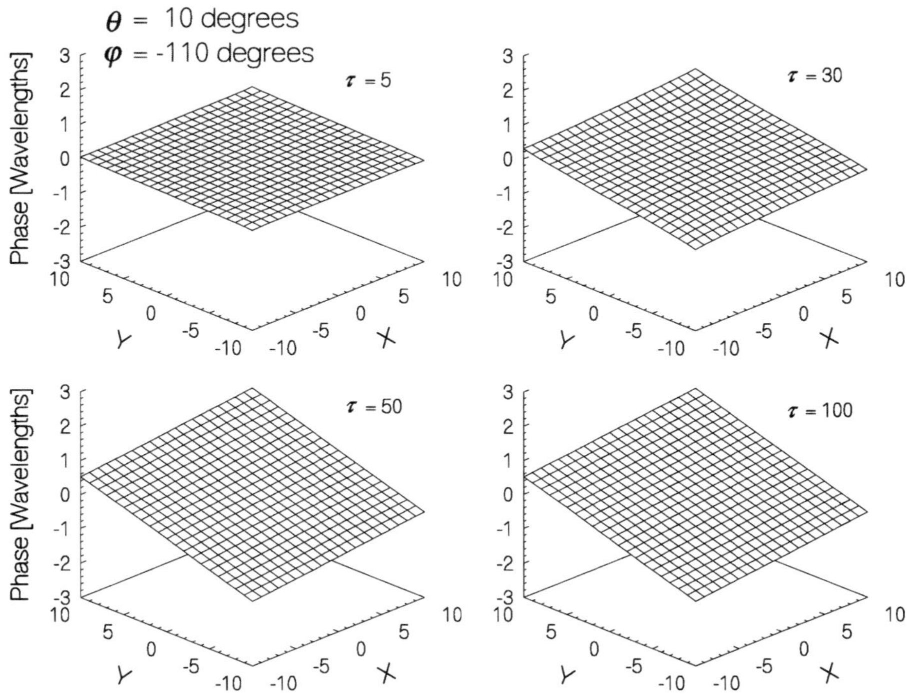

Fig. 4-5. Aperture phase distributions versus time (edge oscillators injection locked for beam-steering). (Reprinted from Ref. 43 with permission, ©2001 IEEE.)

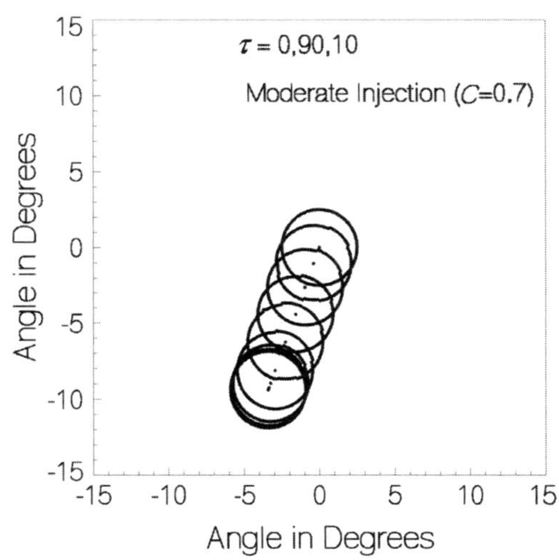

Fig. 4-6. Beam trajectory during the beam-steering transient. (Reprinted from Ref. 43 with permission, ©2001 IEEE.)

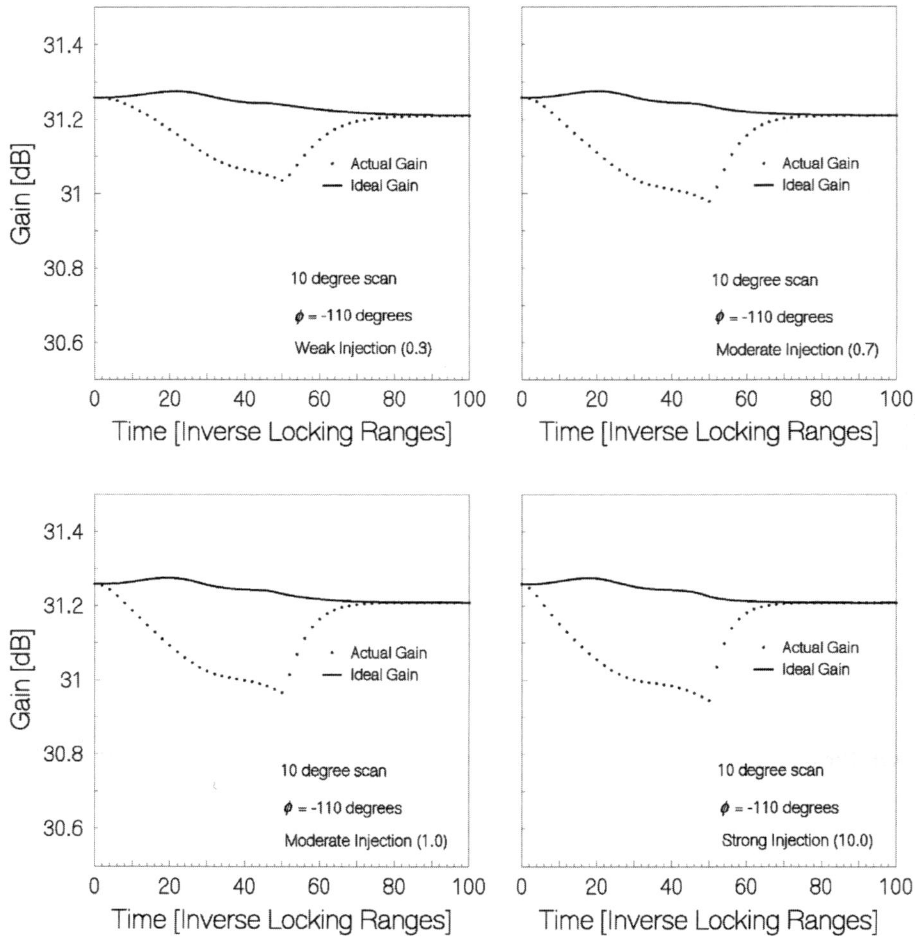

Fig. 4-7. Peak gain dynamics during the beam-steering transient for various injection strengths. (Reprinted from Ref. 43 with permission, ©2001 IEEE.)

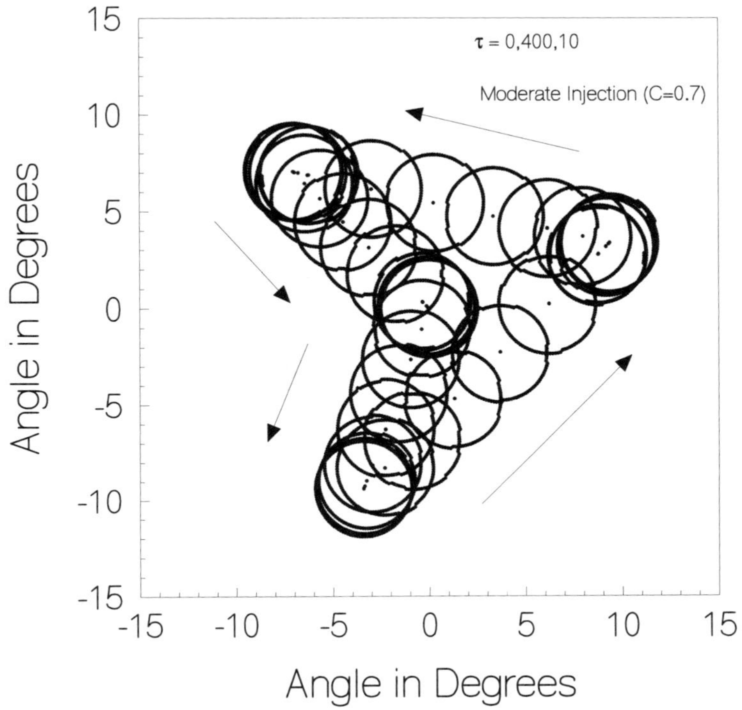

Fig. 4-8. Sequential bream-steering (τ varies from 0 to 400 in increments of 10). (Reprinted from Ref. 43 with permission, ©2001 IEEE.)

Using these coordinates, the discrete model yields the system of differential equations,

$$\begin{aligned}\frac{\partial \varphi_{pq}}{\partial t} &= \omega_{tune,pq} - \omega_{ref} \\ &-\Delta\omega_{lock}\Big[\sin(\varphi_{pq}-\varphi_{p-1,q})+\sin(\varphi_{pq}-\varphi_{p+1,q}) \\ &\quad +\sin(\varphi_{pq}-\varphi_{p+1,q-1})+\sin(\varphi_{pq}-\varphi_{p,q-1}) \\ &\quad +\sin(\varphi_{pq}-\varphi_{p,q+1})+\sin(\varphi_{pq}-\varphi_{p-1,q+1})\Big]\end{aligned} \qquad (4.3\text{-}1)$$

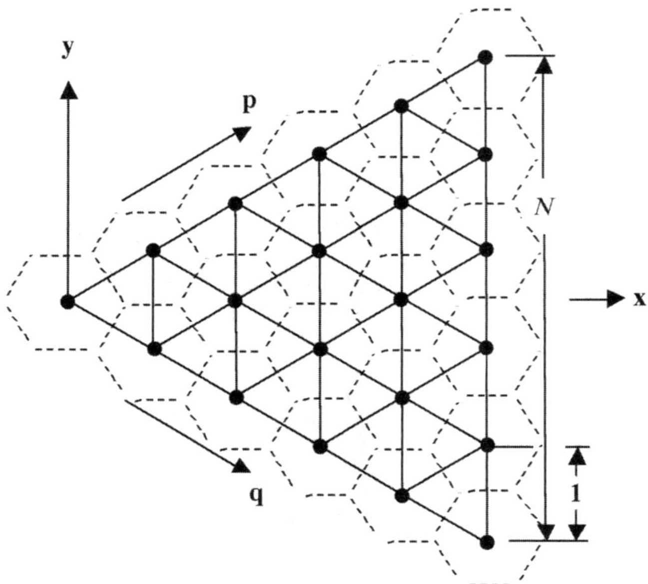

Fig. 4-9. Oscillators coupled on an equilateral triangular lattice. (Reprinted from Ref. 44 with permission, ©2004 IEEE.)

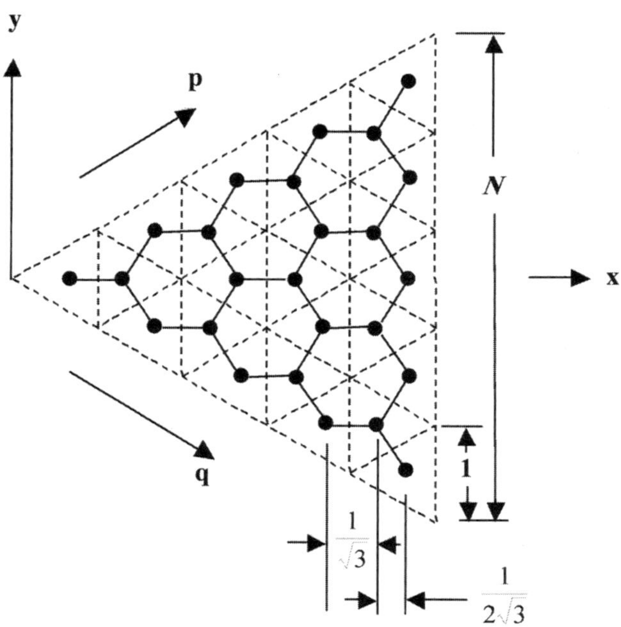

Fig. 4-10. Oscillators coupled on a hexagonal lattice. (Reprinted from Ref. 44 with permission, ©2004 IEEE.)

where we have assumed that the coupling phases are multiples of 2π. We remark that this system of nonlinear equations can be solved numerically to yield the full nonlinear solution for the dynamic behavior of the phase distribution. However, as mentioned earlier, the analytic solution of the linearized formulation provides more insight. Linearizing and expanding in Taylor series to second order leads to

$$2\left(\frac{\partial^2\varphi}{\partial p^2}+\frac{\partial^2\varphi}{\partial p\partial q}+\frac{\partial^2\varphi}{\partial q^2}\right)-\frac{\partial\varphi}{\partial\tau}=-\Delta\Omega_{tune}(p,q) \qquad (4.3\text{-}2)$$

where, as before, $\tau=\Delta\omega_{lock}t$ and $\Delta\Omega_{tune}=(\omega_{tune}-\omega_{ref})/\Delta\omega_{lock}$. Transforming to Cartesian coordinates, x and y, we arrive at

$$\left(\frac{\partial^2\varphi}{\partial x^2}+\frac{\partial^2\varphi}{\partial y^2}\right)-\frac{\partial\varphi}{\partial T_{tri}}=-\frac{1}{\sqrt{3}}\Delta\Omega_{tune} \qquad (4.3\text{-}3)$$

and T_{tri} = (3/2)τ. This equation is very much like Eq. (4.1-2) for the Cartesian case except for the scaling of the time and the detuning.

In the hexagonal case, shown in Fig. 4-10, the unit cells ar triangular and each interior oscillator is coupled to three nearest neighbors [44]. Following a procedure analogous to that presented above for the triangular case leads to

$$\left(\frac{\partial^2\varphi}{\partial x^2}+\frac{\partial^2\varphi}{\partial y^2}\right)-\frac{\partial\varphi}{\partial T_{hex}}=-\sqrt{3}\Delta\Omega_{tune} \qquad (4.3\text{-}4)$$

and T_{hex} = τ/4. Again, this equation is very much like Eq. (4.1-2) for the Cartesian case except for the scaling of the time and the detuning.

Finite arrays using the triangular and hexagonal coupling schemes may be constructed with equilateral triangular boundaries, and the boundary conditions on the phase will again be of the Neumann type. Just as was the case for rectangular arrays, if the oscillators on the triangular boundary are detuned, the steady-state interior phases are governed by Laplace's equation, and planar distributions are an obtainable special case. The desired steady-state solution to produce a beam steered to spherical coordinate angles, (θ_0,φ_0), is

$$\varphi(x,y,\infty)=-\frac{2\pi d}{\lambda}\left[\left(x-\frac{N}{\sqrt{3}}\right)\sin\theta_0\cos\varphi_0+y\sin\theta_0\sin\varphi_0\right] \qquad (4.3\text{-}5)$$

where d is the separation of the radiating elements in the y coordinate for fixed x, and λ is the wavelength. Substituting this desired steady-state phase distribution, Eq. (4.3-5), into the partial differential equations, Eqs. (4.3-3) and (4.3-4), gives the required detuning of the perimeter oscillators.

$$\begin{aligned}
\Delta\Omega_{tune}\big|_{x=y\sqrt{3}} &= -\frac{2\pi d}{\alpha\lambda}\sin\theta_0\cos\left(\varphi_0-\frac{2\pi}{3}\right)\\
\Delta\Omega_{tune}\big|_{x=-y\sqrt{3}} &= -\frac{2\pi d}{\alpha\lambda}\sin\theta_0\cos\left(\varphi_0+\frac{2\pi}{3}\right)\\
\Delta\Omega_{tune}\big|_{x=A\sqrt{3}/2} &= -\frac{2\pi d}{\alpha\lambda}\sin\theta_0\cos(\varphi_0)
\end{aligned} \tag{4.3-6}$$

where A is the length of a side of the array, Nd, and α is $1/\sqrt{3}$ for the triangular coupling and $\sqrt{3}$ for hexagonal coupling. Note that the sum of these detunings is zero for all steering angles so that the ensemble frequency of the array is unchanged.

At this point we note for later reference that it is possible to obtain the needed tuning for a given planar steady-state phase distribution from the full nonlinear formulation Eq. (4.3-1). Inserting the desired phase Eq. (4.3-5) into Eq. (4.3-1) and evaluating on the boundary of the triangle, we obtain

$$\begin{aligned}
\Delta\Omega_{tune}\big|_{x=y\sqrt{3}} &= \frac{1}{\sqrt{3}}\sin\left[\frac{2\pi d}{\alpha\lambda}\sin\theta_0\cos\left(\varphi_0+\frac{\pi}{6}\right)\right]\\
&\quad+\frac{1}{\sqrt{3}}\sin\left[\frac{2\pi d}{\alpha\lambda}\sin\theta_0\cos\left(\varphi_0+\frac{\pi}{2}\right)\right]\\
\Delta\Omega_{tune}\big|_{x=-y\sqrt{3}} &= \frac{1}{\sqrt{3}}\sin\left[\frac{2\pi d}{\alpha\lambda}\sin\theta_0\cos\left(\varphi_0-\frac{\pi}{6}\right)\right]\\
&\quad+\frac{1}{\sqrt{3}}\sin\left[\frac{2\pi d}{\alpha\lambda}\sin\theta_0\cos\left(\varphi_0-\frac{\pi}{2}\right)\right]\\
\Delta\Omega_{tune}\big|_{x=A\sqrt{3}/2} &= \frac{1}{\sqrt{3}}\sin\left[\frac{2\pi d}{\alpha\lambda}\sin\theta_0\cos\left(\varphi_0-\frac{5\pi}{6}\right)\right]\\
&\quad+\frac{1}{\sqrt{3}}\sin\left[\frac{2\pi d}{\alpha\lambda}\sin\theta_0\cos\left(\varphi_0+\frac{5\pi}{6}\right)\right]
\end{aligned} \tag{4.3-7}$$

which reduces to Eq. (4.3-6) for small θ_0—that is, for small inter-oscillator phase differences when the linearization is accurate. Note that the sum of these detunings is exactly zero regardless of steering angle.

We propose that the solutions of the partial differential equations, Eq. (4.3-3) and Eq. (4.3-4), be obtained as series of the eigenfunctions of the differential operators subject to Neumann boundary conditions on the triangular boundary of the arrays. These eigenfunctions have been studied in the context of waveguides of triangular cross section and are thus well known. They are expressed as sums of three products of two of the trigonometric functions, sine and cosine. These eigenfunctions and their useful properties are summarized in the appendix of Ref. 44.

We wish to solve

$$\left(\frac{\partial^2 \varphi}{\partial x^2}+\frac{\partial^2 \varphi}{\partial y^2}\right)-\frac{\partial \varphi}{\partial T}=-\alpha \Delta\Omega_{tune} \tag{4.3-8}$$

with the detuning function given by Eq. (4.3-6), and we will assume that the detuning is a step function in time. Laplace transformation of Eq. (4.3-8) gives

$$\left(\frac{\partial^2 \tilde{\varphi}}{\partial x^2}+\frac{\partial^2 \tilde{\varphi}}{\partial y^2}\right)-s\tilde{\varphi}=-\alpha \Delta\tilde{\Omega}_{tune}=-\frac{\alpha}{s}\Delta\Omega_{tune}(x,y) \tag{4.3-9}$$

The solution will be of the form

$$\varphi(x,y,\tau)=\sum_{mn} C_{mn} H_{mn}^{(i)}(x,y)\left(1-e^{-\sigma_{mn}T}\right)u(T) \tag{4.3-10}$$

where the H functions are the normalized eigenfunctions on the triangle with the superscript denoting even or odd symmetry of the function in y. Thus, the unknown coefficients may be found from the desired steady-state phase distribution, Eq. (4.3-5), by setting T to infinity in Eq. (4.3-10) and setting the resulting sum equal to the steady-state solution. Then, the orthogonality of the eigenfunctions permits us to find the coefficients, C_{mn}. This procedure is completely equivalent to expressing the Green's function as a sum of the eigenfunctions and then integrating the product of the Green's function and the desired steady-state phase distribution as was done in the Cartesian case.

As was done in Ref. 44, we now provide a number of computed examples demonstrating the dynamic behavior obtained via the various formulations of the problem—that is, the nonlinear model, the linearized discrete model, and the continuum model for both the triangular and hexagonal coupling topologies. We begin with triangular coupling with N = 28 resulting in a 435-oscillator array. We completely preclude the appearance of grating lobes in the visible region by selecting the radiating element spacing, d, to be $\lambda/\sqrt{3}$. We note, however, that, because the maximum permissible phase difference between coupled oscillators is 90 deg, the element spacing can be as large as $\lambda\sqrt{3}/2$ without the appearance of visible grating lobes. The array size, A, in the continuum model is taken to be $d\sqrt{(N+1)(N+2)}$ instead of Nd because that makes the array area equal to the sum of the unit cell areas, resulting in a more accurate directivity. Figure 4-11 shows the aperture phase distribution for four instants of time computed using the continuum model with perimeter detuning given by Eq. (4.3-6) and unit step temporal dependence and steady-state beam-pointing angles θ_0, φ_0) = 10 deg, 45 deg). Note the rather severe phase aberration at time equal to 10 inverse locking ranges. Figure 4-12 shows the directivity (gain in the absence of loss) computed by pattern integration as a function of time during the beam-steering transient. The solid curve is the result of planar phase distribution, and "projected aperture loss" is again evident. In the left plot, the continuum result is compared with the full nonlinear solution obtained numerically, and in the right plot the numerical solution of the linearized discrete model is compared with the nonlinear solution. Note that the dip in gain at about 10 inverse locking ranges correlates with the severe aberration at that time in Fig. 4-11. The nonlinear solution used Eq. (4.3-7) while the linear ones used Eq. (4.3-6) as detuning. Because the angle from normal is only 10 deg, the error in the linear approximation of the sine functions is less than 6.5 percent, and the linearized and continuum results agree well with the full nonlinear result taken to be the correct behavior. Figure 4-13 shows the trajectory of the beam peak and 3-dB contour during the beam-steering transient as computed via the three formulations, and, as should be expected for this small steering angle, they agree very well.

Now, if the final beam angle is increased from 10 deg to 25 deg, the error in the linear approximation of the sine function is almost 49 percent, and the discrepancy between the linear and nonlinear results in Fig. 4-14 show the impact of this in that the gain error at the dip is about 2 dB, and the curves are slightly different in shape. However, there is still qualitative agreement between the linear and nonlinear results.

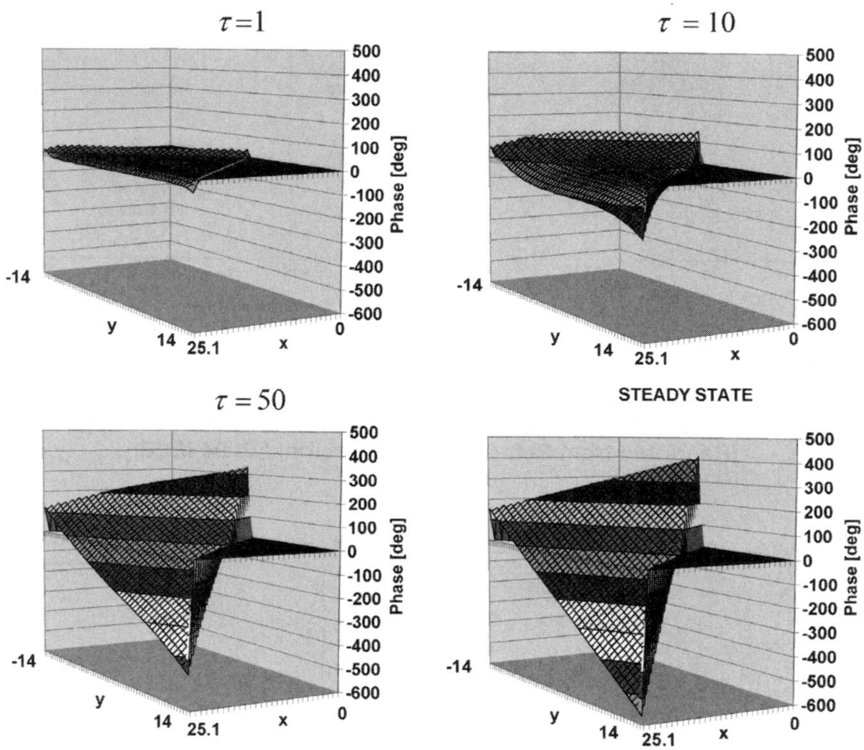

Fig. 4-11. Aperture phase distributions for triangular coupling with (θ_0, φ_0) = (10 deg, 45 deg) for τ values of 1, 10, and 50, plus steady state. (Reprinted from Ref. 44 with permission, ©2004 IEEE.)

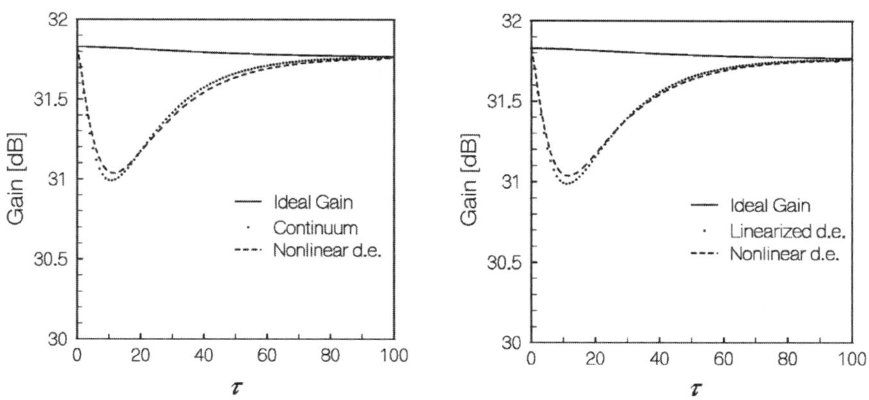

Fig. 4-12. Gain dynamics of gain plotted against percentage of τ for triangular coupling with (θ_0, φ_0) = (10 deg, 45 deg). (Reprinted from Ref. 44 with permission, ©2004 IEEE.)

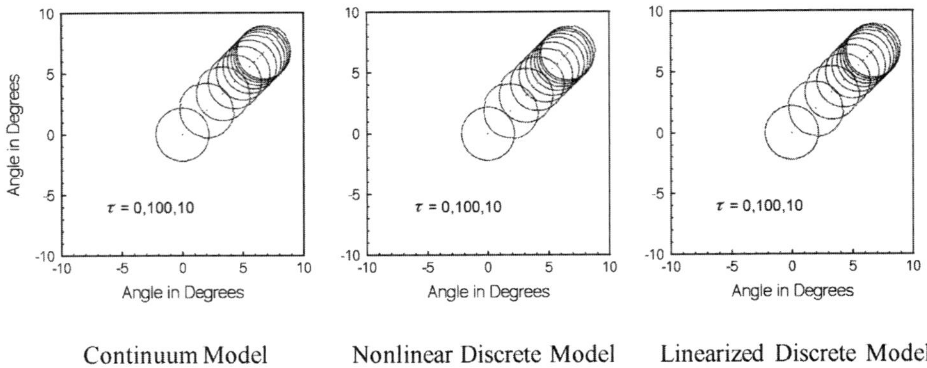

Fig. 4-13. Beam dynamics for triangular coupling with (θ_0, φ_0) = (10 deg, 45 deg). (Reprinted from Ref. 44 with permission, ©2004 IEEE.)

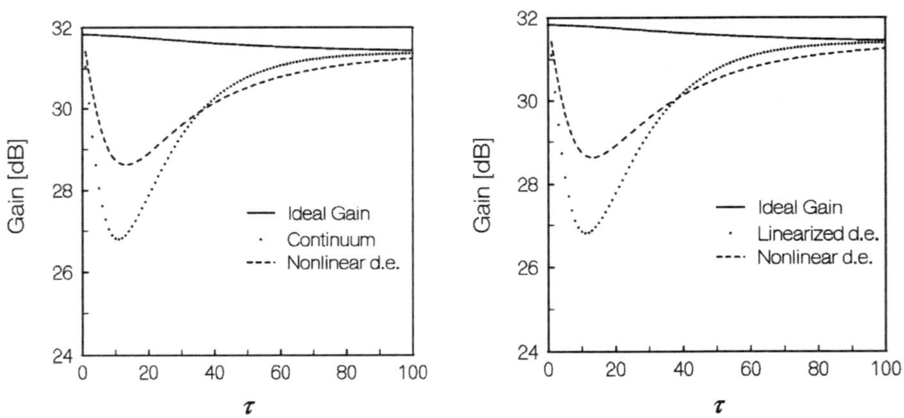

Fig. 4-14. Gain dynamics for triangular coupling with (θ_0, φ_0) = (25 deg, 45 deg). (Reprinted from Ref. 44 with permission, ©2004 IEEE.)

Figure 4-15 shows the trajectory of the beam peak and the 3-dB contour during the beam-steering transient as computed using the three formulations, and the agreement among them is good even though the accuracy of the linearization is questionable.

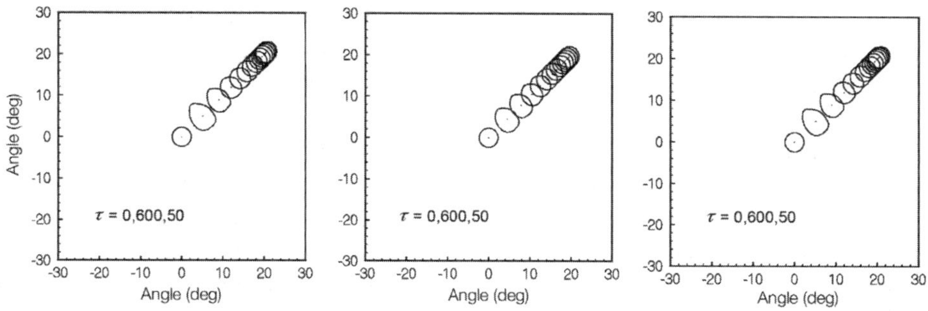

Fig. 4-15. Beam dynamics for triangular coupling with (θ_0, φ_0) = (25 deg, 45 deg). (τ varies from 0 to 600 in increments of 50.) (Reprinted from Ref. 44 with permission, ©2004 IEEE.)

Consider now a hexagonally coupled array in which we again choose d to be $\lambda/\sqrt{3}$. If we choose N to be 28, this hexagonally coupled array will have 784 oscillators. This makes the distance between the corner elements a bit less than in the triangular case for $N = 28$. Choosing the array size, A, to be Nd here makes the array area equal to the sum of the unit cell areas as was done for the triangular coupling example. Assuming that the radiating elements are arranged as in Fig. 4-10, the separation between periodic lines of elements will be $\lambda/2$, and there will be no grating lobes in the visible region. Here again, however, the spacing d can be as large as $\lambda\sqrt{3}/2$ and still not produce grating lobes because the phase differences must be less than 90 deg to maintain lock.

Figure 4-16 shows the aperture phase distribution at a sequence of times for steady-state beam-steering angles of (θ_0,φ_0) = (20 deg, 45 deg). The behavior is very similar to that observed with triangular coupling except for the time scale of the response, which is considerably slower for the hexagonal coupling. Of course, one must remember that this array has many more oscillators than the triangular one. The temporal behavior of the gain of this array is shown in Fig. 4-17 as computed using the full nonlinear model, the linearized model, and the continuum model. Because for this steady-state beam-steering angle, the error in the linear approximation to the sine function is only about 9 percent, these results agree quite well. The dip in the gain at about 60 inverse locking ranges is consistent with the significant phase aberration seen at that time in Fig. 4-16. Finally the trajectories of the beam peak and 3-dB contour as computed using the three formulations are shown in Fig. 4-18 for a steady-state beam position 20 deg from normal to the array. If this angle is increased to 30 deg, the approximation error increases to about 20 percent, and Fig. 4-19 shows the impact of this in terms of the discrepancies between the full nonlinear result taken to be correct and those of the two linearized theories. Finally Fig. 4-20 shows the corresponding beam trajectories.

As pointed out in Ref. 44, it might seem appropriate to correct the detuning needed for a given set of steady-state beam-steering angles as was done for triangular coupling in the manner of (4.3-7), but this is not very effective for reasons that will become clear as we discuss the true steady-state phase distribution for a hexagonally coupled array [45]. For one thing, there is a tendency to choose detunings that do not sum to zero thus producing quadratic phase distributions instead of planar ones. One may artificially impose a zero sum, but the proper choice of detuning still lacks a firm theoretical basis.

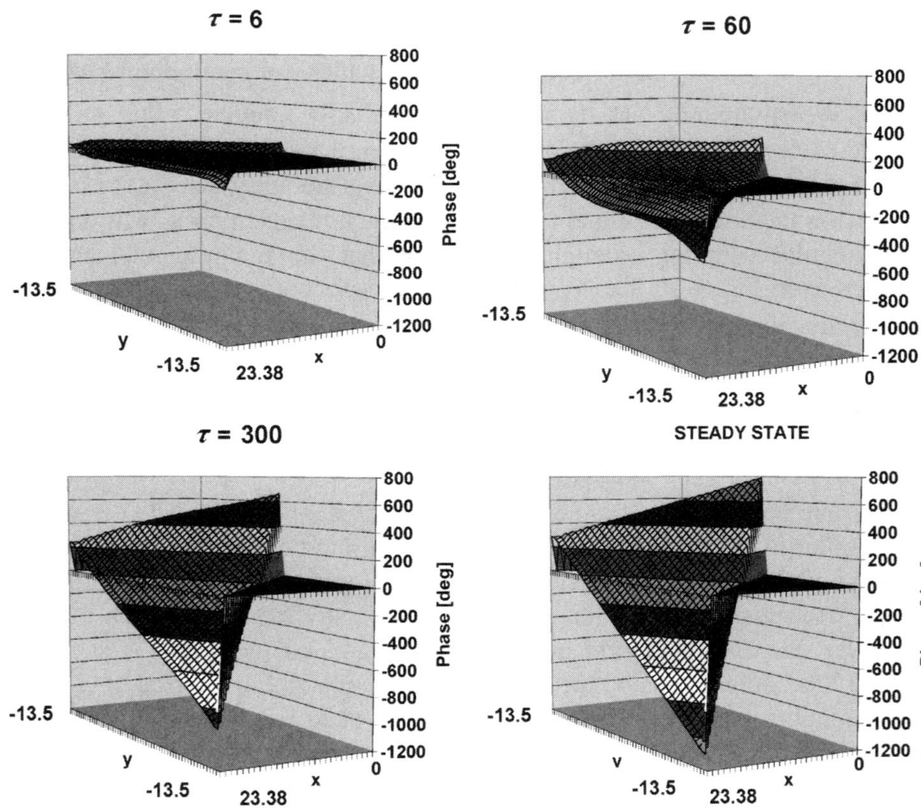

Fig. 4-16. Aperture phase for hexagonal coupling with (θ_0, φ_0) = (20 deg, 45 deg). (Reprinted from Ref. 44 with permission, ©2004 IEEE.)

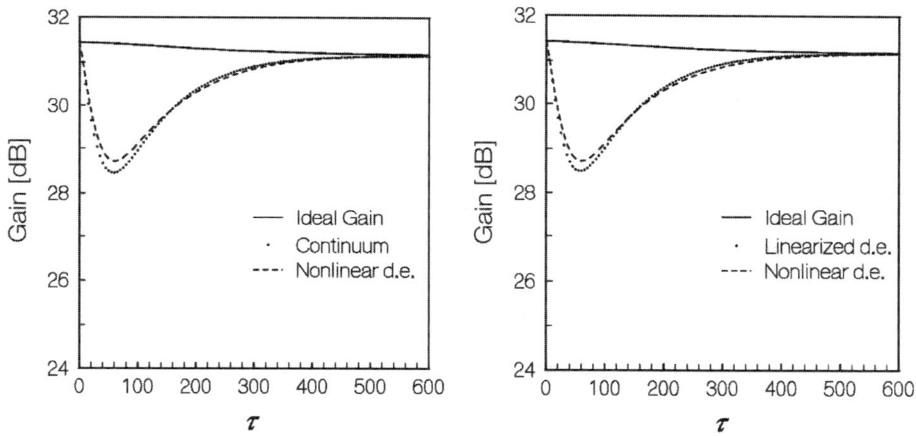

Fig. 4-17. Gain dynamics for hexagonal coupling with (θ_0, φ_0) = (20 deg, 45 deg) (Reprinted from Ref. 44 with permission, ©2004 IEEE.)

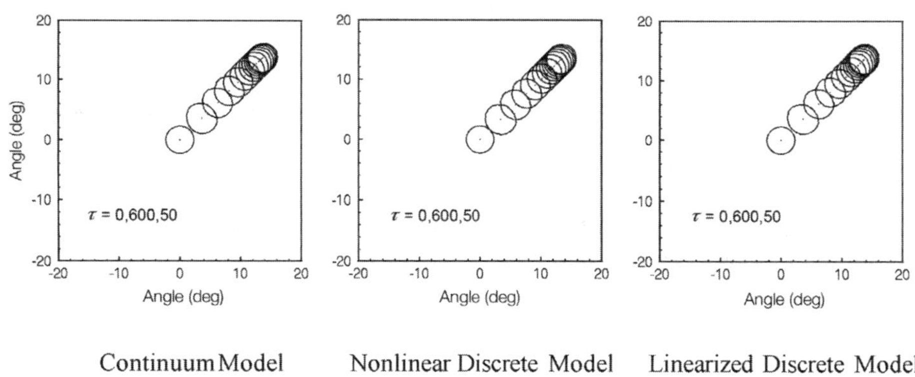

Fig. 4-18. Beam dynamics for hexagonal coupling with (θ_0, φ_0) = (20 deg, 45 deg) (τ varies from 0 to 600 in increments of 50). (Reprinted from Ref. 44 with permission, ©2004 IEEE.)

To obtain the results shown in Figs. 4-19 and 4-20, a correction factor was applied to the three detunings of Eq. (4.3-6). The factor was derived at an azimuth angle of 30 deg because, as will be seen, it is only at 30 deg plus integral multiples of 60 deg that planar phase distributions are rigorous steady-state solutions for the phase distribution. For reference, this correction factor as given in Ref. 44 is

$$Correction\ Factor = \frac{\sin\left[\frac{2\pi d}{\alpha\lambda}\sin\theta_0\cos\left(\frac{\pi}{6}\right)\right]}{\frac{2\pi d}{\alpha\lambda}\sin\theta_0\cos\left(\frac{\pi}{6}\right)} \tag{4.3-11}$$

and this factor is multiplied by each of the edge detunings in Eq. (4.3-6), thus preserving the zero sum.

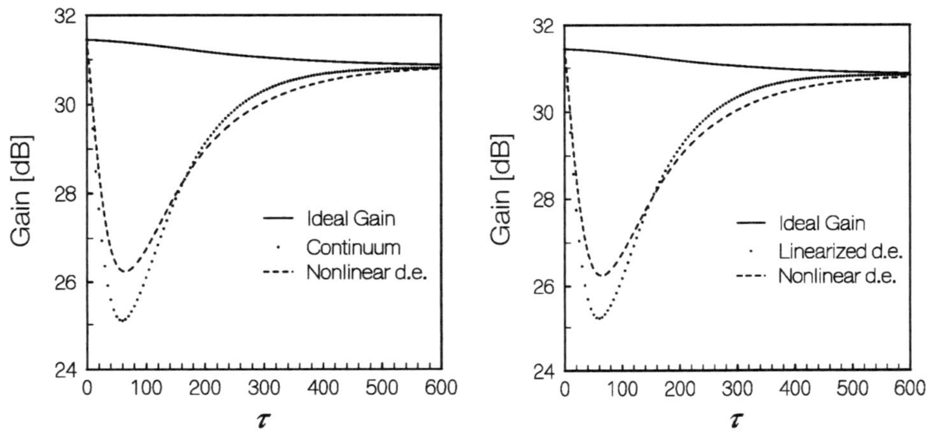

Fig. 4-19. Gain dynamics for hexagonal coupling with (θ_0, φ_0) = (30 deg, 45 deg) (Reprinted from Ref. 44 with permission, ©2004 IEEE.)

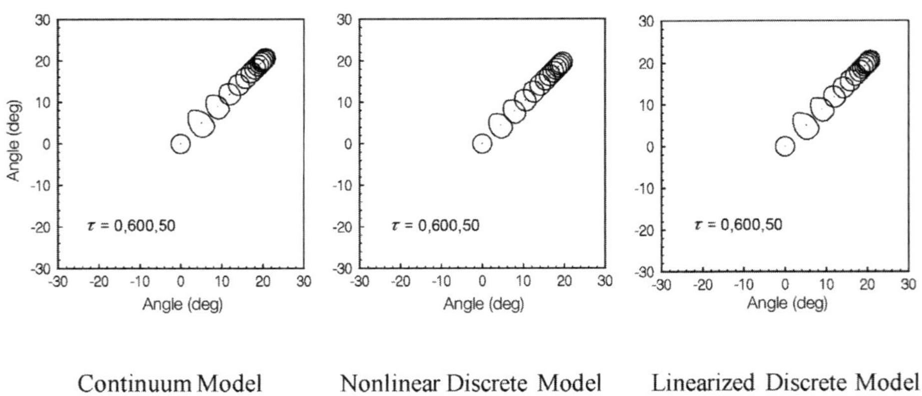

Fig. 4-20. Beam dynamics for hexagonal coupling with (θ_0, φ_0) = (30 deg, 45 deg) (Reprinted from Ref. 44 with permission, ©2004 IEEE.)

Now let us revisit in more detail the matter of the steady-state phase distribution in a hexagonally coupled array. We begin by formulating the full nonlinear set of differential equations for such an array. Following Pogorzelski [45], we write for each oscillator

$$\frac{\partial \varphi_{xy}}{\partial t} = \omega_{tune,xy} - \omega_{ref} - \Delta\omega_{lock} \times \left[\sin(\varphi_{xy} - \varphi_{x-\delta,y}) + \sin(\varphi_{xy} - \varphi_{x+\frac{\delta}{2},y+\frac{\sqrt{3}}{2}\delta}) + \sin(\varphi_{xy} - \varphi_{x+\frac{\delta}{2},y-\frac{\sqrt{3}}{2}\delta}) \right] \tag{4.3-12}$$

in which the coupling phase is assumed to be a multiple of 2π and $\delta = 1/\sqrt{3}$, the spacing between coupled oscillators. We have particular interest in the steady state, so we set the time derivative equal to zero and get

$$\frac{\omega_{tune,xy} - \omega_{ref}}{\Delta\omega_{lock}} = \sin(\varphi_{xy} - \varphi_{x-\delta,y}) + \sin(\varphi_{xy} - \varphi_{x+\frac{\delta}{2},y+\frac{\sqrt{3}}{2}\delta}) + \sin(\varphi_{xy} - \varphi_{x+\frac{\delta}{2},y-\frac{\sqrt{3}}{2}\delta}) \tag{4.3-13}$$

Recall that the desired phase distribution is given by Eq. (4.3-5). Substituting Eq. (4.3-5) into Eq. (4.3-13), we obtain for the nonperimeter oscillators,

$$\sin\left[D\cos(\varphi_0)\right] + \sin\left[D\cos\left(\varphi_0 + \frac{2\pi}{3}\right)\right] + \sin\left[D\cos\left(\varphi_0 - \frac{2\pi}{3}\right)\right] = 0 \tag{4.3-14}$$

where $D = \frac{2\pi d}{\lambda\sqrt{3}} \sin\theta_0$. Now Eq. (4.3-14) can be rewritten in the form

$$\sin\left[\frac{D}{2}\cos(\varphi_0)\right] \sin\left[\frac{D}{2}\cos\left(\varphi_0 + \frac{2\pi}{3}\right)\right] \sin\left[\frac{D}{2}\cos\left(\varphi_0 - \frac{2\pi}{3}\right)\right] = 0 \tag{4.3-15}$$

and it is clear that for small D—that is, small θ_0—this equation holds approximately true. Moreover, it holds exactly true for $\varphi_0 = \frac{\pi}{6} \pm \frac{n\pi}{3}$ for integer values of n. However, it does not hold true for arbitrary D and φ_0. Thus, we conclude that no possible detuning of the perimeter oscillators can result in a planar aperture distribution for azimuth angles other than $\varphi_0 = \frac{\pi}{6} \pm \frac{n\pi}{3}$.

Pogorzelski noted, however, that if one postulates a phase distribution of the form

$$\begin{aligned}\varphi(x,y) &= -\frac{2\pi d}{\lambda} \\ &\times\left[\left(x-\frac{N}{\sqrt{3}}\right)\sin\theta_0\cos\varphi_0 + y\sin\theta_0\sin\varphi_0\right] \pm \Delta\varphi_{xy}\end{aligned} \tag{4.3-16}$$

in which the ambiguous sign denotes alternation from one oscillator to its neighbor, an exact solution for the perimeter detuning that will produce it is possible, provided that $\Delta\varphi_{xy}$ is set to the proper value. Substituting Eq. (4.3-16) into Eq. (4.3-13) yields for the nonperimeter oscillators

$$\begin{aligned}&\sin\left[2\Delta\varphi_{xy} - D\cos(\varphi_0)\right] + \sin\left[2\Delta\varphi_{xy} - D\cos\left(\varphi_0 + \frac{2\pi}{3}\right)\right] \\ &+ \sin\left[2\Delta\varphi_{xy} - D\cos\left(\varphi_0 - \frac{2\pi}{3}\right)\right] = 0\end{aligned} \tag{4.3-17}$$

and solving for $\Delta\varphi_{xy}$, we obtain

$$\Delta\varphi_{xy} = \frac{1}{2}\tan^{-1}\left(\frac{Num}{Den}\right) \tag{4.3-18}$$

where

$$\begin{aligned}Num &= \sin\left[D\cos(\varphi_0)\right] + \sin\left[D\cos\left(\varphi_0 + \frac{2\pi}{3}\right)\right] \\ &+ \sin\left[D\cos\left(\varphi_0 - \frac{2\pi}{3}\right)\right]\end{aligned} \tag{4.3-19}$$

$$Den = \cos\left[D\cos(\varphi_0)\right] + \cos\left[D\cos\left(\varphi_0 + \frac{2\pi}{3}\right)\right] + \cos\left[D\cos\left(\varphi_0 - \frac{2\pi}{3}\right)\right] \tag{4.3-20}$$

Finally, substituting Eq. (4.3-16) with Eq. (4.3-18) into Eq. (4.3-13) yields the perimeter detuning required to produce this nonplanar phase distribution. The result is

$$\begin{aligned}
\Delta\Omega_{tune}\big|_{x=y\sqrt{3}} &= -\sin\left[\frac{2\pi d}{\lambda\sqrt{3}}\sin\theta_0\cos\left(\varphi_0 - \frac{2\pi}{3}\right) - 2\Delta\varphi_{xy}\right] \\
\Delta\Omega_{tune}\big|_{x=-y\sqrt{3}} &= -\sin\left[\frac{2\pi d}{\lambda\sqrt{3}}\sin\theta_0\cos\left(\varphi_0 + \frac{2\pi}{3}\right) - 2\Delta\varphi_{xy}\right] \\
\Delta\Omega_{tune}\big|_{x=(3N-1)\sqrt{3}/6} &= -\sin\left[\frac{2\pi d}{\lambda\sqrt{3}}\sin\theta_0\cos(\varphi_0) - 2\Delta\varphi_{xy}\right]
\end{aligned} \tag{4.3-21}$$

for the edge elements and

$$\begin{aligned}
&\Delta\Omega_{tune}\left(\frac{(3N-1)\sqrt{3}}{6}, \frac{(N-1)}{2}\right) = \\
&\quad -\sin\left[\frac{2\pi d}{\lambda\sqrt{3}}\sin\theta_0\cos\left(\varphi_0 + \frac{2\pi}{3}\right) - 2\Delta\varphi_{xy}\right] \\
&\Delta\Omega_{tune}\left(\frac{(3N-1)\sqrt{3}}{6}, -\frac{(N-1)}{2}\right) = \\
&\quad -\sin\left[\frac{2\pi d}{\lambda\sqrt{3}}\sin\theta_0\cos\left(\varphi_0 - \frac{2\pi}{3}\right) - 2\Delta\varphi_{xy}\right] \\
&\Delta\Omega_{tune}\left(\frac{1}{\sqrt{3}}, 0\right) = -\sin\left[\frac{2\pi d}{\lambda\sqrt{3}}\sin\theta_0\cos(\varphi_0) - 2\Delta\varphi_{xy}\right]
\end{aligned} \tag{4.3-22}$$

for the corner elements. Because of Eq. (4.3-17), the sum of these detunings is always zero, so the ensemble frequency of the array remains constant. Figure 4-21 shows a typical phase distribution for such an array. Figure 4-21a shows the phase distribution, and Fig. 4-21b shows the deviation, ±4.67 deg, from planar.

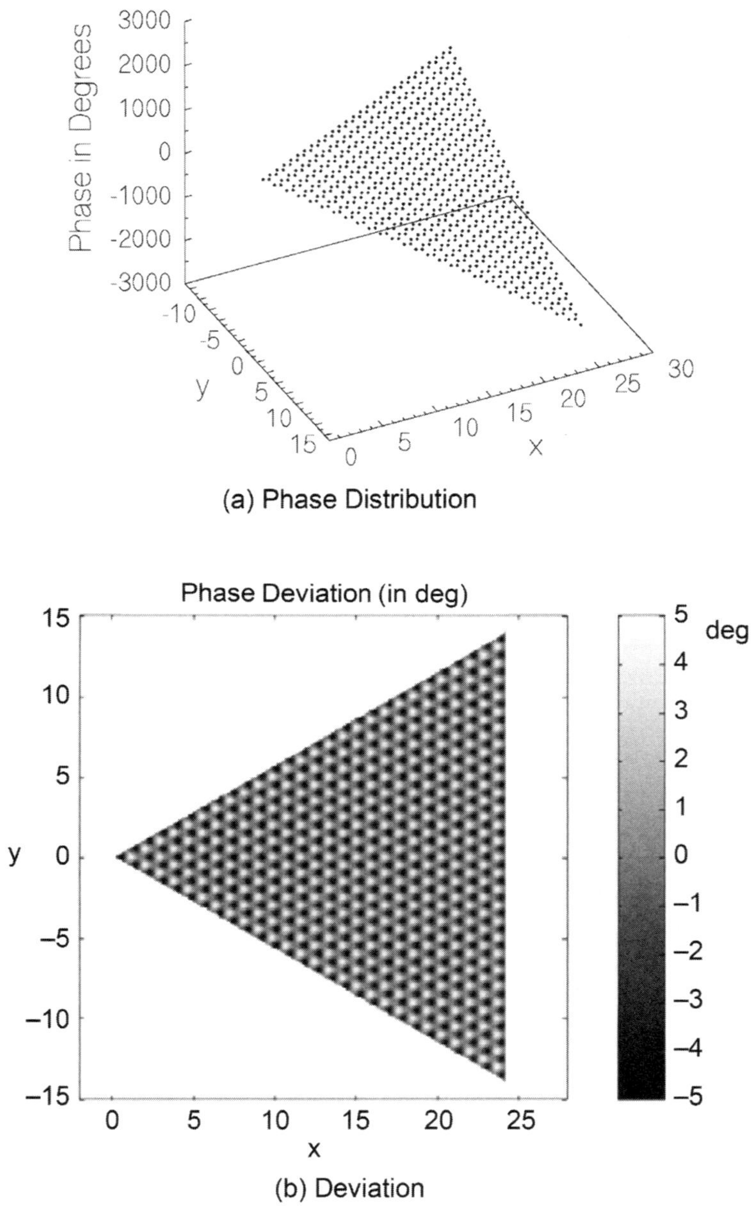

Fig. 4-21. Phase distribution for (θ_0, φ_0) = (22.3 deg, 57 deg) with $d = \lambda$ showing (a) phase distribution and (b) deviation from planar. (Reprinted from Ref. 44 with permission, ©2005 IEEE.)

One would perhaps expect that such a phase aberration would result in a decrease in gain. However, as discussed in detail in Ref. 45, this is typically not the case. The power that would have been radiated by a planar phase distribution is partially shifted by the phase alternation into the invisible region. Thus, this part is not radiated, and the power input to the antenna is decreased by the same amount. As a result, the directivity is unaffected by the phase alternation! The only circumstance resulting in a decrease in the gain is when the combination of element spacing and steering angle results in one or more grating lobes in the visible region. Analytical estimation of this effect is discussed in Ref. 45.

Finally, we remark that a planar steady-state phase distribution is of course attainable if one is willing to detune all of the oscillators in the hexagonally coupled array. This would require that alternate oscillators be detuned in opposite directions in frequency by an amount that depends on the scan angle.

4.4 Conclusion

In this chapter we have discussed a variety of coupling topologies for planar arrays, and we have shown that the continuum model can be used to describe the dynamic behavior of the phase distribution over these arrays. By this means we have demonstrated that beam-steering can be accomplished by detuning the perimeter oscillators or in the Cartesian case by injection locking them to external signals. Beam-steering by external injection in the triangular case was not treated but appears to be possible, though the analysis may become somewhat more challenging.

In the next chapter we point out that all of the preceding results are fundamentally noncausal in that the response begins immediately upon application of the detuning or phase shift of the external locking signal regardless of the physical separation of the cause and effect. A modified formulation is proposed to render the solutions causal.

Chapter 5
Causality and Coupling Delay

In the analysis presented in the preceding chapters, it was tacitly assumed that the coupling was implemented using nondispersive transmission lines characterized by a phase shift of Φ generally taken to be an integral multiple of 2π (plus π in the case of series resonant oscillators). However, the theory made no provision for the transit time through the coupling line. As a result, the solutions were noncausal. That is, each oscillator in the array responded immediately upon changing the tuning of an oscillator or the phase of an injection signal no matter what the distance between the excitation and the response. This is characteristic of the diffusion equation that arises from the continuum model. Heat conduction analyzed in this manner is similarly non-causal. Following Pogorzelski [47], we propose to remedy this situation by explicitly introducing time delay in the coupling. This time delay is determined by the physical length of the line and its propagation velocity.

5.1 Coupling Delay

A nondispersive transmission line introduces a pure time delay in that the signal applied at one end of the line is duplicated at the other end after the delay time. At that point the signal is reflected if the termination is not matched to the line impedance. For our analysis we will assume a matched termination. Now, if the analysis is done via Laplace transformation of the applied signal, the transform of the delayed signal is merely the original transform multiplied by e^{-sd}, where d is the delay time and s is the transform variable conjugate to the time variable.

Suppose now that we envision an array of coupled oscillators and take the reference frequency to be the initial ensemble frequency of the array. We define the coupling phase delay using this reference frequency. That is,

$$\Phi = -\frac{\omega_{ref}\ell}{v_p} = -\omega_{ref}d \tag{5.1-1}$$

where ℓ is the physical length of the line and v_p is the phase velocity. The line length is chosen so that the coupling phase is a multiple of 2π (plus π in the case of series resonant oscillators). Now, using the reference frequency, we define the phase, φ, of the phasor signal voltage, V, by

$$V = A(t)e^{j\theta(t)} \tag{5.1-2}$$

where

$$\theta(t) = \omega_{ref}t + \varphi(t) \tag{5.1-3}$$

Recall from Chapter 1 that V can be written in the form

$$V = e^{j[\theta(t) - j\ln A(t)]} \tag{5.1-4}$$

so that

$$\operatorname{Im}\left[\ln(V)\right] - \omega_{ref}t = \varphi(t) \tag{5.1-5}$$

Crucial to our analysis is the fact that any function of the input signal will be delayed by the nondispersive transmission line in the same manner as the signal itself so that the Laplace transform of any function of the input signal multiplied by e^{-sd} will be the transform of the same function delayed. Thus, we may apply this delay factor to the Laplace transform of $\varphi(t)$ given by Eq. (5.1-5) to obtain the transform of the phase delayed by the coupling line. This forms the basis of our introduction of coupling delay into the analysis of coupled oscillator arrays.

The following question regarding this treatment of time delay was posed by a particularly astute student, so we thought it appropriate to answer it here as you may be similarly puzzled. Slightly paraphrased, the student asked that we consider a linear array in which one of the oscillators is detuned upward, thus changing the ensemble frequency of the array. "Is it not then true," he asked, "that the coupling phase produced by coupling lines of fixed length would be changed and would thus be no longer a multiple of 2π?" To clarify this point, recall that, assuming that the reference frequency is held constant, the solution for the time evolution of the oscillator phases will contain terms linear in time representing the shift in ensemble frequency (as in Eq. (3.1-51)). The slope of this linear dependence relates the time delay to an equivalent phase shift through the transmission line. So, for example, if the phase at the input end of the line is

$$\varphi_{in}(t)=\alpha t \tag{5.1-6}$$

then the phase at the output end of the line is

$$\varphi_{out}(t)=\Phi+\varphi_{in}(t-d)=\Phi+\alpha(t-d)=\varphi_{in}(t)+\Phi-\alpha d \tag{5.1-7}$$

an effective coupling phase delay of $\Phi-\alpha d$. Conversely, due to the linear time dependence, the new ensemble frequency will be

$$\omega_{ens}=\omega_{ref}+\alpha \tag{5.1-8}$$

and the effective coupling phase will be

$$\Phi_{eff}=-\frac{\omega_{ens}\ell}{v_p}=-\omega_{ens}d=-\left(\omega_{ref}+\alpha\right)d=\Phi-\alpha d \tag{5.1-9}$$

So, we conclude that indeed the coupling phase has changed, but that change is embodied in the linear time dependence of the phases arising from the change in ensemble frequency and need not be explicitly imposed on the formulation by a change in the Φ parameter.

5.2 The Discrete Model with Coupling Delay

Returning to the linearized discrete model of a linear array of $(2N + 1)$ oscillators discussed in Section 2.2, we have

$$\frac{d\varphi_i}{dt}=\omega_{0i}-\omega_{ref}+\Delta\omega_{lock}\left(\varphi_{i+1}-2\varphi_i+\varphi_{i-1}\right) \tag{5.2-1}$$

$$\frac{d\varphi_{-N}}{dt}=\omega_{0N}-\omega_{ref}+\Delta\omega_{lock}\left(\varphi_{-N+1}-\varphi_{-N}\right) \tag{5.2-2}$$

$$\frac{d\varphi_{N}}{dt}=\omega_{0N}-\omega_{ref}+\Delta\omega_{lock}\left(\varphi_{N-1}-\varphi_{N}\right) \tag{5.2-3}$$

Laplace transformation with respect to $\tau = \Delta\omega_{lock} t$ results in

$$s\tilde{\varphi}_i = \Delta\tilde{\Omega}_{tune,i} + \left(\tilde{\varphi}_{i+1} - 2\tilde{\varphi}_i + \tilde{\varphi}_{i-1}\right) \tag{5.2-4}$$

$$s\tilde{\varphi}_{-N} = \Delta\tilde{\Omega}_{tune} + \left(\tilde{\varphi}_{-N+1} - \tilde{\varphi}_{-N}\right) \tag{5.2-5}$$

$$s\tilde{\varphi}_N = \Delta\tilde{\Omega}_{tune} + \left(-\tilde{\varphi}_N + \tilde{\varphi}_{N-1}\right) \tag{5.2-6}$$

and, introducing the coupling delay factors for delay of d inverse locking ranges, we have

$$s\tilde{\varphi}_i = \Delta\tilde{\Omega}_{tune,i} + \left(\tilde{\varphi}_{i+1}e^{-sd} - 2\tilde{\varphi}_i + \tilde{\varphi}_{i-1}e^{-sd}\right) \tag{5.2-7}$$

$$s\tilde{\varphi}_{-N} = \Delta\tilde{\Omega}_{tune} + \left(\tilde{\varphi}_{-N+1}e^{-sd} - \tilde{\varphi}_{-N}\right) \tag{5.2-8}$$

$$s\tilde{\varphi}_N = \Delta\tilde{\Omega}_{tune} + \left(-\tilde{\varphi}_N + \tilde{\varphi}_{N-1}e^{-sd}\right) \tag{5.2-9}$$

Rearranging yields

$$\tilde{\varphi}_{i+1}e^{-sd} - (s+2)\tilde{\varphi}_i + \tilde{\varphi}_{i-1}e^{-sd} = -\Delta\tilde{\Omega}_{tune,i} \tag{5.2-10}$$

$$\tilde{\varphi}_{-N+1}e^{-sd} - (s+1)\tilde{\varphi}_{-N} = -\Delta\tilde{\Omega}_{tune,-N} \tag{5.2-11}$$

$$-(s+1)\tilde{\varphi}_N + \tilde{\varphi}_{N-1}e^{-sd} = -\Delta\tilde{\Omega}_{tune,N} \tag{5.2-12}$$

These equations may be written compactly in matrix form as

$$[s[I] - [M]][\tilde{\varphi}] = [\Delta\tilde{\Omega}_{tune}] \tag{5.2-13}$$

in which $[M]$ is given by

$$[M]=\begin{bmatrix} 1 & -e^{-sd} & & & \\ -e^{-sd} & 2 & -e^{-sd} & & \\ & -e^{-sd} & 2 & \ddots & \\ & & \ddots & \ddots & -e^{-sd} \\ & & & -e^{-sd} & 1 \end{bmatrix} \tag{5.2-14}$$

We now have two alternative approaches available for solving this system of linear equations. We can expand the solution as a sum of eigenvectors of the matrix $[I]s - [M]$, or we can solve the system via Cramer's rule. Following Pogorzelski [47], we choose the Cramer's rule approach. The result is

$$\tilde{\varphi}_i = (-1)^{n_> - n_<}\left[\frac{\left(a_0 U_{N+n_<-1} - bU_{N+n_<-2}\right)\left(a_0 U_{N-n_>-1} - bU_{N-n_>-2}\right)}{b\left(a_0^2 U_{2N-1} - 2a_0 bU_{2N-2} + b^2 U_{2N-3}\right)}\right]\Delta\tilde{\Omega}_{tune,j} \tag{5.2-15}$$

where U is the Chebyshev polynomial of the second kind of argument $a/(2b)$, $a_0 = s+1$, $a = s+2$, and $b = -e^{-sd}$. Now, U can be written in the form

$$U_m\left(\frac{a}{2b}\right) = \frac{\sin\left[(m+1)\cos^{-1}\left(\frac{a}{2b}\right)\right]}{\sin\left[\cos^{-1}\left(\frac{a}{2b}\right)\right]} \tag{5.2-16}$$

and defining Q to be

$$Q = -e^{-i\cos^{-1}\left(\frac{a}{2b}\right)} = e^{-\sec h^{-1}\left(-\frac{2b}{a}\right)} = -\frac{a}{2b} - \sqrt{\left(\frac{a}{2b}\right)^2 - 1} \tag{5.2-17}$$

U becomes

$$U_m\left(\frac{a}{2b}\right) = (-1)^{m+1}\frac{(Q)^{-(m+1)} - (Q)^{(m+1)}}{2\sqrt{\left(\frac{a}{2b}\right)^2 - 1}} \tag{5.2-18}$$

Substituting Eq. (5.2-18) into Eq. (5.2-15) yields

$$\tilde{\varphi}_i = \frac{\Delta\tilde{\Omega}_{tune,j}}{\sqrt{a^2-4b^2}}\left[\frac{\left(Q^{n_<}+RQ^{-(2N+1)}Q^{-n_<}\right)\left(Q^{-n_>}+RQ^{-(2N+1)}Q^{n_>}\right)}{\left(1-R^2Q^{-2(2N+1)}\right)}\right] \quad (5.2\text{-}19)$$

in which

$$R=-\frac{Q+b}{1+bQ} \quad (5.2\text{-}20)$$

In Eq. (5.2-19), $n_>$ is the greater of i and j while $n_<$ is the lesser. The form of Eq. (5.2-19) is suggestive of an image series produced by reflections at the ends of the array, where Q plays the role of a "propagator." The series may be obtained by expanding in powers of the reflection coefficient, R. When R is set equal to zero, we obtain the solution for an infinite array,

$$\tilde{\varphi}_i = \frac{\Delta\tilde{\Omega}_{tune,j}}{\sqrt{a^2-4b^2}}Q^{n_<}Q^{-n_>} \quad (5.2\text{-}21)$$

or, using Eq. (5.2-17),

$$\tilde{\varphi}_i = \frac{\Delta\tilde{\Omega}_{tune,j}}{\sqrt{(s+2)^2-4e^{-2sd}}}e^{-|i-j|\operatorname{sech}^{-1}\left(\frac{2e^{-sd}}{s+2}\right)} \quad (5.2\text{-}22)$$

Consider now an example of a 17-element array with coupling delay of two inverse locking ranges (ILRs) and step detuning of the center oscillator by one locking range. The inverse Laplace transform of Eq. (5.2-19) may be easily obtained by expanding it in powers of $-b=e^{-sd}$. Each term of the resulting series will be of the form

$$\frac{C(e^{-sd})^p}{s(s+2)^{p+1}} \quad (5.2\text{-}23)$$

which has a known inverse transform

$$\int_0^{\tau-pd}\frac{C}{p!}\tau'^p e^{-2\tau'}d\tau'\,u(\tau-pd) \quad (5.2\text{-}24)$$

The solution is plotted in Fig. 5-1. This solution exhibits several easily understandable features. First, the center oscillator is the only one detuned, and it is detuned at time zero. Thus, its nearest neighbors on either side do not change phase until one delay time has elapsed, giving the influence of the center oscillator detuning sufficient time to propagate to them. The center oscillator phase evolution continues exponentially and unperturbed until two delay times have elapsed. Then the influence of the phase changes of the neighboring oscillators impact the center oscillator, causing the slope change at time equal to four inverse locking ranges. This multiple reflection-like behavior continues to spread throughout the array creating the ripples in the phase visible in Fig. 5-1. For comparison, the corresponding phase behavior in the absence of coupling delay is shown in Fig. 5-2 and is noncausal.

The preceding discussion indicates that the array behavior will not begin to differ from that of the infinite array until eight delay times have elapsed (16 inverse locking ranges) and that even then the effect will begin with the outermost oscillators. The center oscillator behavior will not differ from that of the infinite array until 16 delay times have elapsed (32 inverse locking ranges). Thus, these plots do not indicate that the array is of finite size. To display finite array effects, we plot similar curves for a seven element array in Fig. 5-3 and Fig. 5-4. In this case the end effects begin to appear in the phase behavior of the end oscillators after three delay times (six inverse locking ranges). This may be seen by comparing the curves for $i = \pm 3$ in Figs. 5-1 and 5-3. Note that they

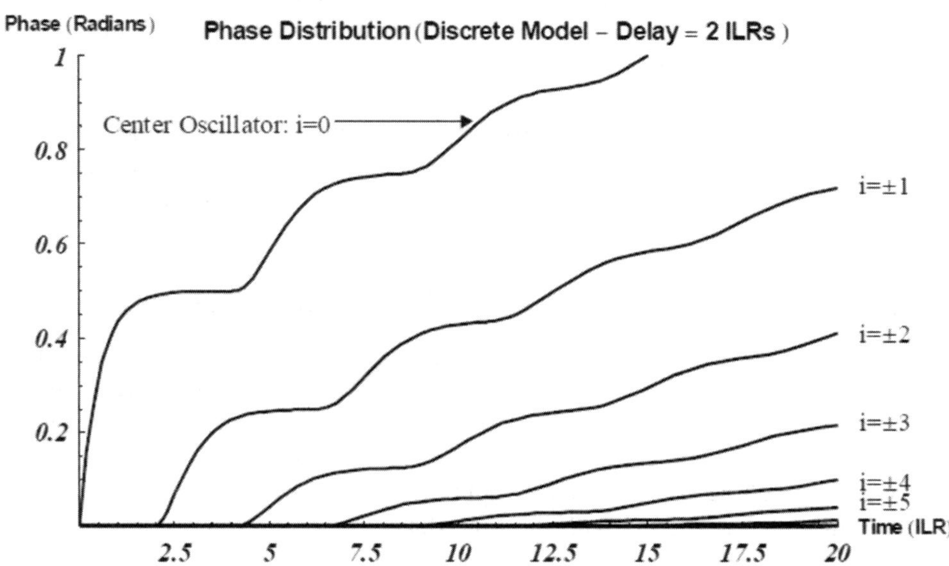

Fig. 5-1. Phase dynamics for a 17-element linear array with two-inverse-locking-range coupling delay. (Reproduced by permission of American Geophysical Union from Ref. 47, ©2008 American Geophysical Union.)

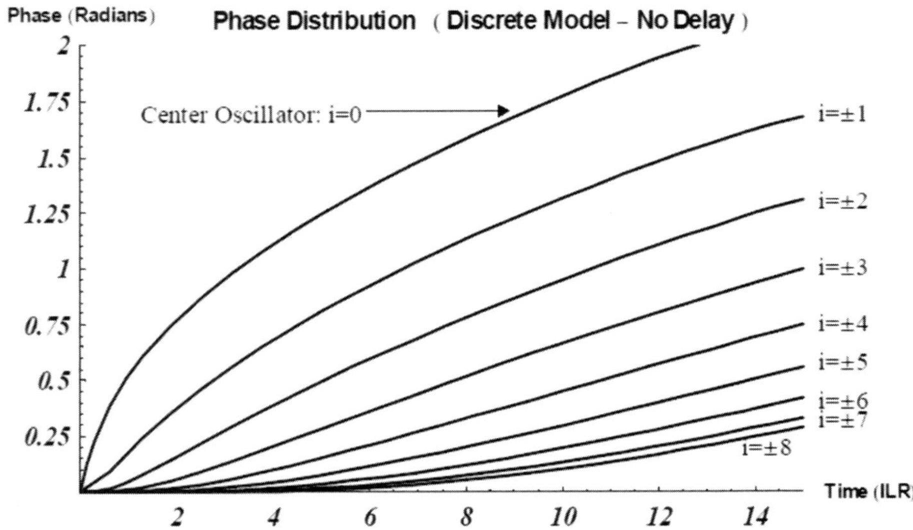

Fig. 5-2. Phase dynamics for a 17-element linear array with no coupling delay. (Reproduced by permission of American Geophysical Union from Ref. 47, ©2008 American Geophysical Union.)

differ only for time greater than six inverse locking ranges. However, the curves in Figs. 5-2 and 5-4 differ for all times because with no coupling delay the end effects begin immediately and, of course, acausally.

Thus, we have shown that the introduction of coupling delay in the linearized discrete model of coupled oscillator arrays eliminates the noncausal nature of the solutions in the absence of coupling delay. We now proceed to apply this approach in the continuum model.

5.3 The Continuum Model with Coupling Delay

In this section we develop a generalization of the continuum model of Section 3.1 that accounts for coupling delay. The causality properties of this generalization will be discussed in terms of the infinite linear array which, of course, behaves identically to a corresponding finite array for times early enough to preclude end effects. Our approach will be that of Section 3.1, where we begin with Eq. (3.1-1) with the time delay, d, inserted in the arguments of the appropriate terms leading to

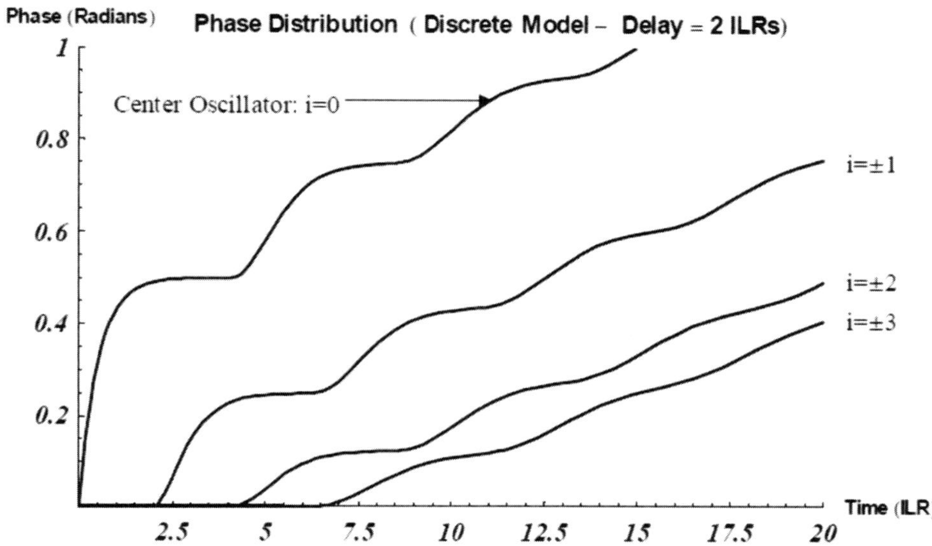

Fig. 5-3. Phase dynamics for a 7-element linear array with two-inverse-locking-range coupling delay. (Reproduced by permission of American Geophysical Union from Ref. 47, ©2008 American Geophysical Union.)

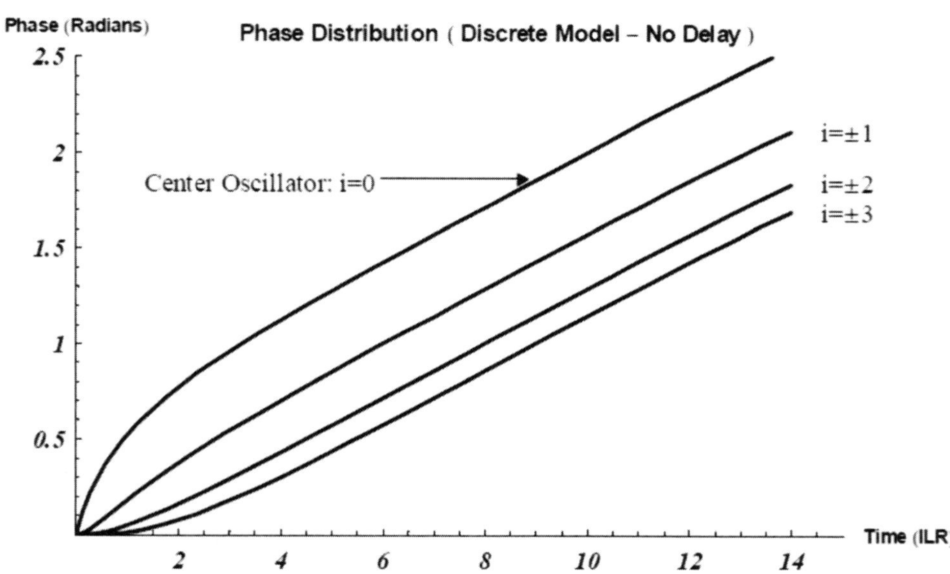

Fig. 5-4. Phase dynamics for a 7-element linear array with no coupling delay. (Reproduced by permission of American Geophysical Union from Ref. 47, ©2008 American Geophysical Union.)

$$\frac{d\varphi(x,t)}{dt} = \omega_0(x) - \omega_{ref} \\ + \Delta\omega_{lock}\left[\varphi(x+\Delta x, t-d) - 2\varphi(x,t) + \varphi(x-\Delta x, t-d)\right] \tag{5.3-1}$$

Introducing the scaled time, τ, and the detuning function, $\Delta\Omega_{tune}$, as before, Laplace transformation leads to

$$s\tilde{\varphi}(x,s) = \Delta\tilde{\Omega}_{tune} \\ +\left[\tilde{\varphi}(x+\Delta x,s)e^{-sd} - 2\tilde{\varphi}(x,t) + \tilde{\varphi}(x-\Delta x,s)e^{-sd}\right] \tag{5.3-2}$$

Then, expanding in Taylor series to second order in Δx, we obtain

$$\frac{d^2\tilde{\varphi}(x,s)}{dx^2} - \left[(s+2)e^{sd} - 2\right]\tilde{\varphi}(x,s) = -\Delta\tilde{\Omega}_{tune}(x,s)e^{sd} \tag{5.3-3}$$

the analog of Eq. (3.1-4). Setting

$$\Delta\tilde{\Omega}_{tune}(x,s) = \frac{1}{s}\delta(x-y) \tag{5.3-4}$$

corresponding to step detuning of the oscillator at $x = y$ at time zero by one locking range, we obtain the Green's function, $\tilde{g}_1$, as the differential equation solution,

$$\tilde{g}_1(s,x,y) = \frac{e^{-|x-y|\sqrt{(s+2)e^{sd}-2}}}{2s\sqrt{(s+2)e^{sd}-2}}e^{sd} \tag{5.3-5}$$

At this point, a serious difficulty is encountered with respect to causality. If one were to compute numerically the inverse Laplace transform integral for Eq. (5.3-5), one would find that the influence of the nearest neighbors of the detuned oscillator begins at time *d*. This violates causality because, as pointed out in Section 5.2, this influence must not begin until time *2d*, the *round-trip* transit time between the detuned oscillator and its neighbors. Following Pogorzelski [48], we begin our study of this apparent paradox by comparing the denominator of Eq. (5.3-5) with that of Eq. (5.2-22) known to be causal. That is, the denominator of Eq. (5.2-22) is

$$\sqrt{(s+2)^2 - 4e^{-2sd}} = \sqrt{(s+2) - 2e^{-sd}}\sqrt{(s+2) + 2e^{-sd}} \tag{5.3-6}$$

while the denominator of Eq. (5.3-5) is

$$\sqrt{(s+2)-2e^{-sd}} \tag{5.3-7}$$

Thus, the two solutions—the causal one Eq. (5.2-22), and the present one, Eq. (5.3-5)—have different branch points in the complex s plane. Solution Eq. (5.3-5) has branch points where $s+2=2e^{-sd}$, whereas the causal solution Eq. (5.2-22) has these plus additional branch points where $s+2=-2e^{-sd}$. Now, computing the inverse Laplace transform via integration on the Bromwich contour will involve deformation of the contour around the branch cuts associated with these branch points. Thus, it becomes clear that the solution Eq. (5.3-5) will be missing the contribution from half of the branch cuts in the causal solution Eq. (5.2-22). As shown in Ref. 48, this is the root of the causality difficulty.

Why do we find ourselves in this situation? Our approach was successful in the absence of coupling delay, but something went wrong when delay was included. This can be understood by looking at the nature of the solutions corresponding to the two sets of branch cuts shown in Fig. 5-5 where the dots correspond to Eq. (5.3-7) and the circles to the remaining branch points of the complete set, Eq. (5.3-6).

In deriving the partial differential equation, Eq. (5.3-3), we have assumed that the solution will be smoothly varying in the interior of the array so that the inter-oscillator phase differences are small, validating the linearization of the sine functions in Adler's formalism. Thus, in the interior of the array where the detuning is zero, the second derivative will be small and

$$\left(s+2\right)e^{sd} \approx 2 \tag{5.3-8}$$

corresponding to the dot branch points in Fig. 5-5. However, we can switch from the dots to the circles by replacing e^{-sd} with $-e^{-sd}$. Doing this in Eq. (5.3-2), we obtain

$$\begin{aligned} s\tilde{\varphi}_1(x,s) &= \Delta\tilde{\Omega}_{tune} \\ &+\left[-\tilde{\varphi}_1(x+\Delta x,s)e^{-sd}-2\tilde{\varphi}_1(x,t)-\tilde{\varphi}_1(x-\Delta x,s)\right] \end{aligned} \tag{5.3-9}$$

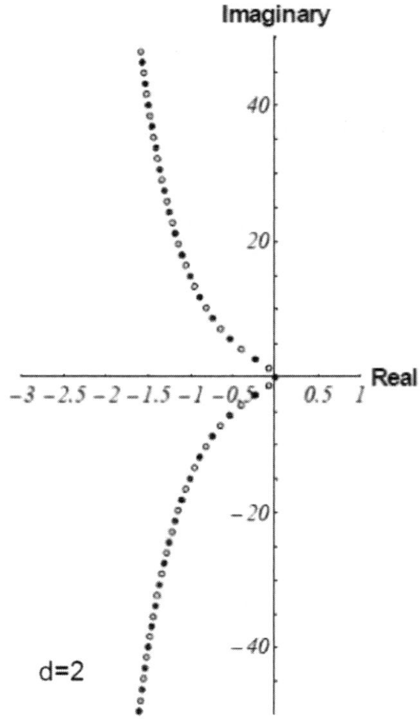

Fig. 5-5. Branch point locations for delay of two inverse locking ranges. (Reproduced by permission of American Geophysical Union from Ref. 48, ©2008 American Geophysical Union.)

so it is evident that φ_1, the solution associated with the circle-type branch points, alternates in sign between adjacent oscillators and is thus clearly not slowly varying. Therefore, we cannot use the Taylor expansion to advantage here. However, if we define

$$\tilde{\varphi}_2(x,s) = \tilde{\varphi}_1(x,t)e^{j\pi x} \tag{5.3-10}$$

then $\tilde{\varphi}_2$ *is* slowly varying even though $\tilde{\varphi}_1$ is not and we may write

$$\begin{aligned}\tilde{\varphi}_2(x,s) &= \Delta\tilde{\Omega}_{tune} \\ &+ \left[\tilde{\varphi}_2(x+\Delta x,s)e^{-sd} - 2\tilde{\varphi}_2(x,t) - \tilde{\varphi}_2(x-\Delta x,s)\right]\end{aligned} \tag{5.3-11}$$

and expand in Taylor series to obtain

$$\frac{d^2\tilde{\varphi}_2(x,s)}{dx^2}+\left[(s+2)e^{sd}+2\right]\tilde{\varphi}_2(x,s)=\Delta\tilde{\Omega}_{tune}(x,s)e^{sd} \tag{5.3-12}$$

corresponding to the circle-type branch points. Thus, it becomes clear that our assumption of slowly varying phase, implicit in the use of the Taylor series, eliminated the solutions associated with the circle-type branch points. The Green's function corresponding to these branch points is

$$\tilde{g}_2(s,x,y)=\frac{-e^{-j|x-y|\sqrt{(s+2)e^{sd}+2}}}{2sj\sqrt{(s+2)e^{sd}+2}}e^{sd+j\pi(x-y)} \tag{5.3-13}$$

and the causal Green's function is a linear combination of Eq. (5.3-5) and Eq. (5.3-13); that is,

$$\tilde{g}(s,x,y)=A\frac{e^{-j|x-y|\sqrt{2-(s+2)e^{sd}}}}{2sj\sqrt{2-(s+2)e^{sd}}}e^{sd}-B\frac{e^{-j|x-y|\sqrt{2+(s+2)e^{sd}}}}{2sj\sqrt{2+(s+2)e^{sd}}}e^{sd-j\pi(x-y)} \tag{5.3-14}$$

where $A + B = 1$ so that the proper detuning function is generated on the right side of the differential equation. It remains to determine A and B.

For large values of s, far from the origin of the s plane, we can obtain a fairly accurate estimate of the branch point locations. These locations are defined by

$$(s+2)=\pm 2e^{-sd} \tag{5.3-15}$$

where the upper sign corresponds to the dots and the lower one to the circles. Inserting $s=\sigma+j\omega$, we obtain

$$(\sigma+2+j\omega)=\pm 2e^{-\sigma d}e^{-j\omega d} \tag{5.3-16}$$

For $|\omega|>>\sigma+2$, we have

$$\omega\approx 2e^{-\sigma d}e^{-j(\omega d\pm\pi/2)} \tag{5.3-17}$$

Thus,

$$\omega_p d \pm \frac{\pi}{2} = 2p\pi \tag{5.3-18}$$

and we have

$$\omega_n \approx \left(2n - \frac{1}{2}\right)\frac{\pi}{d} \tag{5.3-19}$$

for the dots and

$$\omega_m \approx \left(2m + \frac{1}{2}\right)\frac{\pi}{d} \tag{5.3-20}$$

for the circles. Now, from Eq. (5.3-17), we obtain

$$\left|\omega_p\right| \approx 2e^{-\sigma d} \tag{5.3-21}$$

so

$$\sigma_p \approx -\frac{1}{d}\ln\left(\frac{\left|\omega_p\right|}{2}\right) \tag{5.3-22}$$

for $p = m$ or n. Armed with these approximate branch point locations, we are in a position to estimate $\partial g(\tau, y, y)/\partial\tau$, the time derivative of the phase of the detuned oscillator. This will exhibit the temporal discontinuities associated with the arrival of influence from neighboring oscillators and highlight the causal behavior. First, from Eq. (5.3-14), we have

$$s\tilde{g}(s, y, y) = A\frac{e^{sd}}{2j\sqrt{2-(s+2)e^{sd}}} - B\frac{e^{sd}}{2j\sqrt{2+(s+2)e^{sd}}} \tag{5.3-23}$$

Now, envisioning the inverse transform as a sum of branch cut integrals, we recognize that the result will be approximately

$$\frac{\partial g(\tau,y,y)}{\partial \tau} \approx \frac{\alpha}{\pi\sqrt{2d+1}}\left\{ A + 2A\sum_{n=1}^{\infty} \frac{\sin\left[\frac{2n\pi\tau}{d} - \frac{\pi\tau}{2d}\right]}{\left[\frac{n\pi}{d}\right]^{1+\tau/d}}\sqrt{1+\frac{1}{2d}} \right.$$
$$\left. + 2B\sum_{m=0}^{\infty} \frac{\sin\left[\frac{2m\pi\tau}{d} + \frac{\pi\tau}{2d}\right]}{\left[\frac{m\pi}{d}\right]^{1+\tau/d}}\sqrt{1+\frac{1}{2d}} \right\} \tag{5.3-24}$$

Here we have used the *s*'s given by Eq. (5.3-19) through Eq. (5.3-22) and

$$\alpha = \int_0^{\infty} \frac{e^{-u(\tau+d)}}{\sqrt{u}}du = \sqrt{\frac{\pi}{\tau+d}} \tag{5.3-25}$$

The expression on the right side of Eq. (5.3-24) is a Fourier series except for the time dependence of the coefficients. Recall that this series was obtained using the large *s* approximation, so only the high-order terms are accurate. The high-order terms of this series govern the discontinuities in the time dependence. Now, looking at Eq. (5.3-24) for $\tau = d$, we obtain

$$\frac{\partial g(\tau,y,y)}{\partial \tau} \approx \frac{\alpha}{\pi\sqrt{2d+1}}\left\{ A + 2A\sum_{n=1}^{\infty} \frac{\sin\left[\left(2n - \frac{1}{2}\right)\pi\right]}{\left[\frac{n\pi}{d}\right]^{2}}\sqrt{1+\frac{1}{2d}} \right.$$
$$\left. + 2B\sum_{m=0}^{\infty} \frac{\sin\left[\left(2m + \frac{1}{2}\right)\pi\right]}{\left[\frac{m\pi}{d}\right]^{2}}\sqrt{1+\frac{1}{2d}} \right\} \tag{5.3-26}$$

and we see that if $A = B$, the high-order portions of the two series will cancel term by term so that there will be no discontinuity at $\tau = d$. However, at $\tau = 2d$ we have

$$\frac{\partial g(\tau,y,y)}{\partial\tau}\approx\frac{\alpha}{\pi\sqrt{2d+1}}\left\{A+2A\sum_{n=1}^{\infty}\frac{\sin\left[(4n-1)\pi\right]}{\left[\frac{n\pi}{d}\right]^3}\sqrt{1+\frac{1}{2d}}\right.$$
$$\left.+2B\sum_{m=0}^{\infty}\frac{\sin\left[(4m+1)\pi\right]}{\left[\frac{m\pi}{d}\right]^3}\sqrt{1+\frac{1}{2d}}\right\} \tag{5.3-27}$$

and the high-order terms no longer cancel but add. Thus, there *will* be a discontinuity at $\tau = 2d$. This is to be expected because it allows for one round-trip interval to the nearest neighbors from the time when the oscillator is detuned. We conclude that the discontinuities will occur at the proper times for causality to be satisfied only if $A = B$. From this condition and the fact that $A + B = 1$, we determine that both A and B are equal to ½, and from Eq. (5.3-14) the causal Green's function is

$$\tilde{g}(s,x,y)=\frac{e^{-j|x-y|\sqrt{2-(s+2)e^{sd}}}}{4sj\sqrt{2-(s+2)e^{sd}}}e^{sd}-\frac{e^{-j|x-y|\sqrt{2+(s+2)e^{sd}}}}{4sj\sqrt{2+(s+2)e^{sd}}}e^{sd-j\pi(x-y)} \tag{5.3-28}$$

As shown in Ref. 48, a better approximation to the exact discrete model solution may be obtained from the form,

$$\tilde{g}(s,x,y)=C\frac{e^{-j|x-y|C\sqrt{2-(s+2)e^{sd}}}}{4sj\sqrt{2-(s+2)e^{sd}}}e^{sd}-C\frac{e^{-j|x-y|C\sqrt{2+(s+2)e^{sd}}}}{4sj\sqrt{2+(s+2)e^{sd}}}e^{sd-j\pi(x-y)} \tag{5.3-29}$$

with optimal selection of the constant, C. From Ref. 48, the optimal value of C is

$$C = \frac{\pi}{2\sqrt{2}} \tag{5.3-30}$$

The temporal behavior of the phase of each oscillator in the array is most easily seen by plotting the time derivative of the phase because this makes more obvious the times at which the influences from the neighboring oscillators arrive. Thus, in Figs. 5-6 through 5-10, we compare the result of the approximate continuum formula Eq. (5.3-29) in solid lines with that of the discrete model Eq. (5.2-22) in dashed lines considered to be the exact result. The coupling delay in this example is two inverse locking ranges (ILRs). The fine scale wiggles shown in the inset of Fig. 5-6 arise from the truncation of the series of branch cut integrals to a finite number of terms.

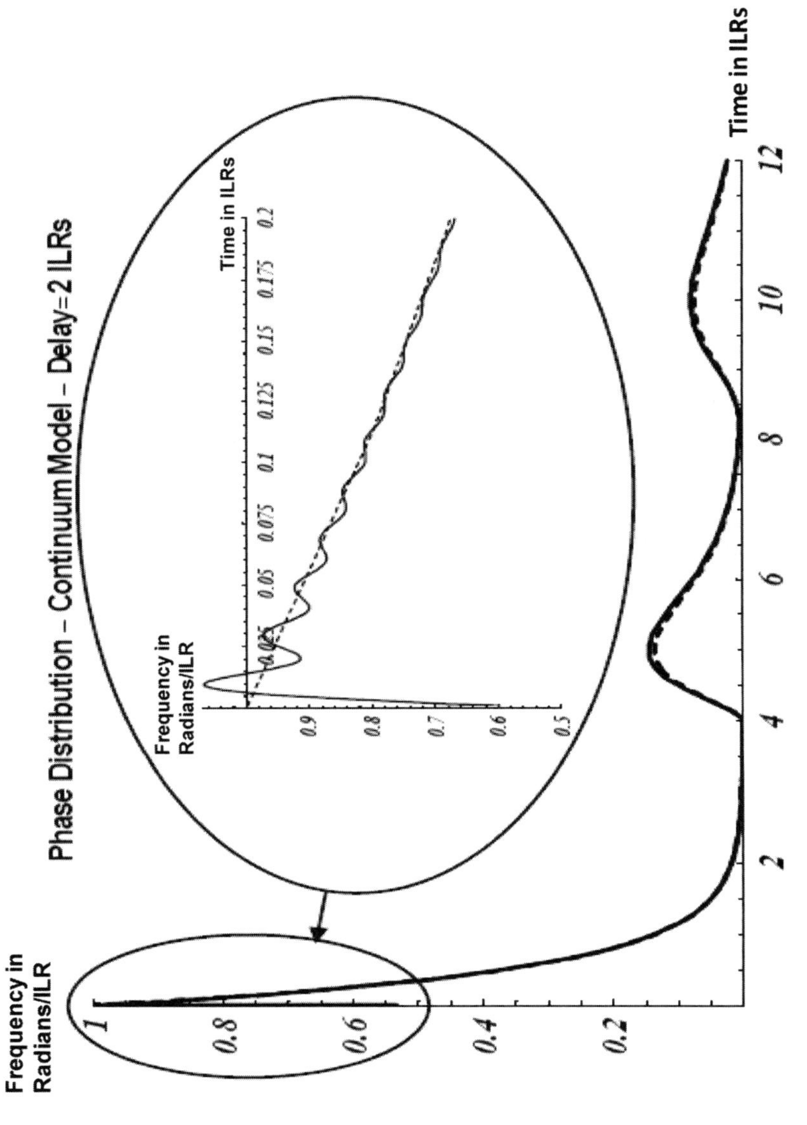

Fig. 5-6. **Time derivative of the phase of the detuned oscillator. (Reproduced by permission of American Geophysical Union from Ref. 48, ©2008 American Geophysical Union.)**

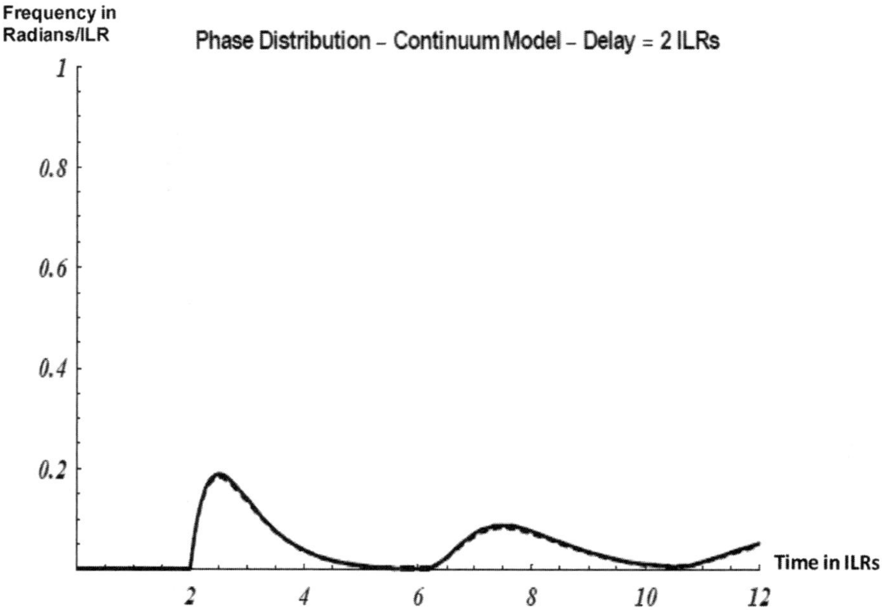

Fig. 5-7. Time derivative of the phase of the nearest neighbors of the detuned oscillator. (Reproduced by permission of American Geophysical Union from Ref. 48, ©2008 American Geophysical Union.)

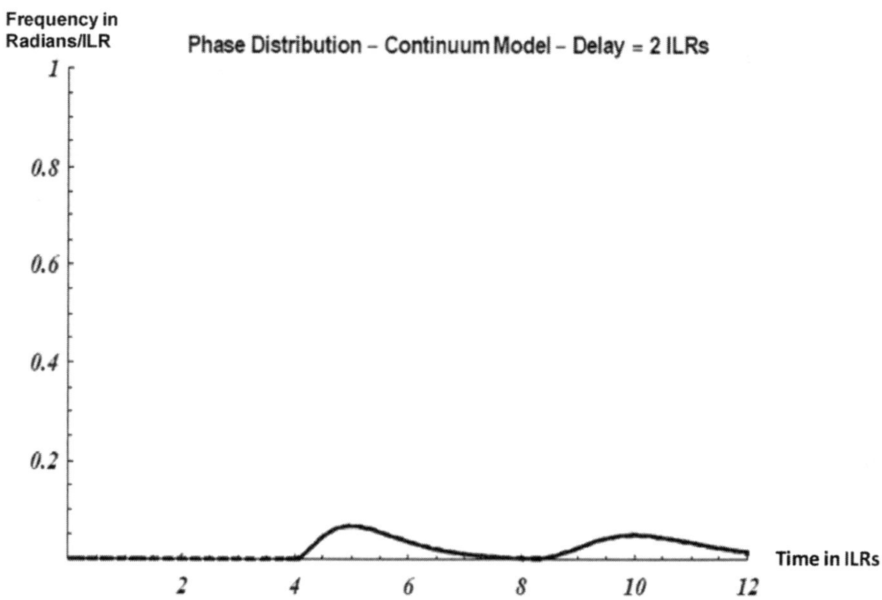

Fig. 5-8. Time derivative of the phase of the second nearest neighbors of the detuned oscillator. (Reproduced by permission of American Geophysical Union from Ref. 48, ©2008 American Geophysical Union.)

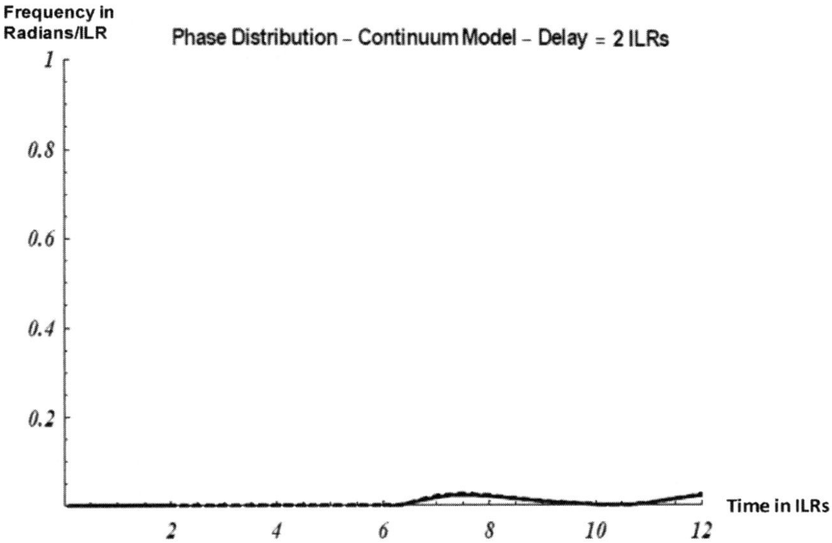

Fig. 5-9. Time derivative of the phase of the third nearest neighbors of the detuned oscillator. (Reproduced by permission of American Geophysical Union from Ref. 48, ©2008 American Geophysical Union.)

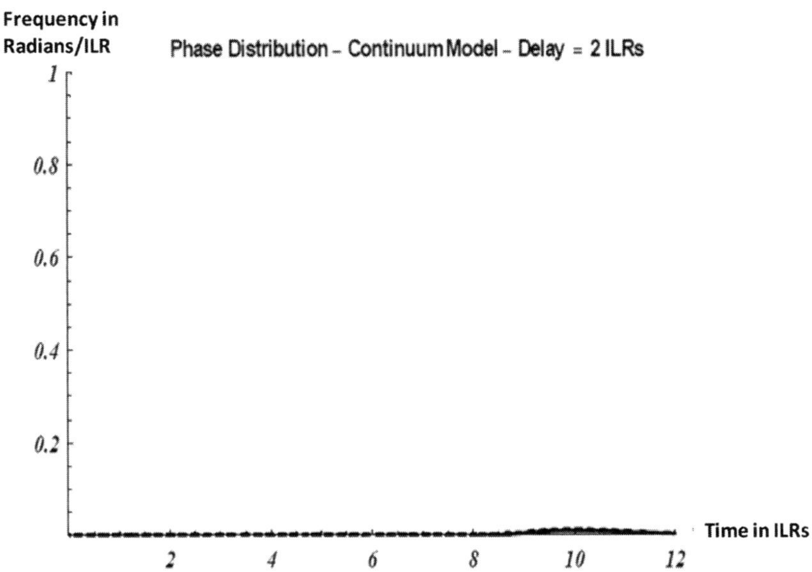

Fig. 5-10. Time derivative of the phase of the fourth nearest neighbors of the detuned oscillator. (Reproduced by permission of American Geophysical Union from Ref. 48, ©2008 American Geophysical Union.)

Notice that the more distant the oscillator from the detuned one, the later the response by exactly two inverse locking ranges (one delay time) per oscillator. Moreover, the influence of the nearest neighbors of the detuned oscillator does not impact that oscillator until four inverse locking ranges (two delay times) have elapsed. Similar delays of two delay times are visible in all of the curves corresponding to round-trip delays between the oscillators. All of these behaviors are consistent with a causal solution.

We return now to the previous analysis of the location of the branch points to highlight two properties that may not have been obvious in the earlier discussion. First, as the delay time is decreased, there is a critical value at which the distribution of the branch points changes character. If the delay time is equal to 0.139232271 inverse locking ranges, the smallest circle-type branch points merge at $\sigma = -9.18224297$. For delays less than that, say for a delay of 0.12 inverse locking ranges, the branch point locations are as shown in Fig. 5-11.

Second, as the delay approaches zero, all of the branch points move to infinity except two: one at the origin and one at –4. Thus, in this zero-delay limit we have from Eq. (5.2-22), taken to be the exact solution, that

$$\tilde{g}(s,x,y)=\frac{e^{-|x-y|\sqrt{s}}}{s\sqrt{s(s+4)}} \tag{5.3-31}$$

which, perhaps surprisingly, does not agree with Eq. (3.1-6). It does agree in the limit of small s, so one can expect that the time functions will agree for late times, but there will be a difference at early times. When $x = y$; that is, for the detuned oscillator, the inverse Laplace transforms of Eqs. (3.1-6) and (5.3-31) can be computed analytically, and we thus obtain from Eq. (5.3-31)

$$g(\tau,y,y)=\tau\, e^{-2\tau}\left[I_0(2\tau)+I_1(2\tau)\right]u(\tau) \tag{5.3-32}$$

where I_n is the Bessel function of imaginary argument and from Eq. (3.1-6), we get

$$g(\tau,y,y)=\sqrt{\frac{\tau}{\pi}}\,u(\tau) \tag{5.3-33}$$

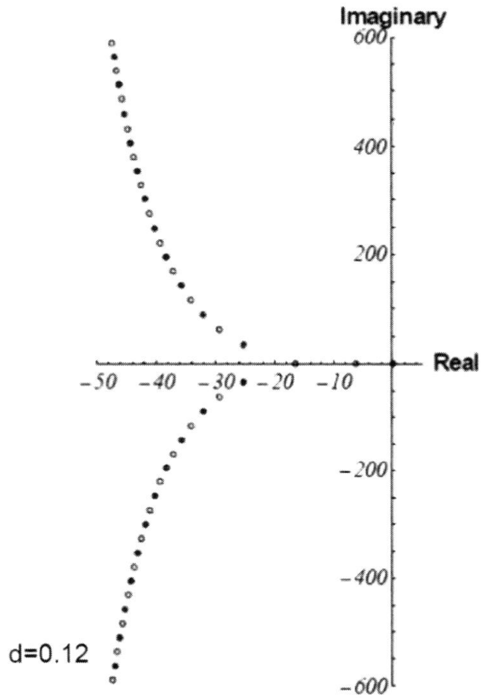

Fig. 5-11. Branch point locations for *d* = 0.12 inverse locking ranges. (Reproduced by permission of American Geophysical Union from Ref. 48, ©2008 American Geophysical Union.)

For comparison, these two functions are plotted in Fig. 5-12. The solid curve is Eq. (5.3-32), and the short dashed curve is Eq. (5.3-33) while the long dashed curve is the difference. Note that, although Eq. (5.3-33) neglects the alternating sign solution, it is nevertheless a very good approximation to the exact solution Eq. (5.3-32).

5.4 Beam Steering in the Continuum Model with Coupling Delay

In this section we apply what we have learned so far regarding the analytical treatment of coupling delay to the analysis of beam-steering of oscillator arrays embodying such delay. We begin with the continuum generalization of the linearized discrete model solution for one detuned oscillator in an infinite array given by Eq. (5.2-22); that is,

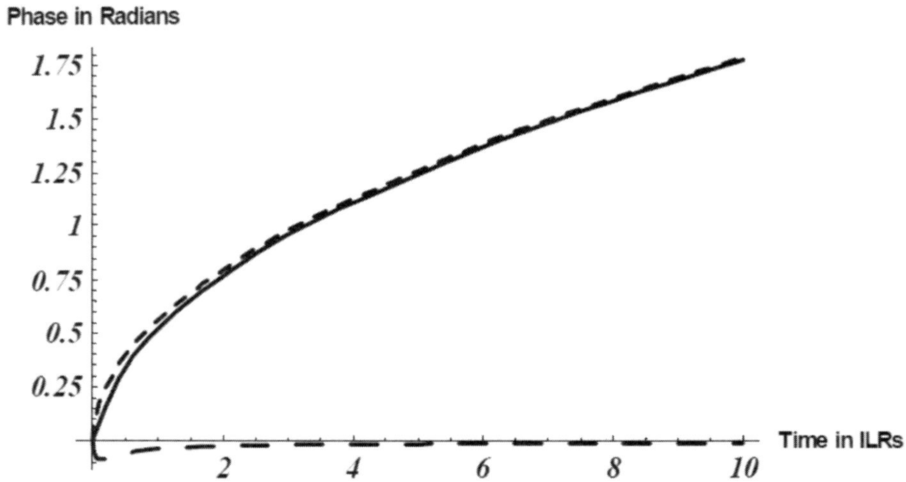

Fig. 5-12. Comparison of the single-branch-point solution (dashed) with the two-branch-point solution (solid). The difference is shown in long dashes. (Reproduced by permission of American Geophysical Union from Ref. 48, ©2008 American Geophysical Union.)

$$\tilde{\varphi}(s,x,y) = \frac{\Delta\tilde{\Omega}_{tune}(y)}{\sqrt{(s+2)^2 - 4e^{-2sd}}} e^{-|x-y| \operatorname{sec} h^{-1}\left(\frac{2e^{-sd}}{s+2}\right)} \tag{5.4-1}$$

Our approach will be to devise a differential equation having Eq. (5.4-1) as its Green's function. When x is not equal to y, this solution satisfies the differential equation,

$$\frac{d^2\tilde{\varphi}}{dx^2} - \left[\operatorname{sec} h^{-1}\left(\frac{2e^{-sd}}{s+2}\right)\right]^2 \tilde{\varphi} = 0 \tag{5.4-2}$$

However, at $x = y$ there will be a discontinuity in the slope of the phase that gives rise to a delta function. Evaluating the magnitude of this slope discontinuity, we determine that the Green's function Eq. (5.4-1) satisfies

$$\frac{d^2\tilde{\varphi}}{dx^2}-\left[\sec h^{-1}\left(\frac{2e^{-sd}}{s+2}\right)\right]^2\tilde{\varphi}$$
$$=-\Delta\tilde{\Omega}_{tune}(y)\frac{2\sec h^{-1}\left(\frac{2e^{-sd}}{s+2}\right)}{\sqrt{(s+2)^2-4e^{-2sd}}}\delta(x-y) \qquad (5.4\text{-}3)$$

We may now follow the procedure of Section 3.1 to express the finite-array Green's function as a sum of the eigenfunctions of the differential operator in this equation. In order to do this, we will need the boundary conditions at the ends of the array. Recall that the reflection coefficient at the array ends was given by Eq. (5.2-20), which is a fairly complicated function of s. However, following Pogorzelski [49], we may simplify matters by assuming the addition of half-length coupling lines at the ends of the array. If this is done, the reflection coefficient becomes unity because the array boundary then becomes an image plane. (See Pogorzelski [47].) A reflection coefficient of unity corresponds to the familiar Neumann condition of zero phase slope. Using this boundary condition, the even and odd normalized eigenfunctions are seen to be

$$\psi_{ne}=\frac{\sqrt{2}}{\sqrt{2N+1}}\cos\left(\frac{2n\pi x}{2N+1}\right)$$
$$\psi_{no}=\frac{\sqrt{2}}{\sqrt{2N+1}}\sin\left(\frac{(2n+1)\pi x}{2N+1}\right) \qquad (5.4\text{-}4)$$

Choosing the detuning time dependence to be a unit step at time zero and following he approach of Section 3.1, the Laplace transform of the phase distribution may be written in terms of the eigenfunctions as

$$\tilde{\varphi}(s,x,y)=\frac{4\sec h^{-1}\left(\dfrac{2e^{-sd}}{s+2}\right)}{(2N+1)s\sqrt{(s+2)^2-4e^{-2sd}}}$$

$$\times\sum_n\left\{\frac{\cos\left(\dfrac{2n\pi y}{2N+1}\right)\cos\left(\dfrac{2n\pi x}{2N+1}\right)}{\left(\dfrac{2n\pi}{2N+1}\right)^2+\left[\sec h^{-1}\left(\dfrac{2e^{-sd}}{s+2}\right)\right]^2}\right. \tag{5.4-5}$$

$$\left.+\frac{\sin\left(\dfrac{(2n+1)\pi y}{2N+1}\right)\sin\left(\dfrac{(2n+1)\pi x}{2N+1}\right)}{\left(\dfrac{(2n+1)\pi}{2N+1}\right)^2+\left[\sec h^{-1}\left(\dfrac{2e^{-sd}}{s+2}\right)\right]^2}\right\}$$

We will obtain the inverse Laplace transform via residue calculus. The poles are determined by

$$\sec h^{-1}\left(\frac{2e^{-sd}}{s+2}\right)=\pm\left(\frac{\ell\pi i}{N+1}\right) \tag{5.4-6}$$

Taking the hyperbolic secant of both sides and then the reciprocal, we obtain the equivalent condition

$$\left(\frac{s+2}{2e^{-sd}}\right)=\cos\left(\frac{\ell\pi}{2N+1}\right) \tag{5.4-7}$$

This equation can be solved in terms of the Lambert W function defined by

$$z=W(z)e^{W(z)} \tag{5.4-8}$$

In terms of this function, the solution of Eq. (5.4-7) is

$$s_{m\ell}=\frac{1}{d}W\left[m,2de^{2d}\cos\left(\frac{\ell\pi}{2N+1}\right)\right]-2 \tag{5.4-9}$$

These pole locations are plotted in Fig. 5-13.

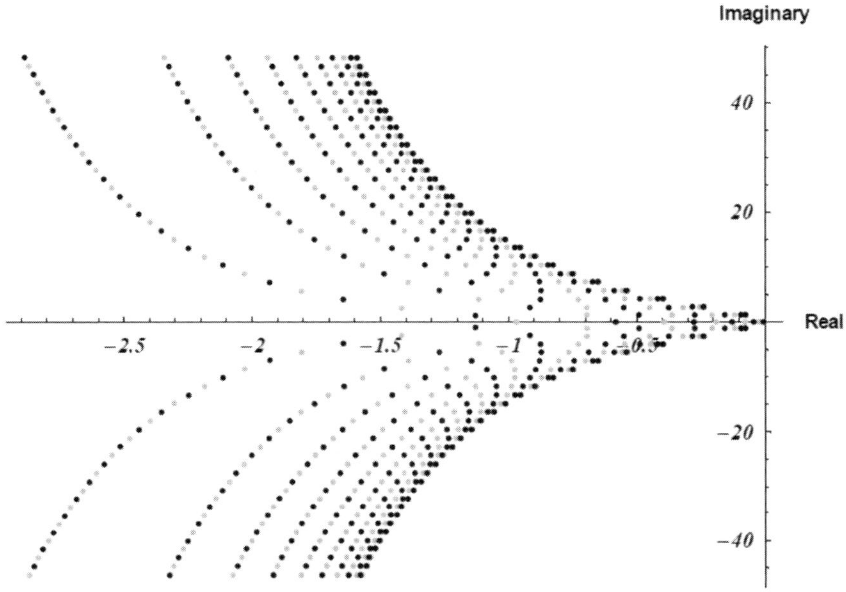

Fig. 5-13. Pole locations for delay of two inverse locking ranges. Black dots denote odd values of ℓ and gray dots denote even values of ℓ. (Reproduced by permission of American Geophysical Union from Ref. 49, ©2008 American Geophysical Union.)

The overall array time constant is determined by the poles closest to the origin. We therefore set about solving Eq. (5.4-9) approximately for small s. To do this we expand the Lambert W function in a Taylor series about $2de^{2d}$.

$$W(z) = W(2de^{2d}) + W'(2de^{2d})\left(z - 2de^{2d}\right) + \cdots \tag{5.4-10}$$

Now, W satisfies the differential equation

$$W'(z) = \frac{W}{z(W+1)} \tag{5.4-11}$$

Thus, the first two terms of the Taylor series yield

$$W(z) = 2d + \frac{z - 2de^{2d}}{e^{2d}(2d+1)} \tag{5.4-12}$$

and, using Eq. (5.4-9), we obtain

$$s_{m\ell} = \frac{z - 2de^{2d}}{e^{2d}(2d+1)} \tag{5.4-13}$$

Recall that z is the argument of the W function, so from Eq. (5.4-9) we have

$$z = 2de^{sd}\cos\left(\frac{\ell\pi}{2N+1}\right) \tag{5.4-14}$$

Substituting Eq. (5.4-14) into Eq. (5.4-13) and setting $m = 0$, we obtain

$$s_{0\ell} = \frac{2de^{sd}\cos\left(\frac{\ell\pi}{2N+1}\right) - 2de^{2d}}{e^{2d}(2d+1)} = \frac{-4}{2d+1}\sin^2\left(\frac{\ell\pi}{2N+1}\right) \tag{5.4-15}$$

The pole at $\ell = 0$ together with the denominator s from the step detuning function produce the double pole at the origin leading to the linear time dependence or shift in ensemble frequency due to the detuning. For the antisymmetric detuning used in beam-steering, the even ℓ poles do not contribute, so the dominant pole is the one for $\ell = 1$ lying on the real axis at

$$s_{01} = \frac{-4}{2d+1}\sin^2\left(\frac{\pi/2}{2N+1}\right) \approx \frac{-1}{2d+1}\left(\frac{\pi}{2N+1}\right)^2 \tag{5.4-16}$$

so the time constant of the array is

$$\tau_c \approx (2d+1)\left(\frac{2N+1}{\pi}\right)^2 \tag{5.4-17}$$

or just $(2d + 1)$ times the time constant without coupling delay. (Compare with Eq. (2.2-40).)

Returning now to Eq. (5.4-5), we form the solution for beam-steering by combining two solutions of the form Eq. (5.4-5), one for detuning of the oscillator at $-N$ and one for detuning of the oscillator at N, at each end of the array.

$$\tilde{\varphi}(s,x,N)-\tilde{\varphi}(s,x,-N)=\frac{8\sec h^{-1}\left(\frac{2e^{-sd}}{s+2}\right)}{(2N+1)s\sqrt{(s+2)^2-4e^{-2sd}}}$$

$$\times\sum_n\frac{\sin\left(\frac{(2n+1)\pi y}{2N+1}\right)\sin\left(\frac{(2n+1)\pi x}{2N+1}\right)}{\left(\frac{(2n+1)\pi}{2N+1}\right)^2+\left[\sec h^{-1}\left(\frac{2e^{-sd}}{s+2}\right)\right]^2} \tag{5.4-18}$$

To obtain the residues, we define $q(s)$ to be the denominator,

$$q(s)=\left(\frac{(2n+1)\pi}{2N+1}\right)^2+\left[\sec h^{-1}\left(\frac{2e^{-sd}}{s+2}\right)\right]^2 \tag{5.4-19}$$

and expand in the Taylor series,

$$q(s)\approx q'(s_{m,2n+1})(s-s_{m,2n+1})+O\left((s-s_{m,2n+1})^2\right) \tag{5.4-20}$$

We thus obtain the aperture phase a function of time in the form

$$\varphi(\tau,x,N)-\varphi(\tau,x,-N)=$$

$$\times\sum_{m,n}\frac{4\left(e^{s_{m,2n+1}\tau}-e^{-s_{m,2n+1}d/4}\right)}{(2N+1)s_{m,2n+1}\left[s_{m,2n+1}d+(2d+1)\right]} \tag{5.4-21}$$

$$\times\sin\left(\frac{(2n+1)\pi y}{2N+1}\right)\sin\left(\frac{(2n+1)\pi x}{2N+1}\right)u(\tau)$$

Recall that at each oscillator, x is an integer and that the phase only has physical meaning at these integral values of x. As a result, the sum on n need only extend from 0 to $N-1$ because for integral x, these terms are equal to those for $n = N + 1$ through $2N$ with the order reversed. Higher-order terms in n only affect the phase values between the oscillators and thus are not relevant.

Typically, the time at which the time function becomes nonzero is determined by when the Bromwich contour used in the inverse Laplace transform integral can be closed in the left half-plane, thus enclosing the poles. Prior to that time, the contour may only be closed in the right half-plane; and since it encloses no poles there, the solution is zero. This is the usual way in which causality enters such analysis. In this case, however, Eq. (5.4-18) reveals that the contour may be closed in the left half-plane beginning at $\tau = -d$. Causality dictates that the solution remain zero until $\tau = 0$. Thus, it turns out that the residue sum remains zero even though the contour is closed in the left half-plane and only becomes nonzero after $\tau = 0$. This is illustrated in Eq. (5.4-21), in which the contour was closed in the left half-plane beginning at $\tau = -d/4$ leading to the $e^{s_{m,2n+1}d/4}$ term. Figure 5-14 shows the resulting solution for each oscillator of a 21-element array with coupling delay of two inverse locking ranges. Causality is obviously satisfied regardless of this unusual closing of the contour. Figure 5-15 shows the same solution extending to later times showing that in steady state the phase increments between oscillators become equal, implying a linear phase progression as needed for beam-steering.

The data in Figs. 5-14 and 5-15 are re-plotted in Figs. 5-16 and 5-17, respectively. Here one may view the aperture phase distribution at all values of time simultaneously.

Finally Fig. 5-18 shows a particular range of time specifically for comparison with Fig. 5-19, which is the same case but with no coupling delay. Pay particular attention to the time scales in these plots.

The point made by comparing Fig. 5-18 with Fig. 5-19 is that the coupling delay of two inverse locking ranges has slowed the response of the array by $2d + 1$ or a factor of five, just as predicted by Eq. (5.4-17).

We now compute the far-zone radiated field when the oscillators in this 21-element array with coupling delay are used to excite the elements of a phased array with half-wavelength element spacing. The result is shown in Fig. 5-20 where we see beam behavior very similar to that of arrays without delay but slower by $(2d + 1)$.

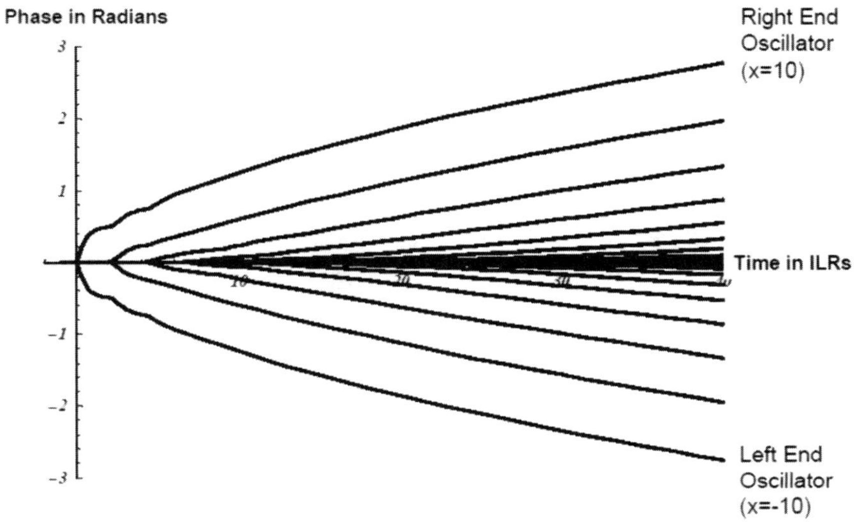

Fig. 5-14. Phase evolution of the oscillators in a 21-element linear array with coupling delay of two inverse locking ranges. (Reproduced by permission of American Geophysical Union from Ref. 49, ©2008 American Geophysical Union.)

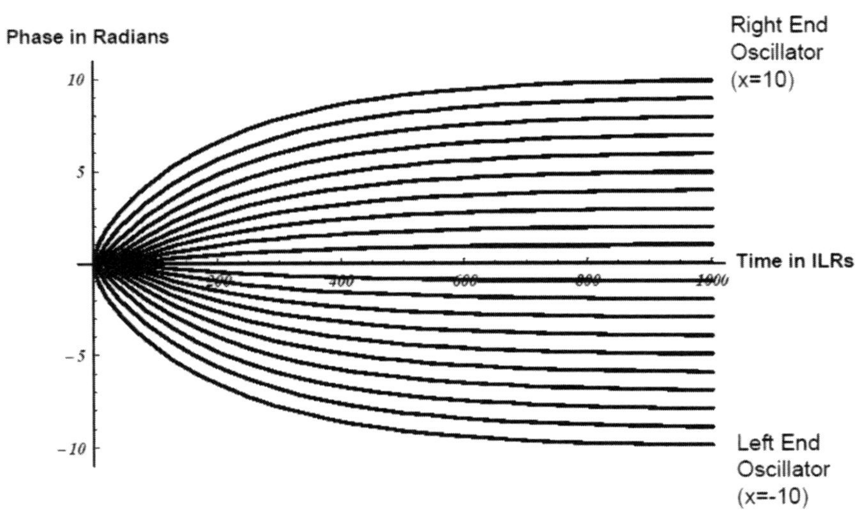

Fig. 5-15. Phase evolution of the oscillators in a 21-element linear array with coupling delay of two inverse locking ranges over a longer duration. (Reproduced by permission of American Geophysical Union from Ref. 49, ©2008 American Geophysical Union.)

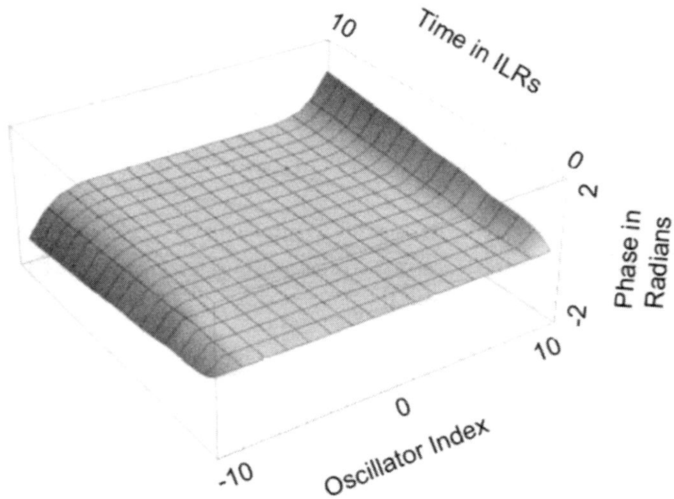

Fig. 5-16. Three-dimensional representation of the phase evolution in a 21-element array at early times. (Reproduced by permission of American Geophysical Union from Ref. 49, ©2008 American Geophysical Union.)

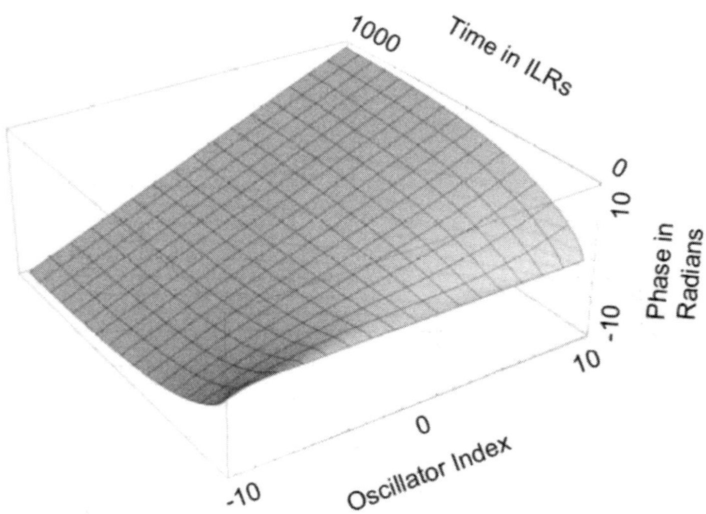

Fig. 5-17. Three-dimensional representation of the phase evolution in a 21-element array at later times. (Reproduced by permission of American Geophysical Union from Ref. 49, ©2008 American Geophysical Union.)

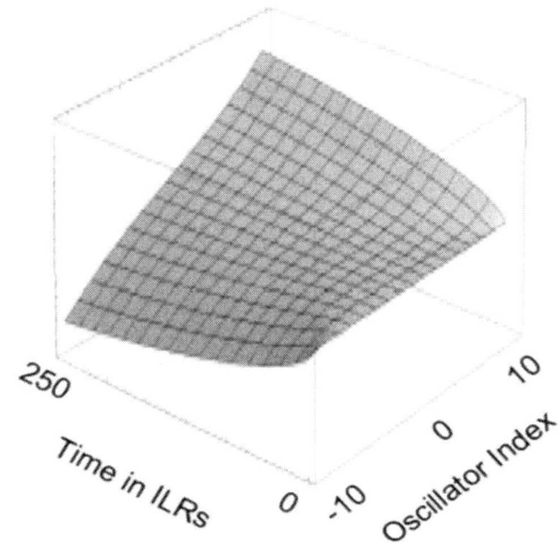

Fig. 5-18. Phase evolution over 250 inverse locking ranges for a 21-element array with coupling delay of two inverse locking ranges. (The vertical scale is from –10 to 10 radians as in Fig. 5-19.) (Reproduced by permission of American Geophysical Union from Ref. 49, ©2008 American Geophysical Union.)

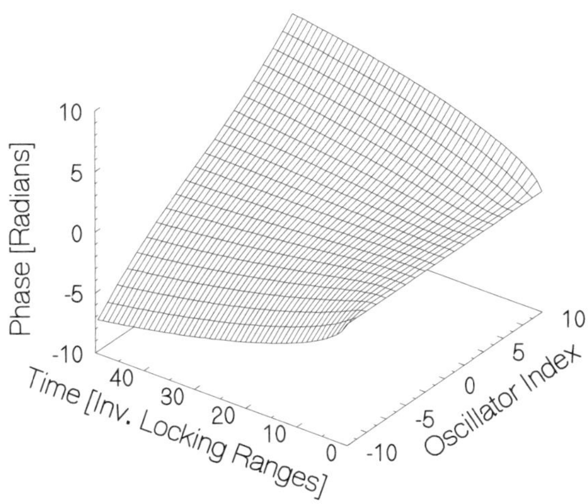

Fig. 5-19. Phase evolution over 250 inverse locking ranges for a 21-element array with no coupling delay. (Reprinted from Ref. 38 with permission, ©2000 IEEE.)

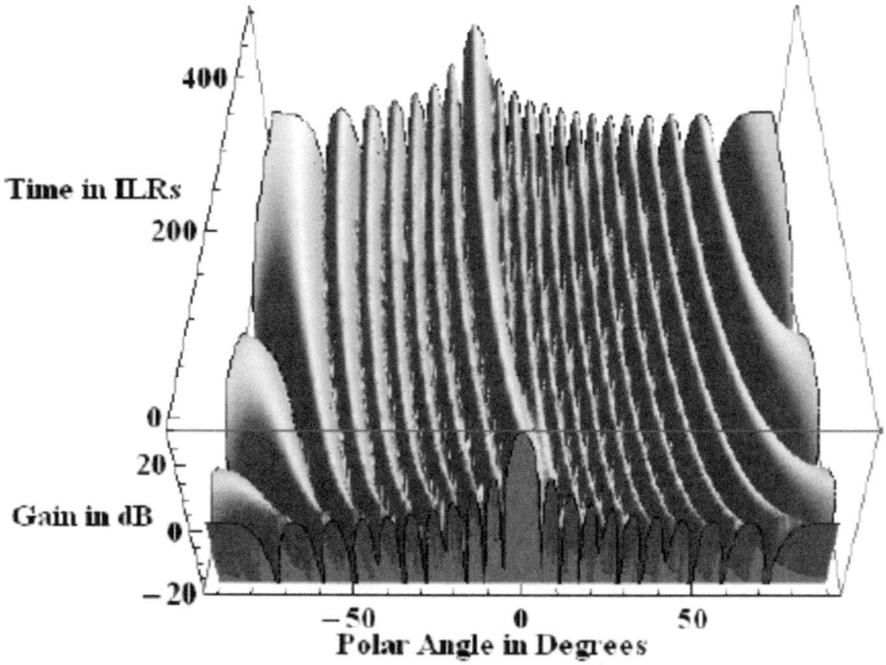

Fig. 5-20. Dynamic behavior of the far-zone radiated field for a 21-element array with coupling delay of two inverse locking ranges. (Reproduced by permission of American Geophysical Union from Ref. 49, ©2008 American Geophysical Union.)

The final value theorem applied to Eq. (5.4-18) gives the steady-state phase distribution as

$$\varphi_{ss}(x) = \frac{4}{(N+1)} \sum_n \frac{\sin\left(\frac{(2n+1)\pi y}{N+1}\right)\sin\left(\frac{(2n+1)\pi x}{N+1}\right)}{\left(\frac{(2n+1)\pi}{N+1}\right)^2} \tag{5.4-22}$$

$$= x \text{ for} -N/2 \le x \le N/2$$

and for half-wavelength spacing of the radiating elements, we find that the steady-state beam position is at

$$\theta = \sin^{-1}\left(\frac{\Delta\varphi}{\pi}\right) = \sin^{-1}\left(\frac{1}{\pi}\right) = 18.56^{\circ} \tag{5.4-23}$$

This one-radian inter-oscillator phase difference stresses the linear approximation a bit in that the error in linearizing the sine function is about 19 percent. However, the dynamic behavior is still qualitatively approximated.

In the above analysis a large number of residues are required for early times and very few are required for late times. However, returning to the discrete model, an alternative formulation is available that provides for more efficient computation for early times. Returning to Eq. (5.2-19) and specializing to the present case of a 21-element array, we have that

$$\begin{aligned} f(s,x) &= \frac{1}{s\sqrt{a^2-4b^2}}\left[\frac{\left(Q^x+Q^{-21}Q^{-x}\right)\left(Q^{-10}+Q^{-21}Q^{10}\right)}{\left(1-Q^{-42}\right)}\right. \\ &\quad \left.-\frac{\left(Q^{-10}+Q^{-21}Q^{10}\right)\left(Q^{-x}+Q^{-21}Q^{x}\right)}{\left(1-Q^{-42}\right)}\right] \\ &= \frac{Q^{10}}{s\sqrt{a^2-4b^2}}\left(\frac{1+Q}{1+Q^{21}}\right)\left(Q^x-Q^{-x}\right) \end{aligned} \tag{5.4-24}$$

where, as before, $a_0 = s+1$, $a = s+2$, and $b = -e^{-sd}$. We now expand this expression in powers of $-b$, and, as before, the inverse Laplace transform of each term in the expansion can be computed analytically. The number of terms required is determined by the time interval over which the response is desired because each term has a delay factor e^{-psd}, where p is the power of $-b$ in the term in the expansion and d is the coupling delay. So, for sufficiently large p, the term will be zero for the interval in question. Thus, in contrast with the eigenfunction expansion, for early times very few terms are required.

This approach was applied to the 21-element array with coupling delay treated earlier, and the results are plotted in Fig. 5-21 for comparison with Fig. 5-14. Interestingly, this power series approach is a bit more flexible in terms of boundary conditions. Recall that without the added half-length coupling lines at each end of the array, the previous method was complicated. Here, however, the use of Eq. (5.2-20) in Eq. (5.2-19) to model an array without the added lines poses no difficulty. The expansion in powers of b proceeds as before and the result is plotted in Fig. 5-22. Notice the difference in the early time ripples due to this alternative boundary condition when compared with Fig. 5-21.

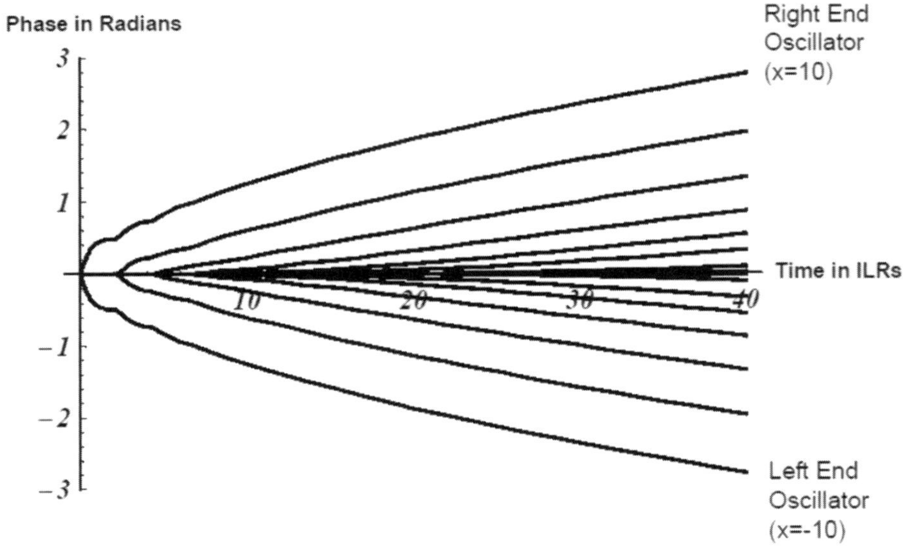

Fig. 5-21. Phase evolution in a 21-element array with coupling delay of two inverse locking ranges via expansion in powers of *b*. (Reproduced by permission of American Geophysical Union from Ref. 49, ©2008 American Geophysical Union.)

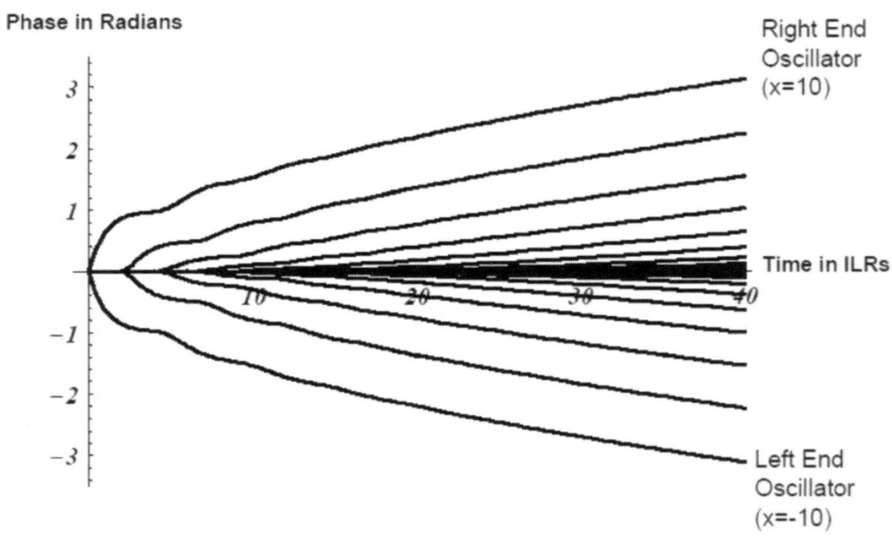

Fig. 5-22. Phase evolution via expansion in powers of *b* for a 21-element array with coupling delay of two inverse locking ranges but without the added half-length coupling lines at the ends. (Reproduced by permission of American Geophysical Union from Ref. 49, ©2008 American Geophysical Union.)

5.5 Conclusion

The primary motivation for this chapter was the issue of causality in coupled oscillator arrays. Because the continuum model leads to a diffusion equation, the response to an excitation always begins immediately regardless of the physical separation of the two. Here, by appropriately introducing a delay factor in the Laplace transforms, we render the solutions causal in that there appears a finite "propagation delay" between the excitation and the response. The result is a more realistic representation of the array response, not to mention some rather interesting inverse Laplace transforms encountered along the way.

Part II: Experimental Work and Applications

Chapter 6
Experimental Validation of the Theory

Much of the very early work in coupled oscillators for phased-array applications involved both theory and experiment. Probably the earliest was the work of Karl Stephan in which he studied a linear array of coupled oscillators for beam-steering of a linear array of radiating elements [1]. In his concept, the phasing was controlled by injection locking the end oscillators to signals whose relative phase was controlled with a phase shifter. While still earlier work was published, Stephan points out that it did not involve mutual injection locking, the defining feature of the concepts treated in this book.

6.1 Linear-Array Experiments

Stephan's pioneering experiment in 1986 involved three very high frequency (VHF) transistor Colpitts oscillators coupled together by a network of lumped elements. The end oscillators were also injected with external signals derived from a master oscillator signal that was split into two signals, one of which was phase-shifted relative to the other by a variable length transmission line (coaxial line stretcher). Stephan measured the oscillator phases with varying amounts of phase shift of the injection signals and verified that the behavior conformed qualitatively to the theory. However, he noted several issues that have persisted throughout the ensuing development of this technology. He noted that manufacturing variation among the oscillators resulting in variation in their free-running frequencies causes nonuniformity in the array behavior and indicated an appreciation of the fact that in his VHF oscillators this can be compensated for with tuning, but that in a monolithic microwave integrated circuit (MMIC) such adjustment would be more difficult. He did some statistical studies of this issue, and such studies were more recently extended by Wang and Pearson [50]. Their approach to mitigating this problem was to design the oscillator to minimize the phase slope of the open-loop gain. The impact of free-running frequency variation on beam pointing was studied by Shen and Pearson [51]. In his early work, Stephan also discussed high-frequency application in spatial power combining and beam-steering, pointing out that there is potential for graceful degradation in the event of oscillator failure. Interestingly, he used a gradual phase shift of the injection signals (as discussed in Section 3.4 above) rather than a step-phase shift in time. The next year, Stephan and Young published theoretical and experimental results concerning two mutually injection-locked oscillators where the coupling was provided by the free-space mutual coupling between the radiating elements excited by them [3]. The coupling was represented by a two-by-two admittance matrix, and the stability of the two modes, even and odd, for the system was treated. The radiation patterns as a function of coupling phase—that is, element separation—showed behavior consistent with the analysis.

Three years later, Robert York, then a student working under Professor Richard Compton at Cornell University, published the results of a study of power combining in mutually injection-locked Gunn-diode oscillators arranged in a four-by-four planar array with Cartesian coupling [4]. Beam steering, however, was not considered; probably at least partly motivated by the publication by James Mink, the emphasis was on power combining [8]. Shortly thereafter, York and Compton published a description of mode locking in arrays of coupled oscillators [5]. They also described excitation of a linear array of radiating elements with a set of mode-locked oscillators [6]. The experiment described involved three Gunn-diode oscillators at 11 GHz. Although beam-steering is discussed, the concept is not the usual phased-array approach to scanning. Rather, mode locking is used to obtain a train of pulses that continuously scan at a rate determined by the spacing of the spectral lines of the periodic pulse train. (See Section 6.5.)

Later in 1992, Hall and Haskins described a two-oscillator element designed for implementation of Stephan's external locking scheme for beam-steering [52]. A four-element array of 2.28-GHz elements was constructed, and beam-steering to 40 deg from normal was demonstrated.

A turning point was reached in 1993 with the publications by York and his student Peter Liao, in a special issue of the *IEEE Transactions on Microwave Theory and Techniques* edited by Mink and Rutledge on Quasi-Optical Techniques [28, 53]. The reported analysis and experiment marked the first application of mutually injection-locked oscillator arrays to beam-steering via detuning of the end oscillators and no external injection. Liao and York constructed a four-element linear array of 10-GHz field effect transistor (FET) oscillators driving radiating elements that also served as resonators for the oscillators. Inter-oscillator coupling was due to mutual coupling among these radiating elements and fell in the weak-coupling regime. Beam-steering to 15 deg from normal was achieved, and the theoretical and experimental results agreed quite well.

In the same special issue discussed above, Nogi et al. described analysis and experimental work with strongly coupled Gunn-diode oscillators at 12.45 GHz [17]. They showed analytically that the array could oscillate in a number of modes, only one of which had constant amplitude across the array as might be desirable in excitation of a phased-array antenna, and they suggested that a resistor at the center of each coupling line would favor this desired mode and suppress the others.

By 1994 the use of coupled oscillators to excite phased-array antenna elements and steer the radiated beam had become a vital and growing area of research. Liao and York reported a six-element microstrip patch array at 4 GHz that could steer to 40 deg from normal [54]. This array did not depend on mutual coupling between the radiating elements for coupling. Rather, the coupling was achieved with transmission lines connected between neighboring patches. This was the first attempt at decoupling the oscillator array design from the radiating aperture design. Later the same year a similar five-element array was reported that steered from –30 to +40 deg from normal. The associated theoretical treatment was based on a general admittance matrix description of the coupling network [33]. In 1997, a similar coupling scheme was used by Ispir et al. in demonstrating the first planar array steered via detuning of the edge oscillators [55]. The array was three elements by three elements and coupled in a Cartesian topology. Thus, all but one element are edge elements. The beam was scanned 10 deg in the E plane and 15 deg in the H plane. Experiments with and without half-length coupling lines at the ends of a linear array were conducted, and it was found that the scan range was larger with the added lines. Kagawa et al. demonstrated beam-steering in arrays with two and three circularly polarized elements [56], and Ispir et al. experimented with unidirectional coupling in a three-element array [55]. They showed that extended inter-oscillator phase range could be had by switching between two different values of coupling phase while steering via detuning the free running frequencies [57]. A very nice compendium of the work prior to 1997 is provided by Lynch et al. in the book by York and Popovic on power combining [58].

During the development of these oscillator arrays, there was a concern about the phase noise of the oscillators and the possibility that arrays of many oscillators of wide-locking range might have prohibitively large phase noise. As a result, an injection signal from an external stable oscillator might be needed. This was investigated by Chang et al. [59]. Using a five-element array at 8.5 GHz, they demonstrated reduction of the array phase noise to the level of the injection source phase noise near the carrier frequency.

In August of 2000, Pogorzelski et al. reported a seven-element linear array feeding a radiating aperture consisting of seven microstrip patches at 2.5 GHz [60]. Commercial MMIC voltage-controlled oscillators [Pacific Monolithics PM 2503] were used. This array is shown in Fig. 6-1. The coupling lines are visible on the circuit board as are the shorting bars used to adjust the coupling phase.

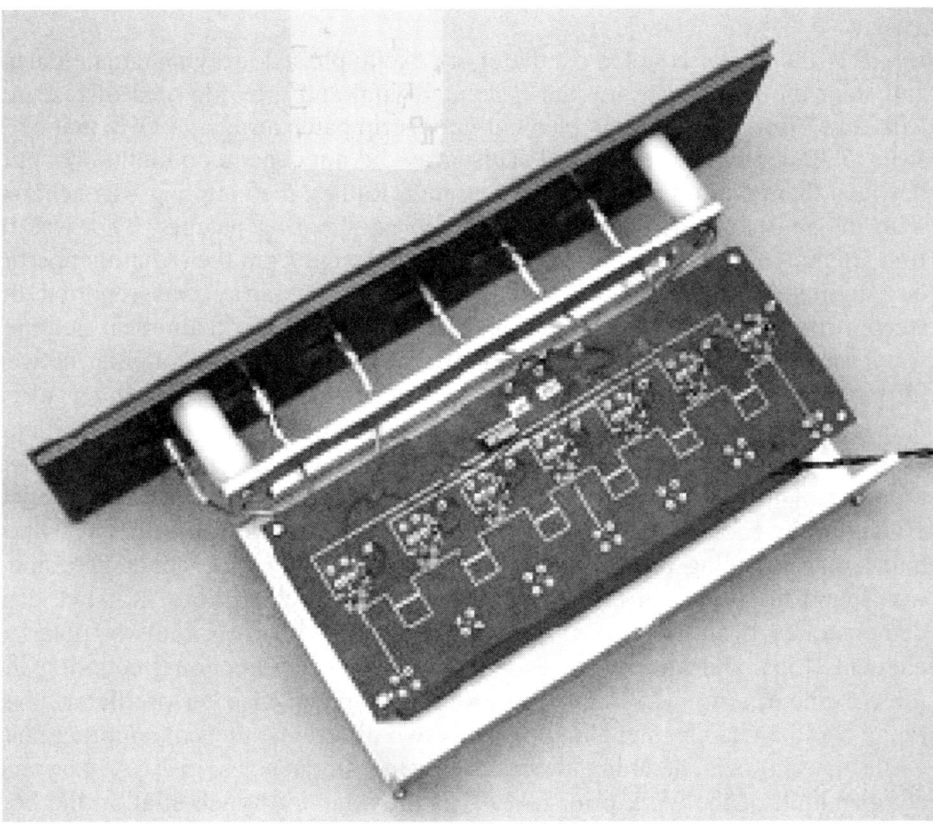

Fig. 6-1. Seven-element linear array. (Reprinted from Ref. 60 with permission, ©2001 IEEE.)

Initially, tuning of the oscillators was accomplished by using a network analyzer to measure the phase difference between adjacent oscillators, one pair at a time, and adjusting the free running frequencies to achieve the desired phase distribution. This process was impractically time-consuming, so a multichannel phase comparator system was devised that measured the phase differences between all adjacent oscillators simultaneously. This system consisted of a set of quadrature hybrid couplers and mixers arranged as shown in Fig. 6-2. The oscillator output signals from neighboring oscillators were mixed, and the lower hybrid frequency was at zero frequency. This DC output voltage was taken to be a measure of the relative phase of the oscillator signals. The hybrid couplers introduced a 90-deg phase shift in one of the signals so that zero output voltage from the mixer corresponded to zero relative phase. The output voltages from the mixers were then integrated from the center outward using a virtual instrument implemented in LabView™ to produce a graphical representation of the aperture phase distribution as shown in Fig. 6-3. The mixer outputs are shown in the bar graph, and the phase distribution is shown in the line graph below.

This seven-element array was evaluated on an antenna measurement range, and the patterns were compared with predictions for both unscanned and scanned beams. The results are shown in Fig. 6-4.

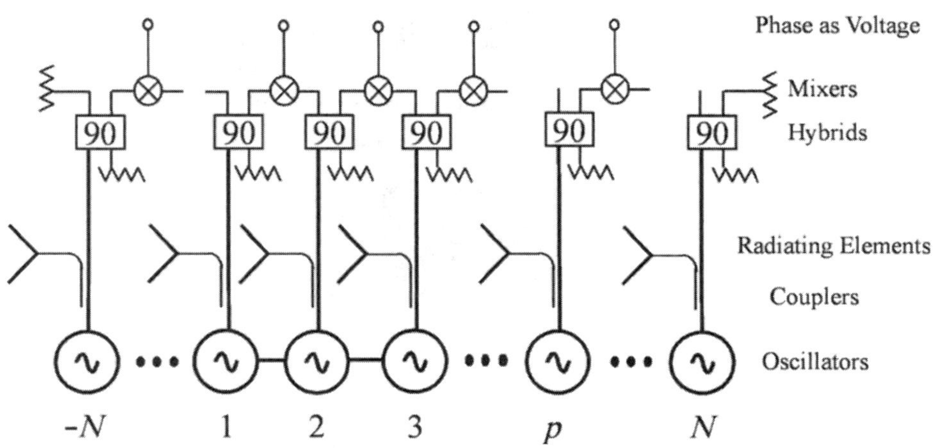

Fig. 6-2. Phase measurement system. (Reprinted from Ref. 61 with permission, ©2000 IEEE.)

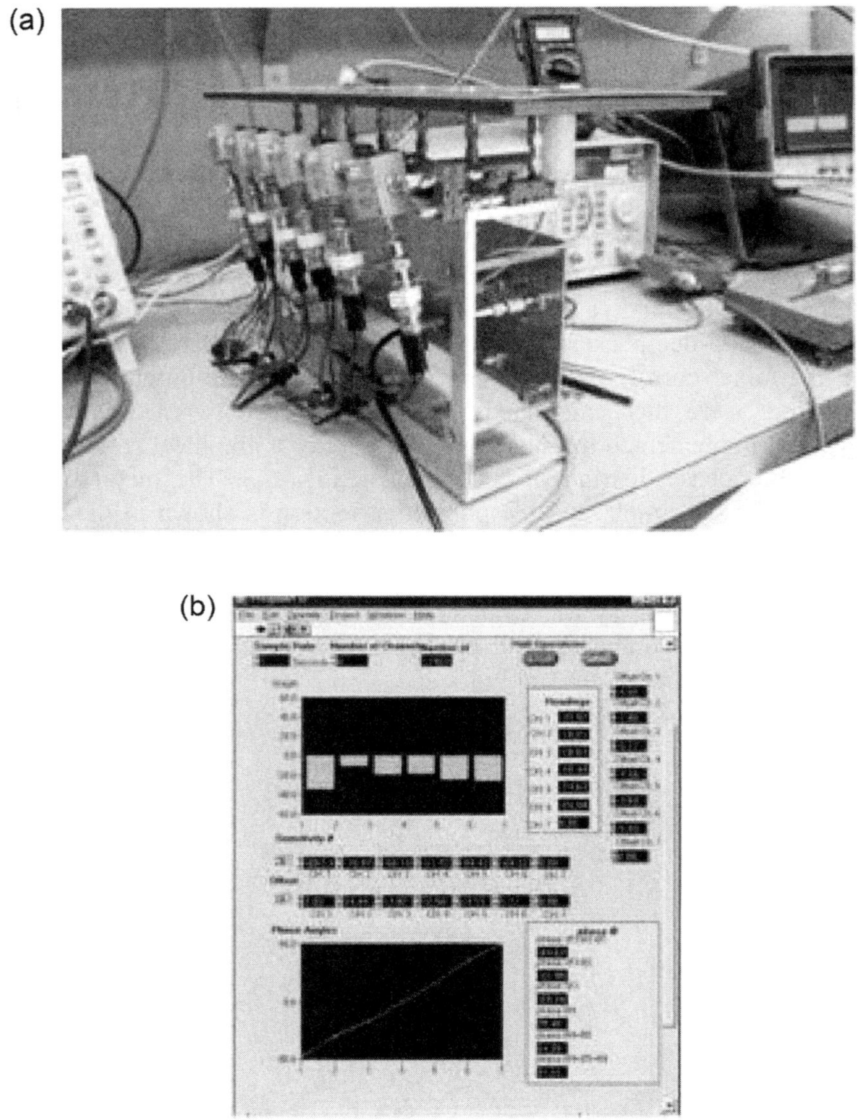

Fig. 6-3. Virtual instrument showing (a) multichannel phase comparator connected to the seven-element array and (b) a screen capture of the virtual instrument computer display. (Reprinted from Ref. 62 with permission, ©2006 IEEE.)

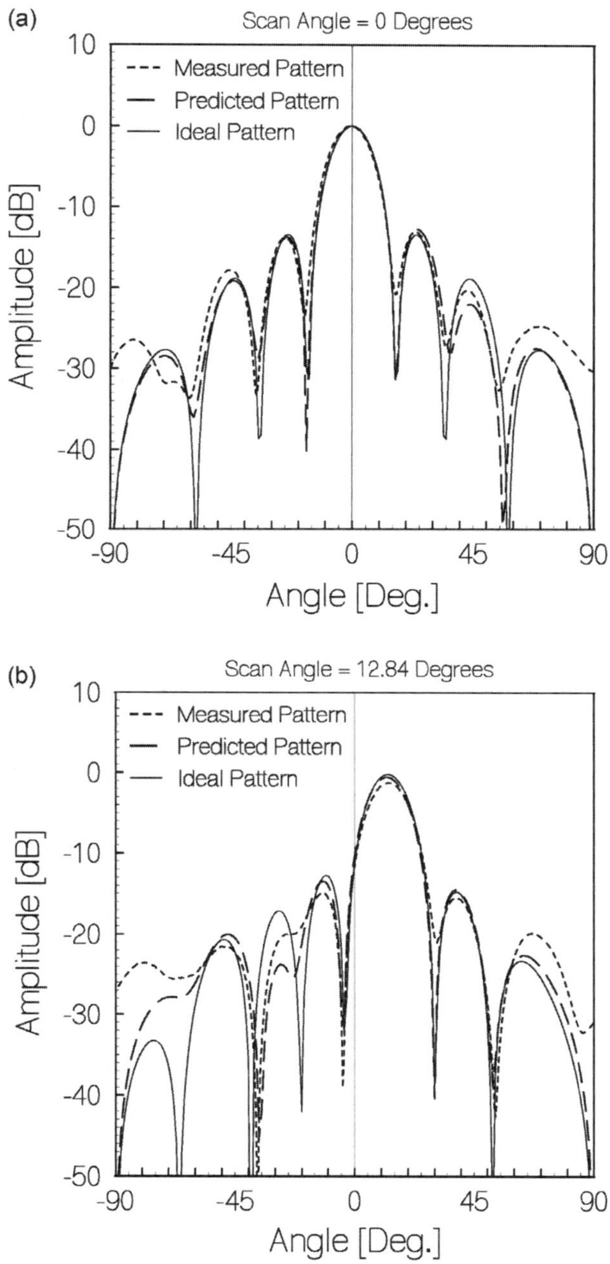

Fig. 6-4. Unscanned and scanned beams from the seven-element array with amplitude plotted against angle for unscanned and scanned beams at (a) 0-deg scan angle and (b) 12.84-deg scan angle. (Reprinted from Ref. 60 with permission, ©2000 IEEE.)

In Section 3.1 we obtained the transient behavior of the oscillator phases under step detuning of one oscillator. The seven-element array described above was also used to experimentally confirm that result [63]. One oscillator was repeatedly detuned by applying a tuning signal from a pulse generator, and the repeated transient behavior was recorded digitally using a multichannel oscilloscope. The results, both experimental and theoretical, for detuning one end oscillator are shown in Fig. 6-5. Figure 6-6 shows a similar comparison when one interior oscillator is step-detuned.

Recently, preliminary results on coupled oscillator arrays implemented using substrate-integrated waveguide (SIW) technology have been reported. SIW technology allows for compact, low-cost, lightweight, and high-performance implementation of microwave active and passive circuits, including active antennas and coupled oscillator arrays. SIW structures were initially proposed in the mid-1990s [64]. SIWs are waveguide-like structures fabricated by using two periodic rows of metallic vias or slots connecting the top and bottom ground planes of a dielectric substrate. An SIW-cavity-backed coupled-oscillator antenna array, shown in Fig. 6-7, was proposed by Giuppi et al. [65].

SIW strucutres share advantages of both microstrip and waveguide technology [66]. Similarly to planar structures, SIWs are compact, light weight, and cost effective due to the fact that they can be easily fabricated on single substrates using conventional fabrication techniques such as the ones used for their planar counterparts. Similarly to waveguide structures, they exhibit increased shielding, low loss, high quality factor, and high power-handling capability. Finally, they allow for high integration by implementing multilayer architectures. SIW technology allows for compact, low-cost implementation of coupled-oscillator arrays, suitable for large array configurations.

Giuppi et al. demonstrated a single-substrate implementation of a cavity-backed coupled-oscillator antenna array [65]. A two-element slot-array prototype that was implemented is shown in Fig. 6-7. Cavity-backed antennas have received interest due to attractive properties such as isolation, reduction of backward radiation, and surface-wave suppression [67].

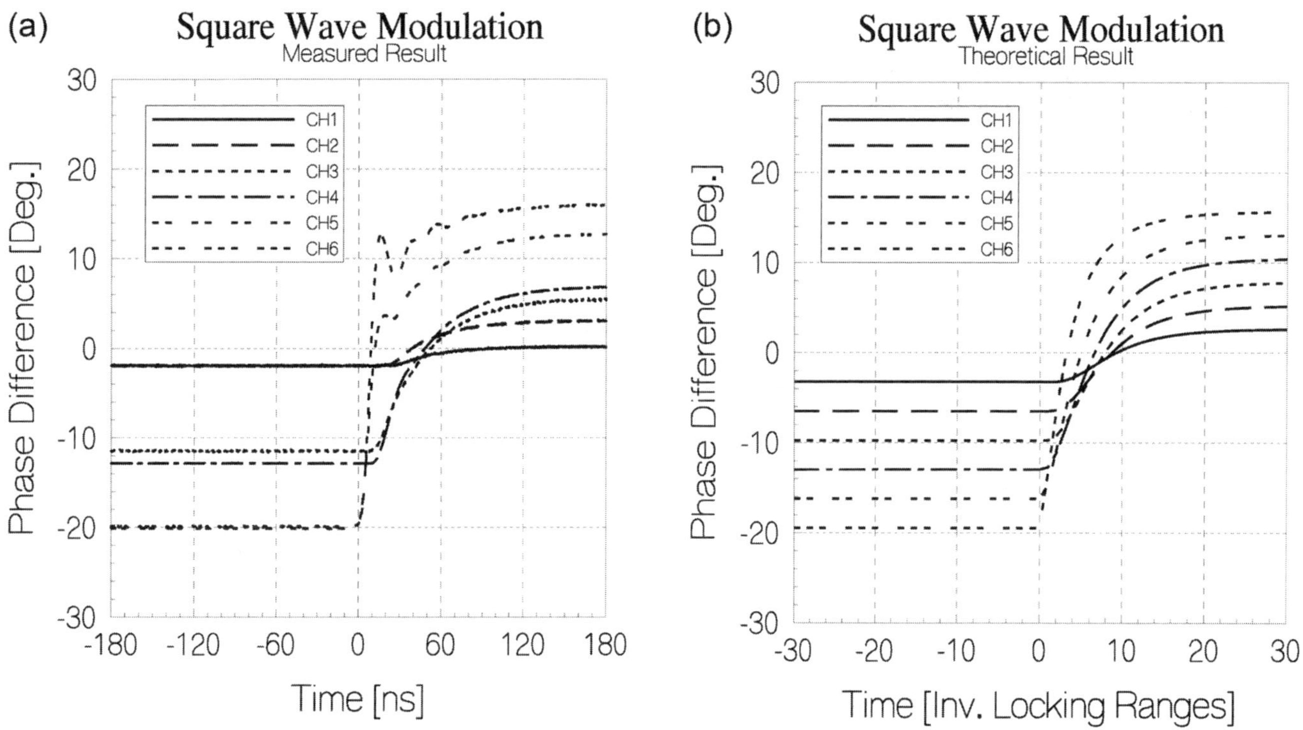

Fig. 6-5. E-plane transient response of the seven-oscillator array with one end oscillator step-detuned for (a) experimental results and (b) theoretical results. (Reprinted from Ref. 63 with permission, ©2002 IEEE.)

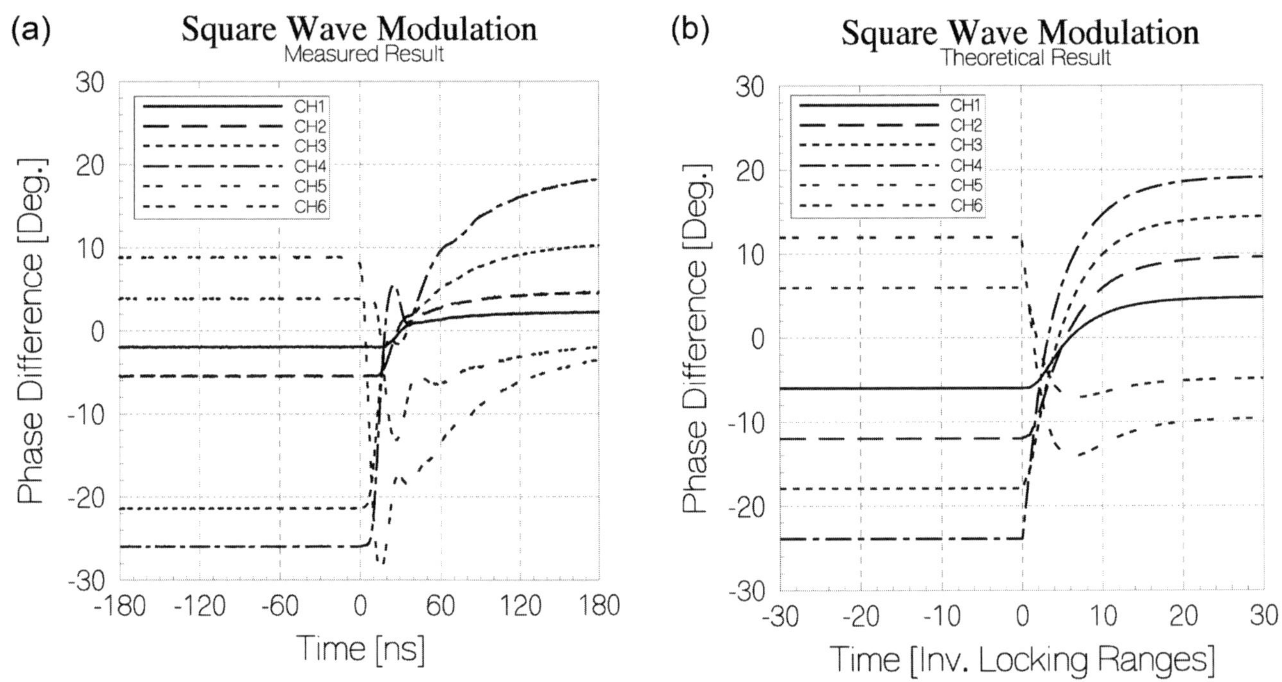

Fig. 6-6. H-plane transient response of the seven oscillator array with one interior oscillator step-detuned for (a) experimental results and (b) theoretical results. (Reprinted from Ref. 63 with permission, ©2002 IEEE.)

In the work of Ref. 65, SIW technology was used to fabricate cavity-backed slot-antenna oscillators and additionally control the coupling among the oscillator elements. The effect of the coupling aperture size A on the coupling strength is shown in Fig. 6-8. It was found that a double aperture symmetrically placed around the middle of the cavity wall, such as the one used in Fig. 6-7, leads to a smoother variation of the coupling factor as a function of the aperture size, compared to a single aperture at the center of the cavity wall, and therefore is less sensitive to fabrication tolerances.

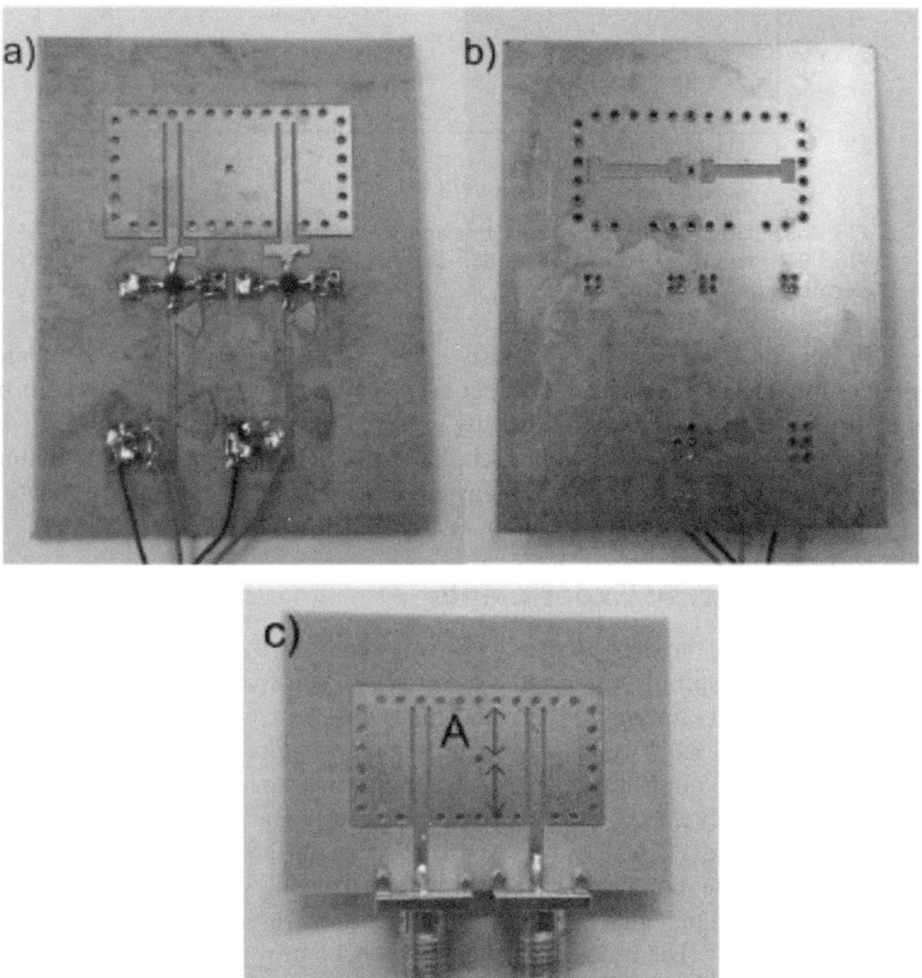

Fig. 6-7. Single-substrate two-element cavity-backed coupled oscillator antenna array in SIW technology: (a) top (active circuit) side, (b) bottom (antenna) side, (c) passive antenna array. Reprinted with permission from Ref. 65, copyright EurAAP 2010; used with permission.

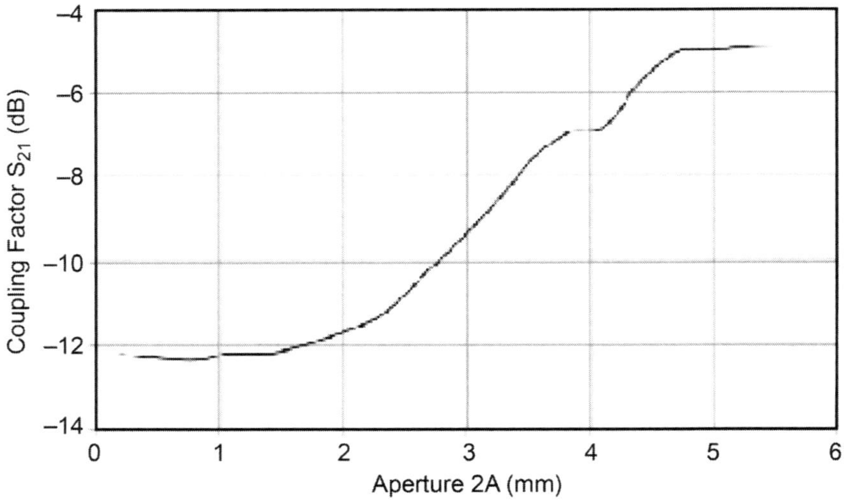

Fig. 6-8. Coupling factor versus the aperture size between the cavities of Fig. 6-7. Reprinted from Ref. 65, copyright EurAAP 2010; used with permission.

Finally, it is possible to control the oscillation frequency of the active oscillator antenna by varying the resonance frequency of the cavity. This is achieved by removing one of the vias from the cavity wall and placing a varactor diode in the cavity providing a capacitance between the top cavity conductor and the bottom ground conductor. Using this topology, a frequency tuning of approximately 2 percent was demonstrated by Giuppi et al. [68] (Fig. 6-9).

6.2 Planar-Array Experiments

A planar three-by-three oscillator array was reported by Pogorzelski in 2000 [61]. Recall that Ispir also reported a three-by-three array in 1997 using coaxial transmission line coupling between the radiating elements [55]. However, the 2000 array by Pogorzelski had no radiating aperture Its purpose was to demonstrate phase control by perimeter detuning via a phase diagnostic system similar to that developed for the seven-element linear array. Basically, the linear array diagnostic system was "woven" through the planar array one row at a time, and the computer-based virtual instrument was reprogrammed to display the computed phase values in a planar representation. This array is shown in Fig. 6-10. The precision potentiometers control the tuning bias of each oscillator.

The phase distributions over the array with various oscillator detuning distributions are shown in Fig. 6-11.

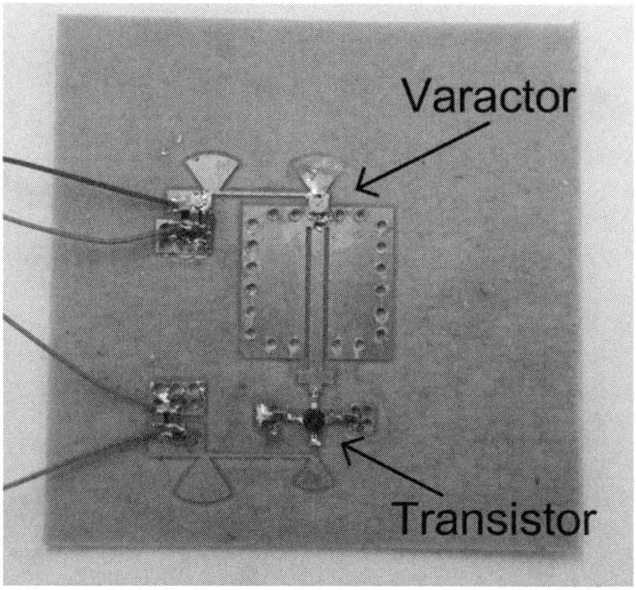

Fig. 6-9. SIW cavity-backed active-oscillator slot antenna with frequency tuning capability. (Reprinted with permission from Ref. 68, IET.)

Fig. 6-10. Nine-element planar oscillator array. (Reprinted from Ref. 62 with permission, ©2006 IEEE.)

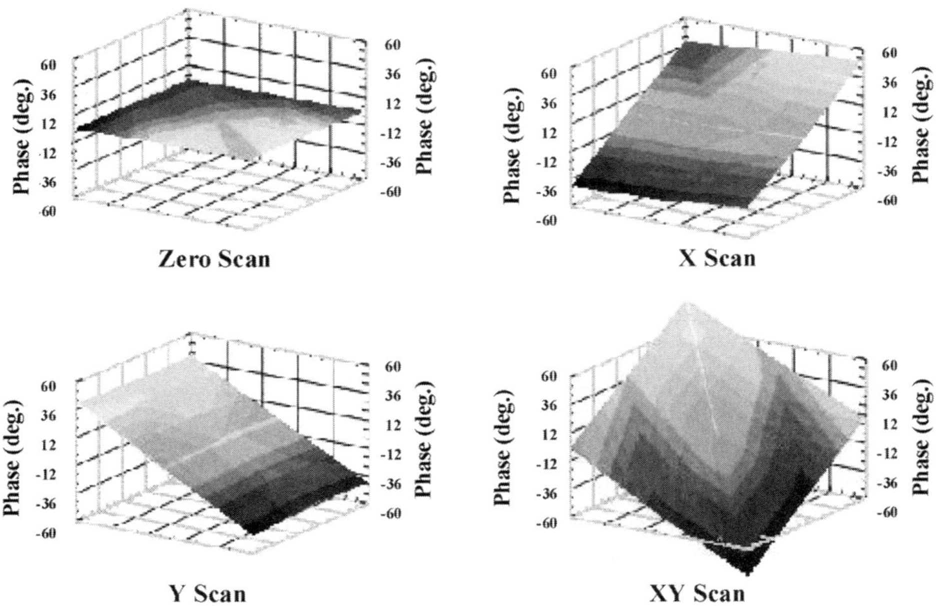

Fig. 6-11. Screen captures of phase distributions for beam-steering in a planar array. (Note: The tonal scale redundantly duplicates the phase vertical scale). (Reprinted from Ref. 61 with permission, ©2000, IEEE.)

As discussed earlier, the maximum phase difference between adjacent oscillators is 90 deg, at which value the oscillators lose lock. This limits the attainable beam-steering angle from normal. However, as described by York and Itoh, this range may be extended by frequency multiplication, which also multiplies the phase excursion [40]. To demonstrate this technique, frequency triplers were added to the above nine-element array, and the resulting 8.4-GHz output was used to drive a nine-element microstrip patch array. The aperture phase was inferred by tripling the outputs of the mixers in the slightly more sophisticated diagnostic system shown in Fig. 6-12, and the resulting far-zone radiation patterns were measured on an antenna range. Note that, unlike its predecessor, this diagnostic system uses attenuators to equalize the input amplitudes at the mixers. The measurement setup is shown in Fig. 6-13 wherein the coaxial line stretchers equalize the phases of the transmission lines to the radiating elements. The resulting measurements are shown in Fig. 6-14b where the "X's" label a pattern that corresponds to steering 90 deg from normal, which is not achievable without frequency multiplication. Note that this pattern is symmetric, indicating that the main lobe points 90 deg to the left while the similar looking lobe on the right is a grating lobe. The element pattern is shown as the dashed curve and indicates why all of the patterns have a null 90 deg from normal.

In 2005, Pogorzelski reported construction and demonstration of a five-by-five element planar array using using the same S-band MMIC oscillators used in the earlier seven-element linear array and microstrip patch radiating elements [70]. This MMIC contained a buffer amplifier at its output, thus isolating the oscillators and patches and completely separating the coupled oscillator array design from the radiating aperture design. In this array, the oscillators were located on one side of a Duroid™ board, and the patches were located on the other side, with a coaxial pin connecting each oscillator output to the corresponding patch. The phase-measurement system was mounted on a phenolic board for physical support and connected to the oscillators via stripline couplers obviating the need for direct physical connection and rendering the measurement system removable. The assembled array and phase measurement system is shown in Fig. 6-15. The Duroid™ circuit board is located between the aluminum plate and the phenolic board.

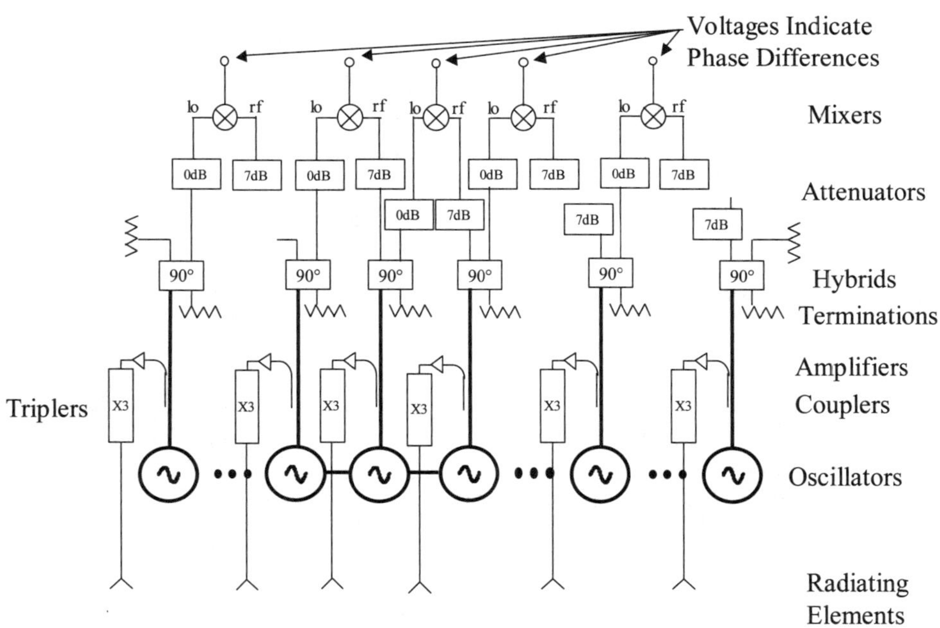

Fig. 6-12. Phase measurement system for the frequency-tripled array. (Reprinted from Ref. 69 with permission, ©2004 IEEE.)

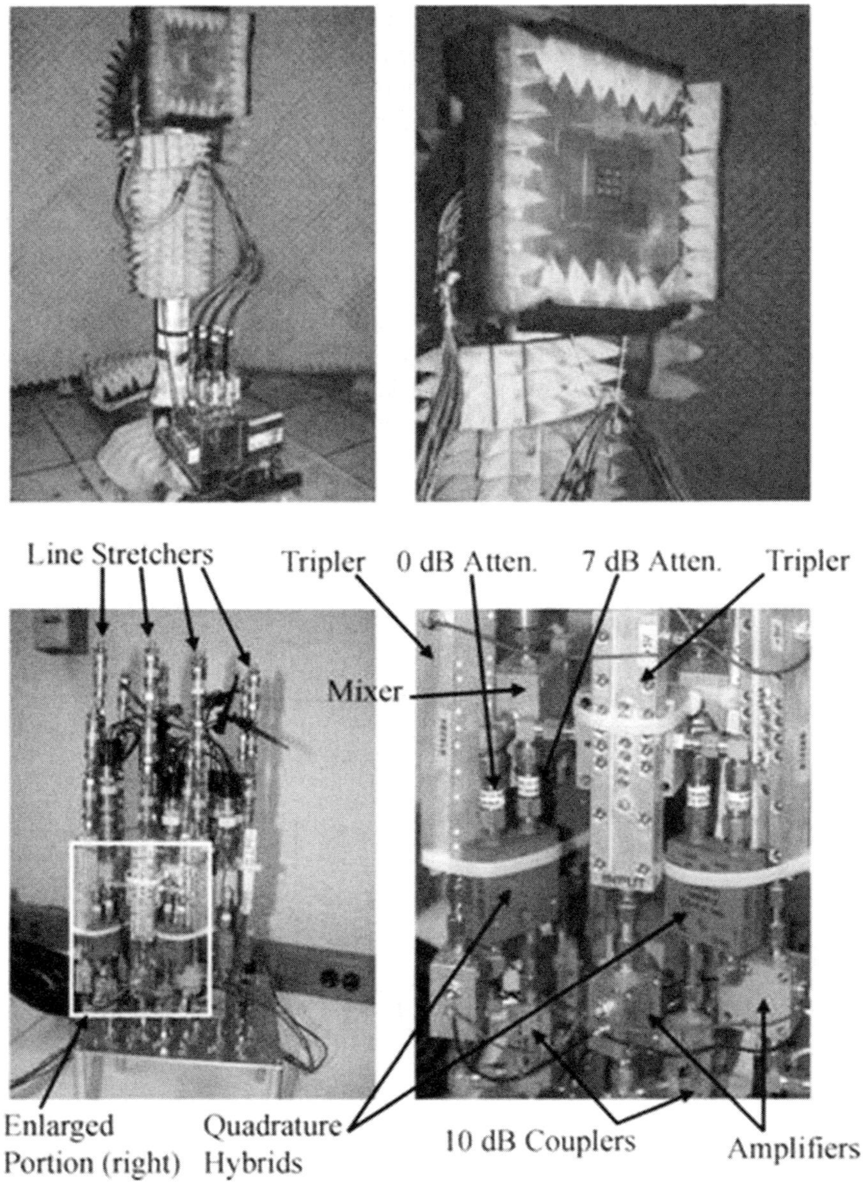

Fig. 6-13. Experimental setup for evaluation of the nine-element frequency-tripled planar array. The white box of the lower left picture is enlarged in the lower right picture. (Reprinted from Ref. 69 with permission, ©2004 IEEE.)

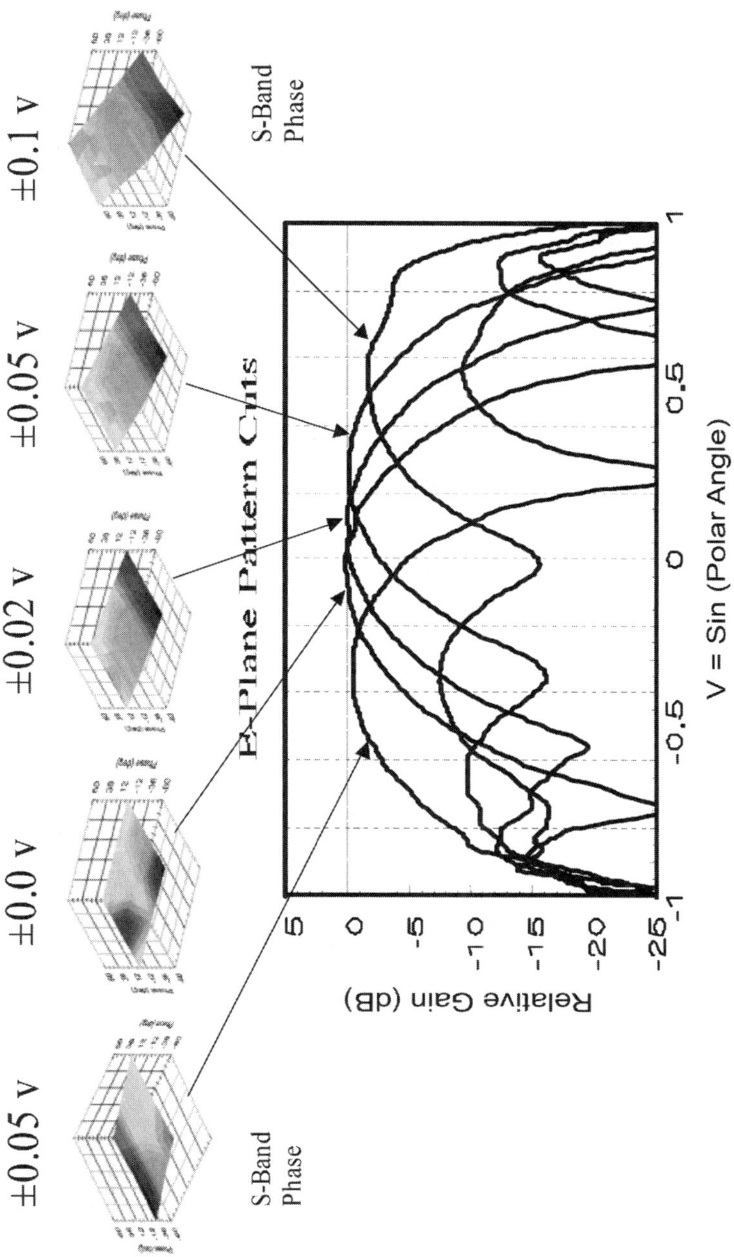

Fig. 6-14(a). Measured phase distributions and antenna patterns of the nine-element frequency-tripled planar array showing E-plane pattern cuts. The vertical axis of the small phase surface plots ranges from –60 to +60 degrees of phase, and the tonal scale reduntantly duplicates this same information. (Reprinted from Ref. 69 with permission, ©2004 IEEE.)

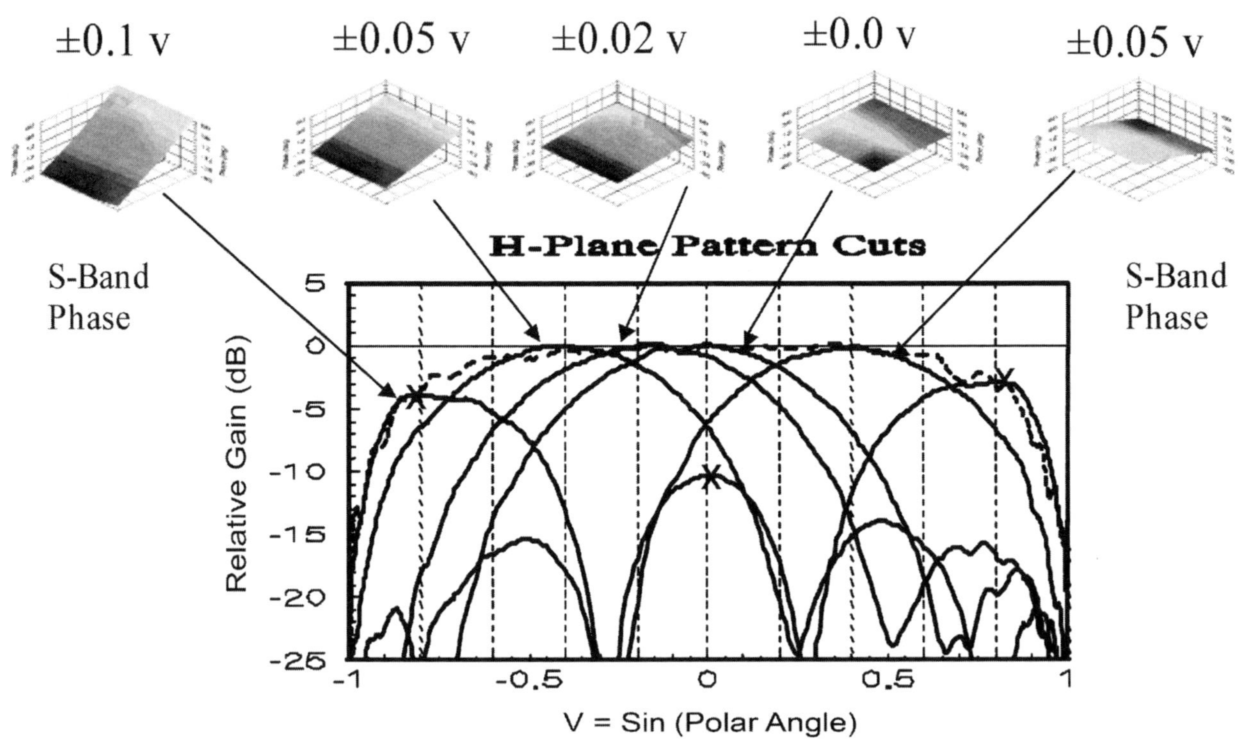

Fig. 6-14(b). Measured phase distributions and antenna patterns of the nine-element frequency-tripled planar array showing H-plane pattern cuts. The vertical axis of the small phase surface plots ranges from –60 to +60 degrees of phase, and the tonal scale reduntantly duplicates this same information. (Reprinted from Ref. 69 with permission, ©2004 IEEE.)

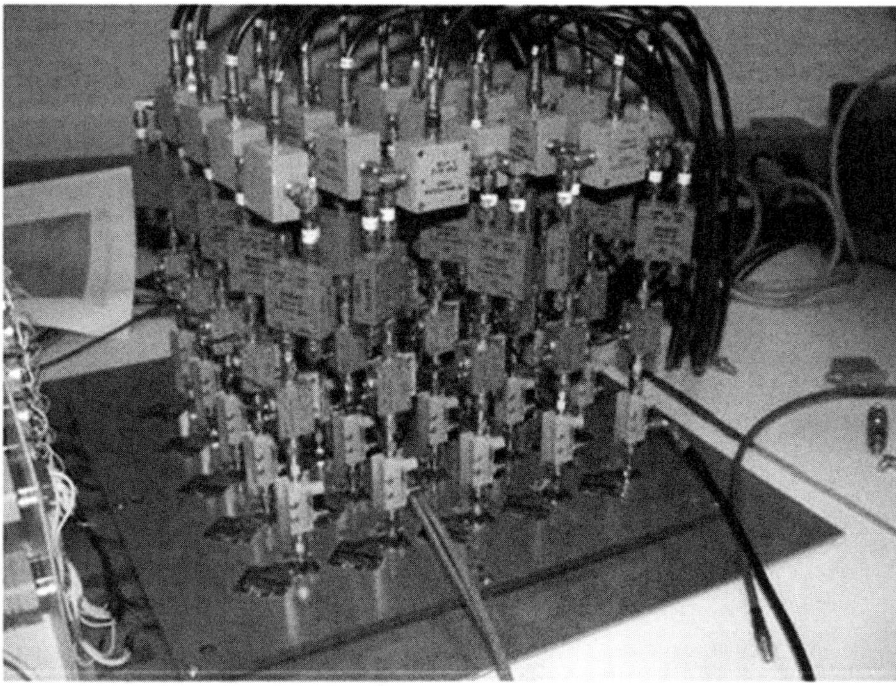

Fig. 6-15. Five-by-five S-band array and phase-measurement system. (Reprinted from Ref. 62 with permission, ©2006 IEEE.)

The Duroid™ board and aluminum plate attached to the measurement fixture are shown in Fig. 6-16. The array of potentiometers controls the power supplied to each oscillator and were used to approach an untapered aperture distribution for the array. Fig. 6-17 shows the assembled array on the measurement range and Fig. 6-18 shows an example measured result for aperture phase distribution and steered beam.

A similar five-by-five element array was reported by Heath et al. [71] of the Georgia Tech Research Institute (GTRI) using a new phase comparator chip that rendered the phase measurement system much more compact. In fact it was integrated with the oscillator array shown in Fig. 6-19. The GTRI group used a LabView™-based display sytem very similar to that of Pogorzelski [70]. However, unlike Pogorzelski, GTRI also used a LabView™ user interace for tuning the oscillators (Fig. 6-20.) This GTRI array was similar to the three-by-three frequency tripled array discussed earlier in that the oscillators were connected to the radiating aperture via cables visible in Fig. 6-21 showing the near-field test setup at GTRI. However, it did not use frequency multiplication as did the three-by-three array. Figure 6-22 displays an example near-field measurement transformed to the far zone showing the beam steered to –20 deg.

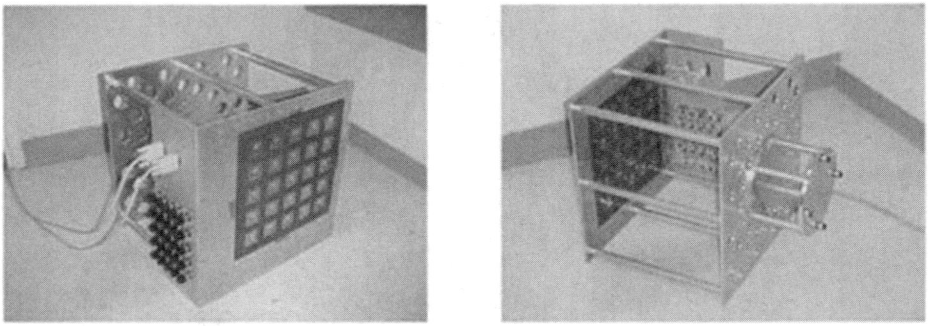

Fig. 6-16. Five-by-five oscillator array and radiating aperture on the measurement fixture. (Reprinted from Ref. 62 with permission, ©2006 IEEE.)

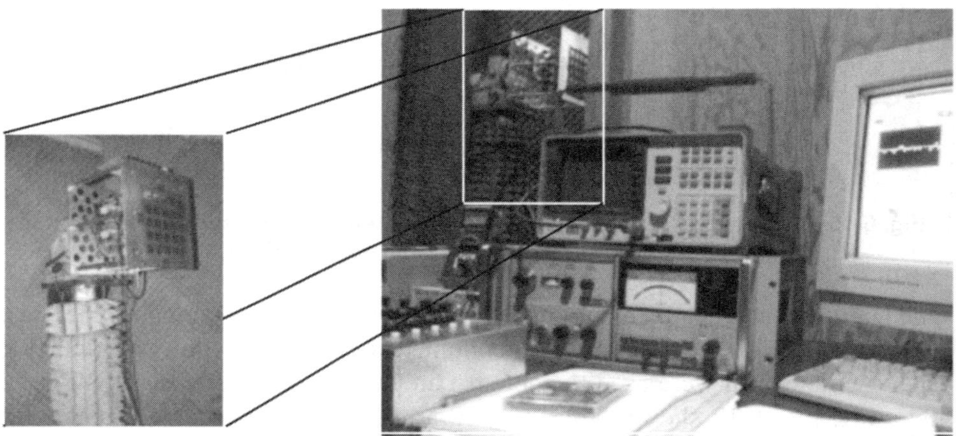

Fig. 6-17. Five-by-five array on the measurement range. (Reprinted from Ref. 62 with permission, ©2006 IEEE.)

The ability to electronically control the phase differences among the synchronized elements of a coupled oscillator array by varying the free-running (uncoupled) frequency of each element has been used by Yen and Chu in order to simultaneously scan and control the polarization of a linear antenna array [72].

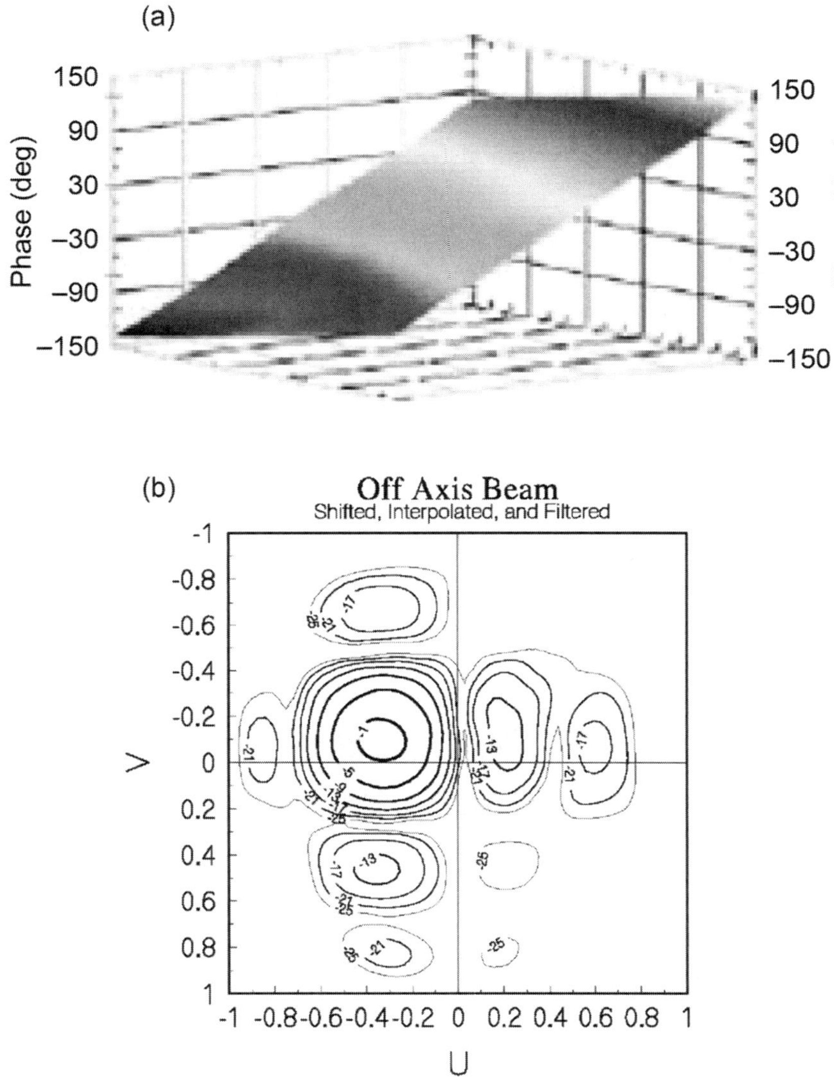

Fig. 6-18. Five-by-five array: (a) Measured aperture phase and (b) steered beam. (Note: The tonal scale redundantly duplicates the phase information in (a).) (Reprinted from Ref. 62 with permission, ©2006 IEEE.)

The block diagram of the proposed architecture is shown in Fig. 6-23. A linear dual linearly polarized patch antenna array of N elements is used as the radiating structure. A two-dimensional coupled oscillator array consisting of two rows of N oscillators is connected to the $2N$ antenna ports. The phase difference between the oscillator elements within each row is used to generate a progressive phase distribution and steer the main beam of the array. The phase difference between the two rows controls the phase difference between the two orthogonal polarization states and therefore the polarization of the array. Frequency doublers are used at each oscillator output in order to produce a stable phase variation of up to 360 deg. A prototype consisting of a four-element antenna array was successfully demonstrated [72] (Fig. 6-24).

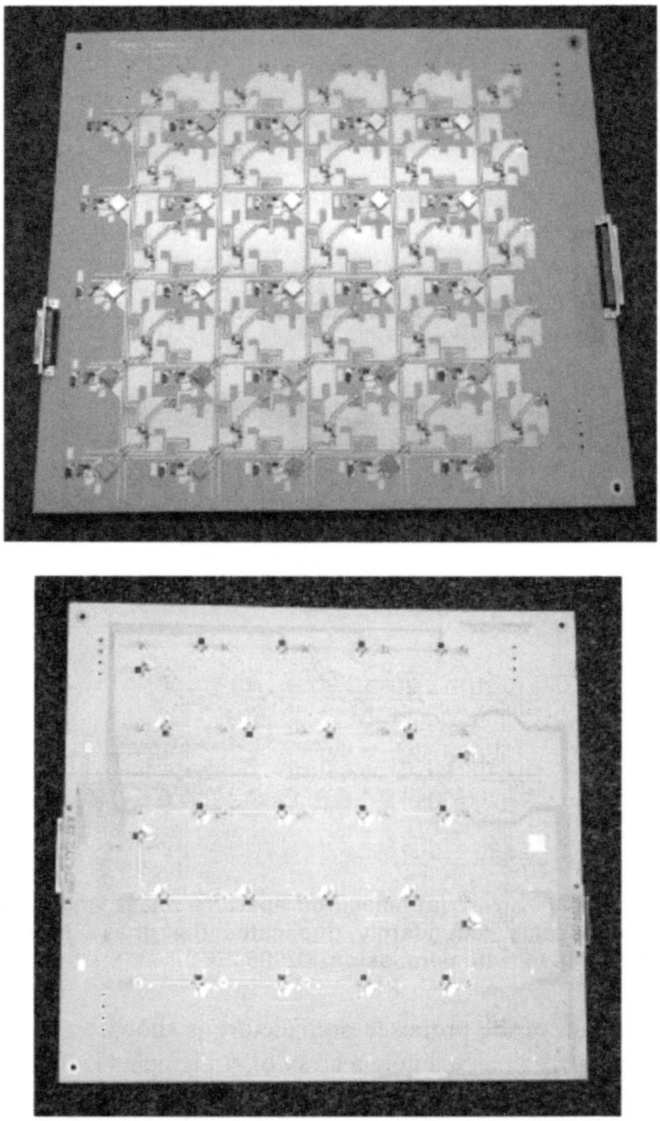

Fig. 6-19. Both sides of GTRI five-by-five oscillator array board. (Reprinted from Ref. 71 with permission, ©2005 IEEE.)

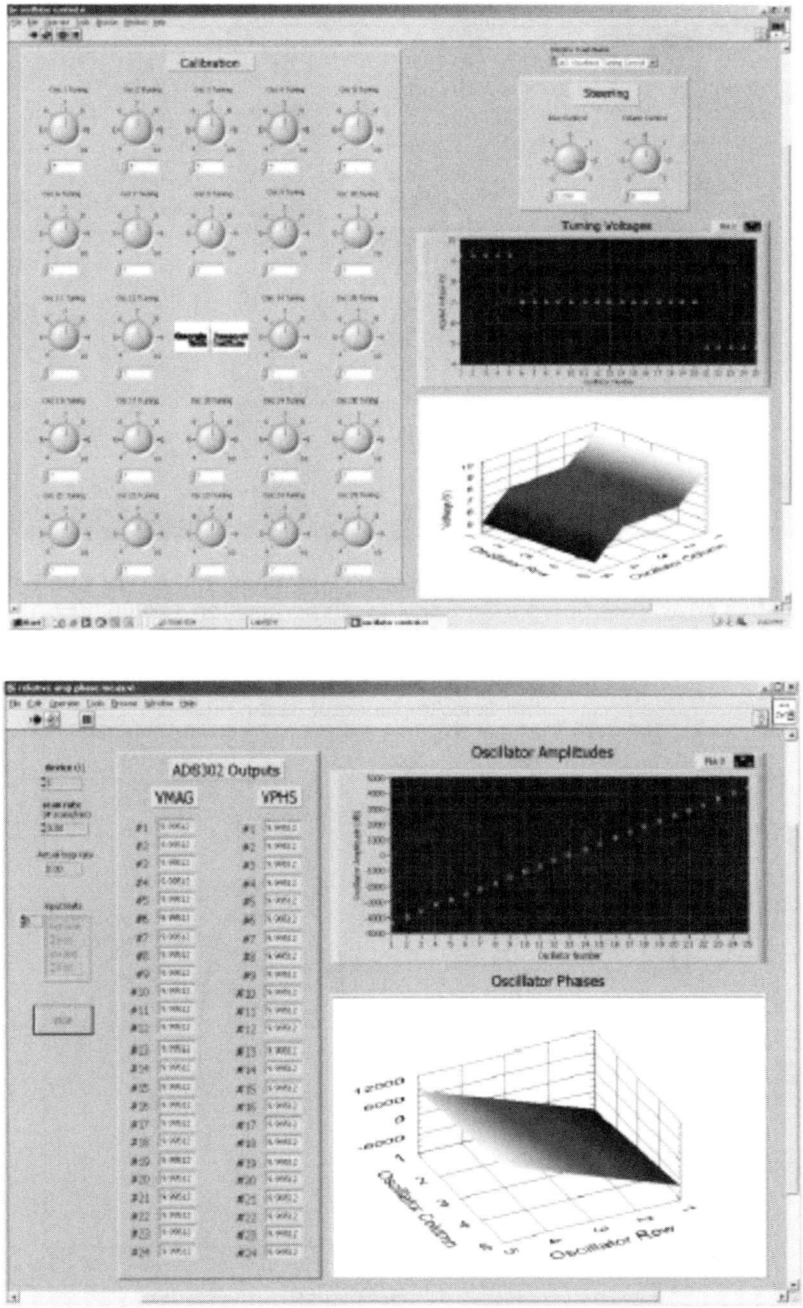

Fig. 6-20. Screen captures of the GTRI LabView virtual instrument for tuning (above) and phase measurement (below). (Reprinted from Ref. 71 with permission, ©2005 IEEE.)

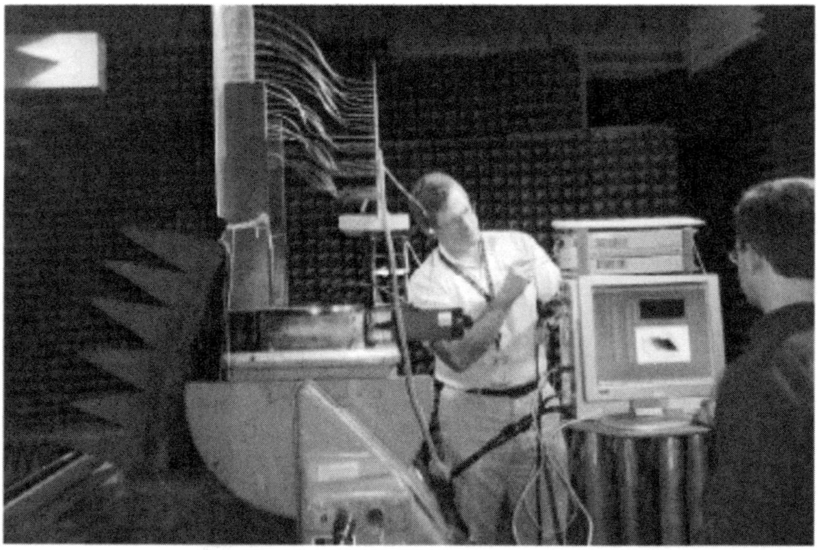

Fig. 6-21. The GTRI test setup for near-field measurement of the five-by-five array. (Reprinted from Ref. 71 with permission, ©2005 IEEE.)

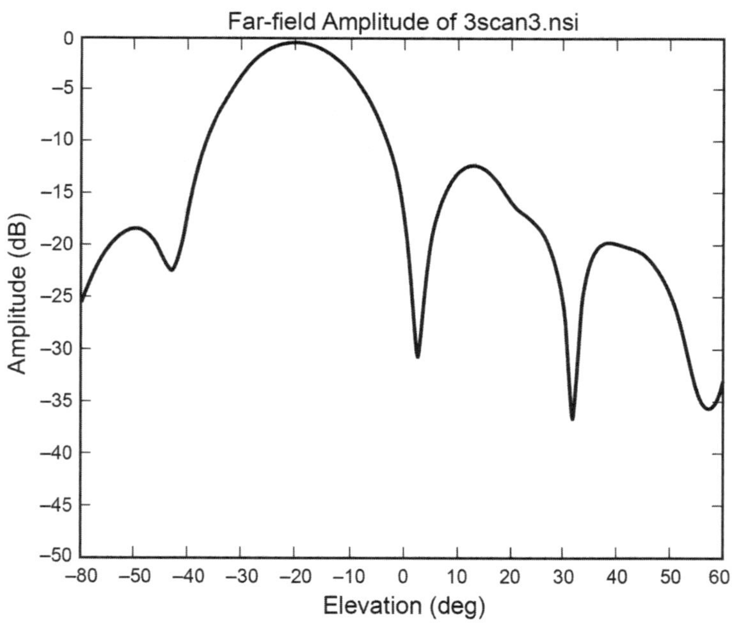

Fig. 6-22. Far zone steered beam of the GTRI five-by-five array. (Reprinted from Ref. 71 with permission, ©2005 IEEE.)

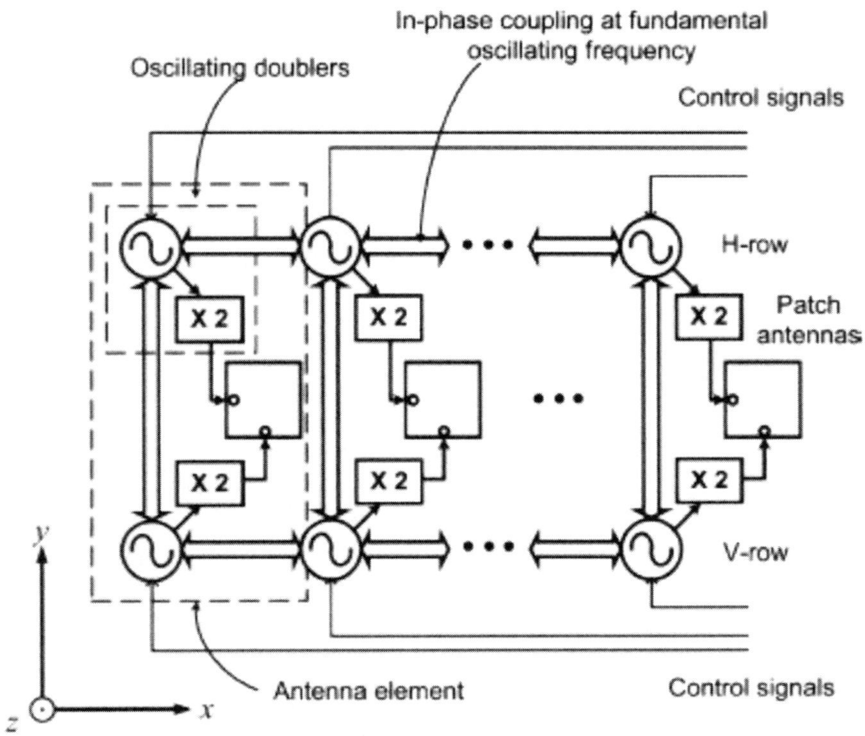

Fig. 6-23. Polarization agile, beam-scanning coupled oscillator antenna array architecture. (Reprinted from Ref. 71 with permission, ©2005 IEEE.)

6.3 Receive-Array Experiments

As early as 1995 it was suggested by Cao and York that oscillator arrays could be used to steer the beam of a receiving antenna [73]. The concept is illustrated in Fig. 6-25. Basically, the oscillators are used as local oscillators to down convert the signals received by each of the elements in antenna aperture. The phasing of the local oscillator signals may then be adjusted to cancel the phasing of the element signals due to the angle of arrival of the incident wave. Thus, the phasing of the oscillators as determined by detuning of the end oscillators may be said to steer the receive beam.

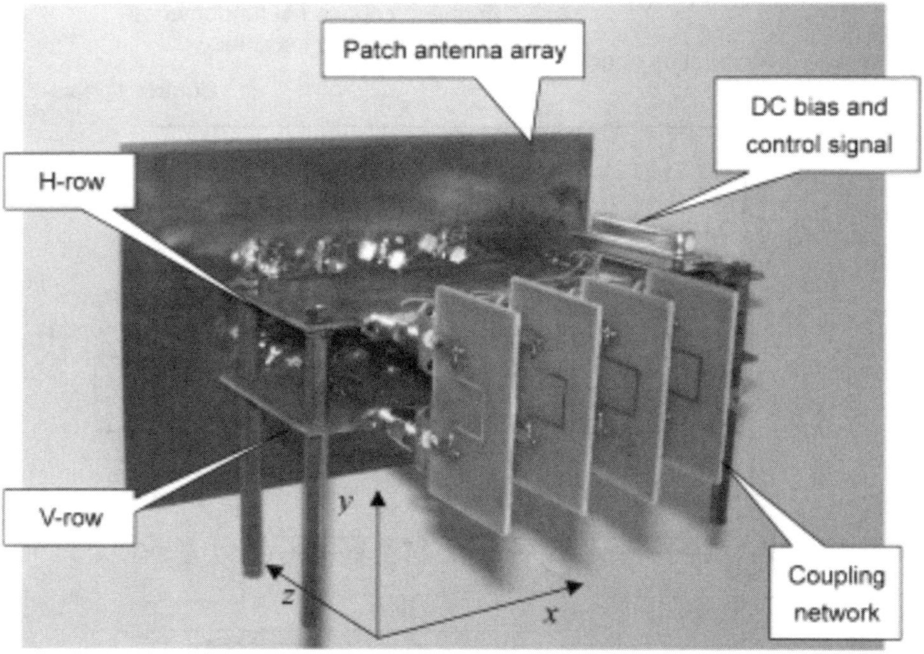

Fig. 6-24. Photo of a four-element prototype of the polarization agile, beam-scanning coupled oscillator antenna array (H-row = horizontal row and V-row = vertical row). (Reprinted from Ref. 72 with permission, ©2005 IEEE.)

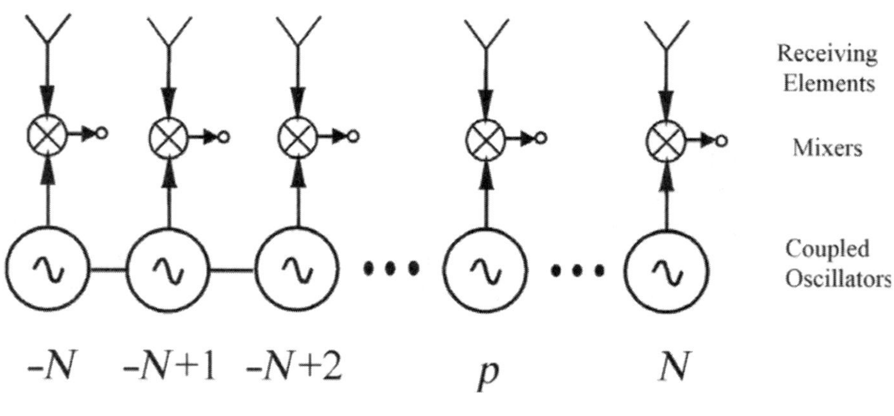

Fig. 6-25. Receive concept using coupled oscillators.

This concept was demonstrated experimentally by Pogorzelski and Chiha [74] using a 15-element array of L-band oscillators [Modco CM1398MST] coupled in a linear configuration. In the absence of a receiving aperture, the received signals were simulated using a 16-way power divider. These 1.95-GHz output signals from the power divider were, of course, in-phase. They were mixed with the 1.265-GHz outputs of the linear-array oscillators producing intermediate frequency signals at 685 MHz. These intermediate frequency signals were then combined using another 16-way power divider in reverse. The testbed setup is shown in Fig. 6-26 together with a close-up of one of the oscillator circuits. By using only every other oscillator in the array, the maximum phase difference between adjacent local oscillator signals was extended to 180 deg. Thus, only eight signals are combined. The combined output at 685 MHz is plotted versus beam-steering angle in Fig. 6-27. The solid line is the theoretically predicted result. The phase distributions across the array corresponding to points A and B are shown in Fig. 6-28.

This apparatus was also used to demonstrate a very interesting scheme patented by Kott for the reduction of sidelobes [75]. Kott proposed the placement of an additional element at each end of an array positioned and excited so as to provide an interferometer pattern with null spacing matching the null spacing of the sidelobes of the array. Then by properly combining the interferometer signal with the array signal, entire regions of sidelobes could be canceled. It turns out that the receive-array testbed described above provides just the proper phasing of the end elements to achieve this cancelation [76] The concept is shown in Fig. 6-29. The attenuators at each end oscillator permit proper weighting of the interferometer signal relative to the receive array signal to achieve cancelation. Figure 6-30 shows the output of the intermediate frequency combiner versus beam-steering angle for the center elements (solid), the interferometer pattern of the end elements (short dashes), and the coherent combination of the two (long dashes), showing that the left sidelobe has been removed while the right one has been enhanced. The beamforming capabilities of coupled oscillator arrays are studied in more detail in Chapter 9.

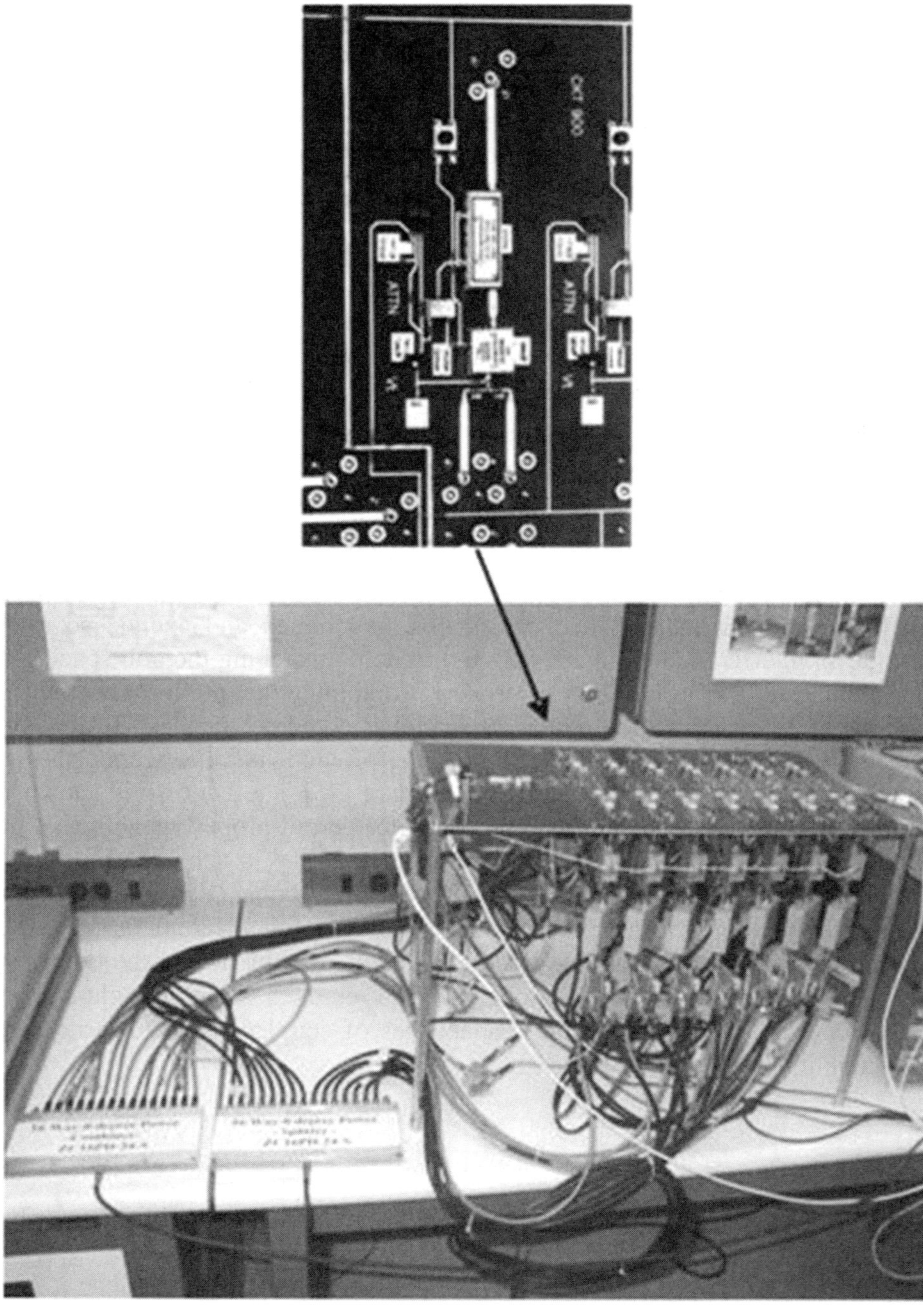

Fig. 6-26. L-band receive-array testbed with close-up of one oscillator. (Reprinted from Ref. 62 with permission, ©2006 IEEE.)

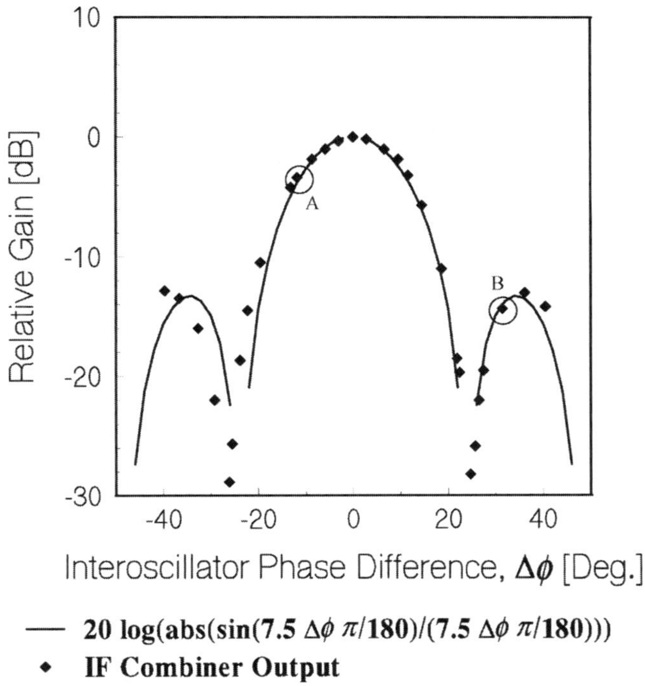

Fig. 6-27. Combined intermediate frequency signal versus receive beam-steering angle for the 15-oscillator L-band array. (Reprinted from Ref. 62 with permission, ©2006 IEEE.)

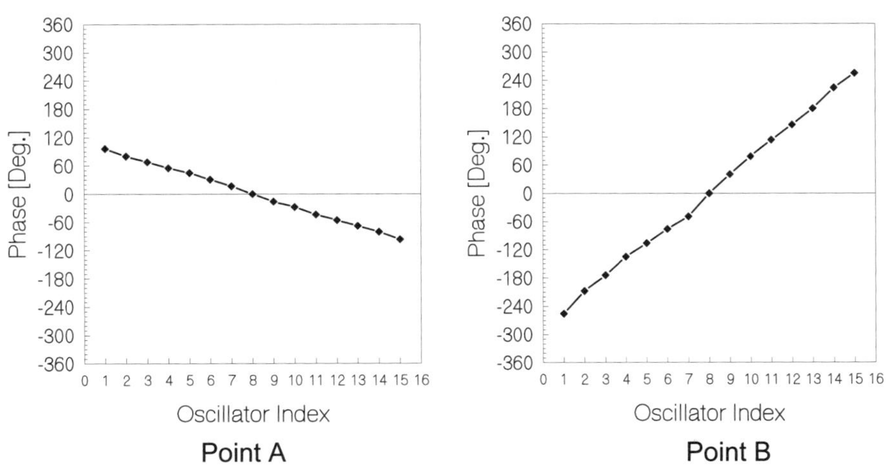

Fig. 6-28. Phase distributions corresponding to the two indicated points A and B in Fig. 6-27. (Reprinted from Ref. 62 with permission, ©2006 IEEE.)

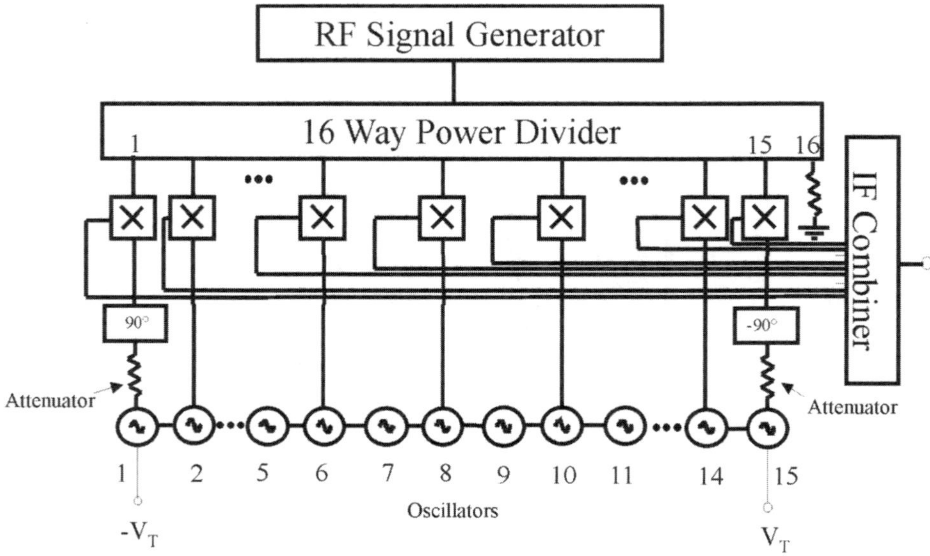

Fig. 6-29. Circuit arrangement to implement the Kott scheme using the 15-element receive-array testbed. (Reprinted from Ref. 62 with permission, ©2006 IEEE.)

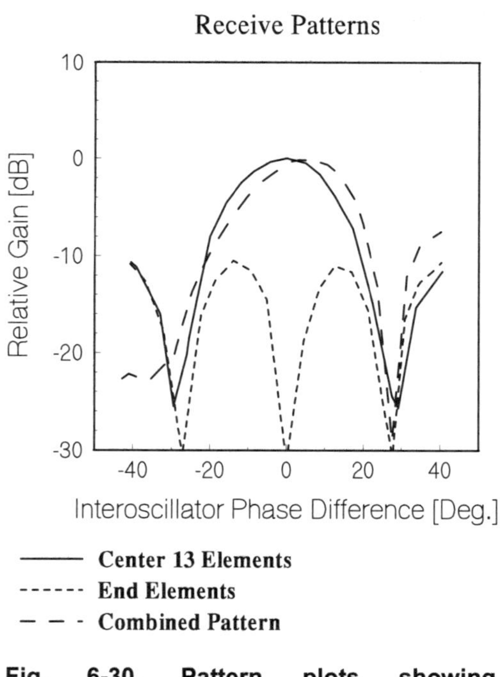

Fig. 6-30. Pattern plots showing cancelation of the left sidelobe. (Reprinted from Ref. 62 with permission, ©2006 IEEE.)

It should be recognized that using the same oscillator array for both transmit and receive poses a certain complexity. This is true because in the proposed receive array, the oscillator signals are used to *down* convert the signal, so it is the *lower* hybrid frequency of the mixers that is combined to produce the output. Thus the oscillator phases are *subtracted* from the phases of the signals received at the array elements. The result is that the receive phasing of the oscillators is the *conjugate* of the phasing required to transmit a beam in the same direction by using the oscillator signals to excite the elements. If instead the *upper* hybrid frequency were combined on receive, this would not be the case, but the combining for receive would then be done at approximately twice the oscillator frequency rather than at a low intermediate frequency.

In closing this section on receive arrays it should be noted that an array of self-oscillating mixers, when properly coupled, is synchronized in frequency forming a coupled oscillator array. Coupled self-oscillating mixer (SOM) arrays have been used in retro-directive array applications such as the work by Shiroma et al. [77]. Additionally, the use of coupled SOM arrays in receive phased-array applications has been investigated by Sanagi et al. [78] and Ver Hoeye et al. [79]. The array topology proposed in Ref. 78 is shown in Fig. 6-31.

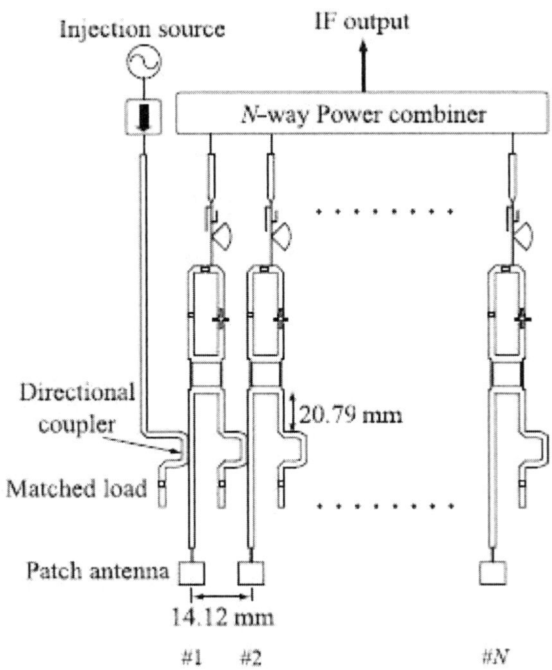

Fig. 6-31. Coupled self-oscillating mixer array. (Copyright ©2005 IEICE [78].)

The SOM element consists of an amplifier and a branch line coupler providing the necessary feedback to obtain the oscillation. A varactor diode in the feedback loop is used to provide frequency tuning of the individual SOM element. A linear array is considered where the various elements are unilaterally coupled using directional couplers. The RF signal at 17.1 GHz is mixed with the second harmonic of the SOM circuit at 8.5 GHz in order to obtain an IF output at 100 MHz. The SOM array is synchronized to an external injection signal coupled to one of the edge array elements. Subharmonic mixing using the second harmonic component results in phase tuning capability of 360 deg. The IF outputs of the SOM array elements are combined using a power combiner. Small prototypes of two and three elements were used to demonstrate the beam-steering capabilities of the proposed architecture. The antenna elements are placed 14.12 millimeters (mm) apart, which corresponds to approximately $0.8\lambda_0$ at 17.1 GHz. As a result, the maximum beam scanning angle that can be achieved by this topology is 38.4 deg. Measured radiation patterns of a three-element array are shown in Fig. 6-32.

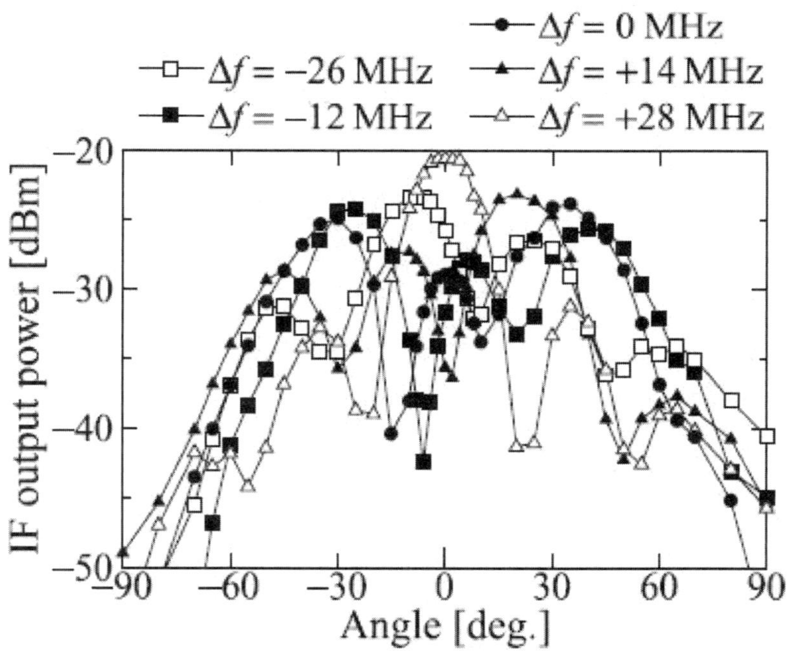

Fig. 6-32. Measured radiation patterns of three-element-coupled SOM array. (Copyright ©2005 IEICE [78].)

Radiation patterns were obtained for different (free-running) frequency difference values between the array elements. Variation of the free-running frequency of the synchronized array elements results in variation of the relative phase among the elements. The frequency difference between successive array elements is experimentally mapped to the inter-element phase difference according to Fig. 6-33. As an example, a frequency difference of 28 MHz corresponds to the in-phase state leading to a radiation pattern with a main beam along the broadside direction.

Finally, a four-element receive SOM array was demonstrated by Ver Hoeye et al. in Ref. 79. The circuit topology is shown in Fig. 6-34, followed by the implemented prototype in Fig. 6-35. Each array element is an SOM circuit designed by the authors in Ref. 80 and described in Section 8.7. An input RF signal of 11.25 GHz is mixed with the third harmonic of the oscillator at 3.25 GHz, producing an IF output at 1.5 GHz. Using the third harmonic in the mixing product allows for a theoretical phase-tuning range of 540 deg for an individual externally injection locked SOM element. In the proposed circuit topology, an external injection locking signal is applied to all SOM elements using a Wilkinson power-divider network. The SOM elements are not coupled to each other directly; therefore, the array topology can be visualized as a star network where the external injection signal synchronizes all array elements. Each oscillator is connected to a patch anrtenna and the array outputs are combined using an IF Wilkinson combiner network. A measured beam-scanning range of 23.5 deg was reported.

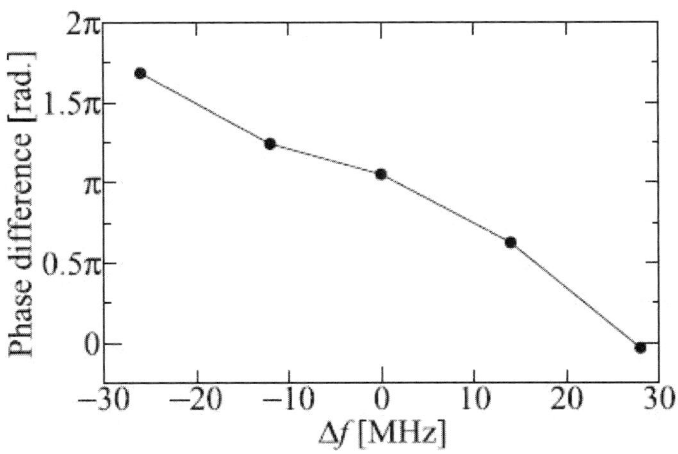

Fig. 6-33. Array element phase difference versus frequency detuning. (Copyright ©2005 IEICE [78].)

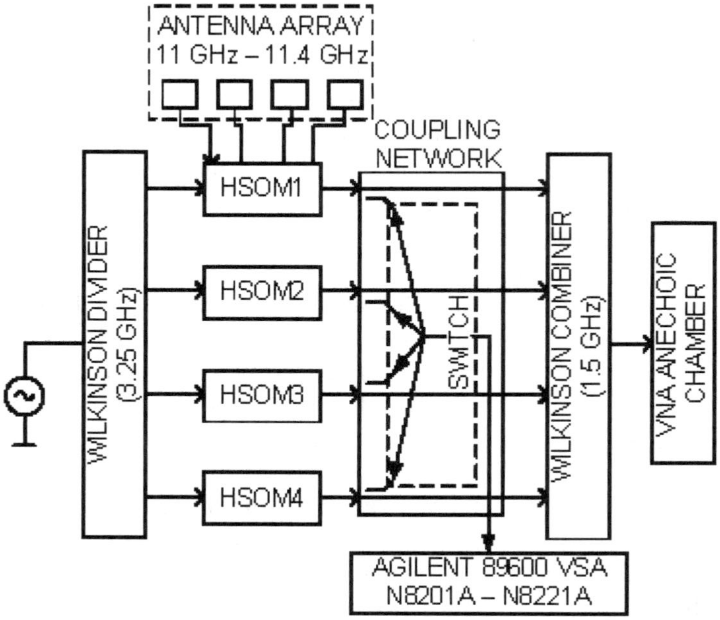

Fig. 6-34. Four-element coupled SOM array block diagram (HSOM = harmonic self-oscillating mixer; VNA = vector network analyzer; VSA = vector signal analyzer). (Adapted from and used with permission from Ref. 79, ©2009 IEEE.)

6.4 Phase Noise

Throughout the early development of coupled oscillator arrays there was a concern about stability in terms of phase noise. It was recognized that phase control can be enhanced by designing oscillators to have wide locking range because by this means the phase change for a given change in VCO tuning bias is reduced. However, associated with this wider locking range will be lower oscillator *Q* and an increase in phase noise. Thus, means were sought to mitigate this situation. For example, Chang et al. were able to double the locking range of a VCO while reducing the phase noise below that expected for such a wide locking range by means of an amplified feedback path [81]. Zheng et al. reduced the phase noise of an individual oscillator by coupling it to a resonant cavity [82]. Similarly, Colwell and Pearson achieved enhanced locking range via passive feedback [83].

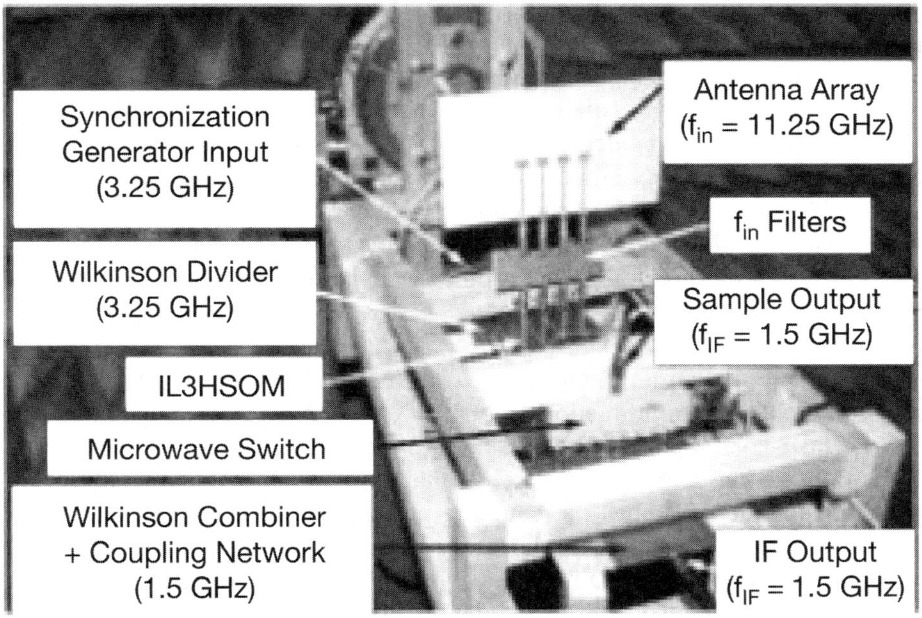

Fig. 6-35. Four-element coupled SOM array implementation. (Reprinted with permission from Ref. 79, ©2009 IEEE.)

Aside from the internal design of the oscillators, phase noise is inherently reduced via the coupling and mutual injection locking of a number of oscillators countering the phase noise associated with wide locking range. This phenomenon was studied by Chang, et al., who verified theoretically and experimentally that mutually injection locking N oscillators results in an N-fold reduction in phase noise relative to that of a single one of the oscillators by itself [84]. (See Fig. 6-36.) This happens because the noise signals of the oscillators are incoherent, whereas the carriers are coherent by virtue of the locking. Thus, the carrier voltages add, resulting in output power N^2 times that of a single oscillator, whereas the noise powers add, resulting in noise power only N times that of a single oscillator. They also showed that the noise increases near the edges of the locking range and reported that no corresponding reduction in phase noise results if the inter-oscillator coupling is unidirectional. An overview of phase noise analysis of coupled-oscillator arrays is presented in Section 7.10.

Recently, a significant decrease in phase noise with increasing coupling strength into the strong coupling regime was reported by Seetharam and Pearson [19]. Interestingly, the behavior of coupled oscillators has been proposed as an alternate means of measuring phase noise rather than the use of a delay line discriminator [85].

A common method of reducing phase noise is injection locking with a signal from a quiet (stable) external oscillator. As mentioned earlier in Section 6.1, this approach has been investigated in the context of arrays of coupled oscillators by Chang et al. [59]. They theoretically investigated injection of one or of all of the oscillators and experimented with injection of the center oscillator of a five-element array of X-band metal semiconductor field-effect transistor (MESFET) VCOs. They found that near the carrier frequency the noise is reduced to the level of the injection signal, while far from the carrier frequency it reverts to that of the array without external injection. The experimental results are shown in Fig. 6-37.

Dussopt and Laheurte designed a four-element array in a two-by-two configuration using unidirectional coupling to produce circular polarization at 4 GHz [86]. They reported that this configuration produced the expected factor-of-four reduction in phase noise, but that with unidirectional coupling, this reduction is independent of the coupling phase [87]. Yang et al. fabricated and tested a similar four-element ring array of linearly polarized elements using bidirectional coupling via lumped capacitors to produce circular polarization at 4.4 GHz [88]. They also experimented with an external injection locking signal.

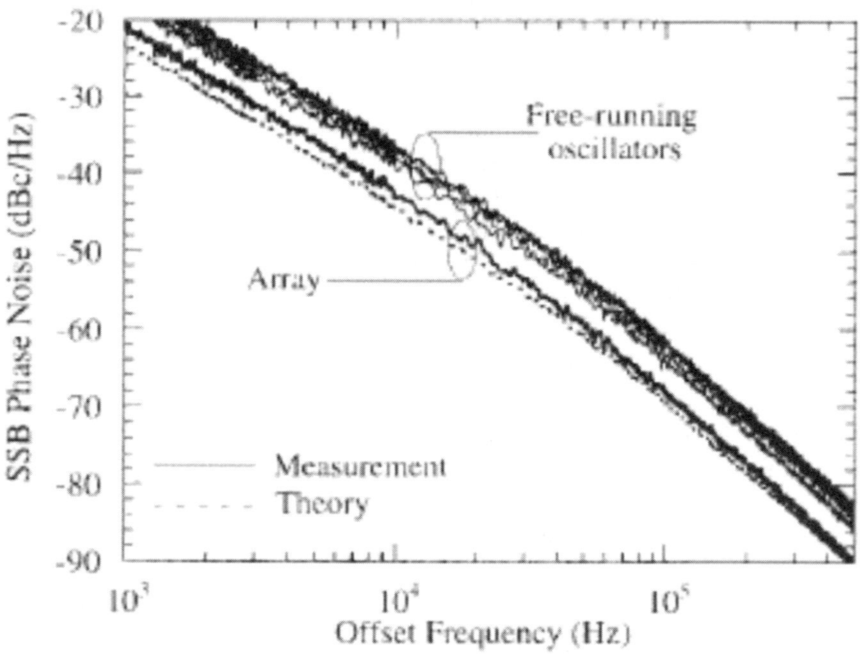

Fig. 6-36. Phase noise of coupled oscillators. (Reprinted with permission from Ref. 84, ©1997 IEEE.)

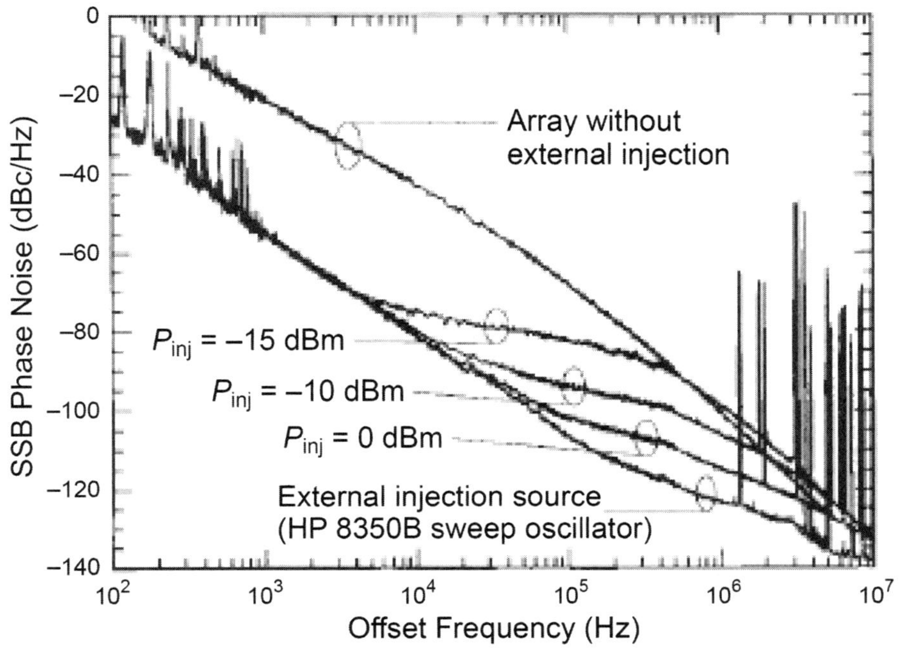

Fig. 6-37. Phase noise of externally injected coupled oscillators. (Reprinted with permission from Ref. 59, ©1997 IEEE.)

6.5 The Unlocked State

In the early days of research in microwave coupled oscillators for antenna applications, when spatial power combining was the primary objective, York and Compton observed a phenomenon closely related to mode locking in lasers [5]. The laser cavity supports a large number of modes of oscillation equally spaced in resonant frequency. By modulating a parameter such as the cavity Q at a frequency equal to the mode spacing, these *modes* can be coupled so that the phases of the oscillations become coherent. Under such conditions the sum of the modal signals form a Fourier series, and the laser output becomes a periodic sequence of equal-amplitude pulses. The energy in each pulse is proportional to the square of the number of modes summed because the combining is coherent. York and Compton showed that a similar effect occurs in a coupled oscillator array if the mutually injection-locked state discussed here in connection with beam-steering is *avoided*. In their array, the oscillators were tuned to a set of equally spaced frequencies separated by more than the locking range. Thus, the spectrum of the resulting spatially combined signal consists of a finite number of equally spaced spectral lines, one for each oscillator. The spectral lines are evenly spaced and tend to remain that way. This may be understood on an intuitive level by recalling that the spectrum of an injected but unlocked oscillator has the form shown in Fig. 1-2. In the limit of injection frequency far from the free-running frequency of the oscillator, the line spacing of the unlocked spectrum is approximately equal to the difference between the injection frequency and the free-running frequency. That is,

$$\sqrt{\Delta\omega_{inj}^2 - \Delta\omega_{lock}^2} \approx \Delta\omega_{inj} \tag{6.5-1}$$

Thus, because the injection signals come from the nearest neighbors, this means that the line spacing of the unlocked spectra is approximately equal to the difference in the free-running frequencies of the neighboring oscillators. In effect then, the oscillators each lock to a line of the unlocked spectrum of their neighbors, and the line spacing of the array becomes uniform. The stability of such mode-locked states has been studied in some detail by Lynch and York [89]. Note that as the differences in the tuning of the neighboring oscillators approach the locking range, the approximation Eq. (6.5-1) fails, the line spacing approaches zero, and the array becomes mutually injection locked, producing a monochromatic output. Maintenance of the mode-locked condition requires that mutual injection locking be avoided. As described in Section 1.4, the locking range can be controlled by adjusting the coupling phase, and in fact, if the coupling phase is 90 deg, the locking range becomes zero, and mutual injection locking is precluded. Thus, from a mode-locking perspective, a 90-deg coupling phase is to be preferred as noted by Lynch and York [89, 90].

One may view the finite line spectrum of the combined output as an infinite line spectrum filtered by a bandpass filter passing only the lines corresponding to the range of oscillator tunings. From Fourier theory, the corresponding time function will be an infinite sequence of equally spaced pulses whose shape is the inverse Fourier transform of the filter bandpass characteristic. For example, if the filter is a square pulse in frequency, the temporal pulses will be sinc functions. York and Compton demonstrated this with an array of three oscillators [5].

A few months later, York and Compton published additional results showing that, when a mode-locked array of oscillators is used to feed a linear array of radiating elements, the resulting beam scans as a function of time [6]. This is a consequence of the fact that the radiating elements are fed with slightly differing frequencies. The frequency differences may be viewed as relative phases changing linearly with time. Thus, the inter-element phasing of the array of elements changes linearly with time, resulting in a beam that scans with time. The repetition rate of the scan is just the period of the pulse output of the array, and at any given angle in the far-zone pattern, the received signal will repeat temporally with this same period as the beam repeatedly scans past that angle.

More recently, the unlocked state of such arrays has been studied as a generator of a chaotic output signal. The array is controlled by modulation of the coupling parameters with the objective of embedding information in the transitioning of the signal between the various unstable periodic orbits [91].

As indicated by York and Itoh [40], all of the phenomena observed for coupled voltage controlled oscillators (VCOs) may also be produced by coupled phase-locked loops (PLLs); one merely has more design flexibility when using PLLs. Section 7.12 contains an introduction to the analysis of coupled phase-locked loops. These principles were demonstrated in a two-element array by Martinez and Compton [92]. This also holds true for mode-locked arrays [93].

6.6 Conclusion

In this chapter we have outlined the experimental work leading to the current level of understanding of the design and fabrication of coupled-oscillator arrays and associated radiating apertures and their performance characteristics. Of course the work has continued as we write, and much of the most recent work severely taxes the capabilities of the linear approximation in explaining the results. Thus, the current trend favors full nonlinear design and analysis. While more complex, such an approach more accurately describes the expected behavior and permits exploitation of the nonlinear effects. These are aspects discussed in Part III of this book.

Part III: Nonlinear Behavior

Chapter 7

Perturbation Models for Stability, Phase Noise, and Modulation

The complex dynamics of coupled-oscillator arrays lead to the existence of a multitude of steady-state solutions. In addition to finding or selecting a desired steady-state solution, one further needs to guarantee its stability. In this section, perturbation methods are described that allow the designer to examine both the existence and the local stability of the various steady-state solutions of coupled oscillator arrays. An introduction to stability analysis of nonlinear dynamical systems is presented [94], followed by its application to coupled oscillator systems [95, 96].

The perturbation nature of noise, leads to phase-noise analysis methods that are closely related to the formulation used in the stability analysis. Analytical models are presented that demonstrate the attractive properties of coupled inter-injection locked oscillator systems, among them improved phase-noise performance compared to single elements [97].

A straightforward application of coupled-oscillator arrays has been in power-combining arrays where, by controlling the phase shift within an array of synchronized oscillator elements, one can direct the radiated beam toward a desired direction taking advantage of free-space power combining and eliminating the use of lossy power-combining networks. The simple topologies associated with such arrays have led to their consideration in communication system applications where one introduces modulation into the oscillator signals [98, 99]. Thus, methods to introduce modulation in such arrays are presented.

These architectures are distinguished from mixer-oscillator arrays where the modulation is not applied in the oscillator signal. Finally, an introduction to the analysis of coupled phase-locked loops is provided.

7.1 Preliminaries of Dynamical Systems

We have demonstrated in Part I of this book that coupled-oscillator arrays are able to synchronize in frequency while maintaining a fixed distribution of the relative phases between their elements and that, despite the complex nature of their dynamics, there are simple methods to control the phase relationships among the array elements, which require a small number of control parameters. It was also demonstrated that as the number of elements increases, there exist many different synchronized solutions, with different ensemble frequency values and different phase distributions. In order to be able to study the behavior of the various solutions as selected parameters of the array are varied, we must first provide a theoretical framework from nonlinear dynamical system theory. This will allow us to classify the types of the solutions and the phenomena that lead to creation or elimination of solutions as well as to changes in the solution stability.

In this section, principles of stability analysis of nonlinear dynamical systems are presented. The theory can be found in standard literature on dynamical systems [100, 101] and nonlinear differential equations [94].

Following Parker and Chua [101] an autonomous continuous time dynamical system is described by the system of differential equations

$$\dot{\boldsymbol{x}} = \boldsymbol{f}(\boldsymbol{x}) \qquad (7.1\text{-}1)$$

where the N-dimensional vector $\boldsymbol{x} \in \mathbb{R}^N$ contains the state variables of the system, and $\boldsymbol{f}(\boldsymbol{x})\colon \mathbb{R}^N \rightarrow \mathbb{R}^N$ is the vector field describing the dynamics of the system. The order of system Eq. (7.1-1) is N. An initial condition $\boldsymbol{x}(\boldsymbol{t_0}) = \boldsymbol{x_0}$ is assumed, where typically $t_0 = 0$ is set, since the vector field does not depend explicitly on time.

In contrast, a nonautonomous continuous time dynamical system is described by a system of equations of the form

$$\dot{\boldsymbol{x}} = \boldsymbol{f}(\boldsymbol{x}, \boldsymbol{t}) \qquad (7.1\text{-}2)$$

where the vector field depends explicitly on time. A nonautonomous system with period T can be expressed in the format of Eq. (7.1-1) by extending the state vector by one more dimension θ defined by $\dot{\theta} = 2\pi/T$, with $\theta(t_0) = 2\pi t_0/T$. In the following a dynamical system defined by Eq. (7.1-1) will be considered. A solution of Eq. (7.1-1) for a given initial condition is called a trajectory or orbit.

A free-running oscillator and a coupled-oscillator array are autonomous dynamical systems. They become nonautonomous when an external injection source is present.

A steady state is the asymptotic behavior of a dynamical system governed by (7.1-1) when $t \to \infty$, when the transient behavior has decayed to zero. A steady state is also called a limit set. The mathematical definition of a limit set includes the asymptotic behavior of a dynamical system both as time progresses forward ($t \to +\infty$) and backward ($t \to -\infty$), distinguishing between ω-limit sets and α-limit sets, respectively.

Steady states can be classified into four different types, equilibrium points, periodic solutions, quasi-periodic solutions, and chaotic solutions.

Equilibrium points x_0 correspond to the solution of

$$\boldsymbol{f}(\boldsymbol{x}_0) = \boldsymbol{0} \tag{7.1-3}$$

An equilibrium point is the DC solution of an oscillator circuit.

A periodic solution $x_0(t)$, is a solution of Eq. (7.1-1) that has a minimum period T, such as

$$\boldsymbol{x}_0(t+T) = \boldsymbol{x}_0(t) \tag{7.1-4}$$

for every t. A periodic solution of an autonomous system is also called a limit cycle.

A quasi-periodic solution is a solution that is equal to a countable sum of periodic solutions with noncommensurate periods, in other words:

$$\boldsymbol{x}_0(\boldsymbol{t}) = \sum_{i=1}^{M} \boldsymbol{h}_i(\boldsymbol{t}) \tag{7.1-5}$$

where $\boldsymbol{h}_i(t)$ are periodic solutions with minimum period T_i. The various frequencies $f_i = 1/T_i$ form a linearly independent set of dimension p with $1 < p \leq M$.

Finally, any bounded steady-state behavior that cannot be classified in one of the previous types is a chaotic steady state.

An N-dimensional dynamical system can be efficiently analyzed using a Poincaré map [101, 102]. The Poincare map is a transform that maps an N-dimensional continuous system to an $(N-1)$ dimensional discrete time system. This is illustrated in Fig. 7-1. Let us consider a continuous-time autonomous dynamical system described by Eq. (7.1-1), which has a periodic solution denoted by L. At some point $\boldsymbol{x}_0$ of the periodic orbit, we define locally a cross-section Σ that is a surface of dimension $N-1$ intersecting L at a non-zero angle. The periodic orbit returns to the point $\boldsymbol{x}_0$ on the cross-section Σ every T second. The sequence of points on the cross section defines a discrete time system, which is equivalent to the original continuous-time system. Furthermore, the limit cycle of an N-dimensional dynamical system is represented by a point on an (N–1)-dimensional surface. The reader is referred to the literature for a precise mathematical definition of a Poincare map for both cases of an autonomous and a nonautonomous system [101, 102]. It should be noted that the computation of a Poincaré map for autonomous systems is complicated by the fact that the period of the limit cycle is not known a priori; whereas in the case of nonautonomous systems, the sampling period T is known in advance due to the explicit dependence of Eq. (7.1-2) in time.

Having provided the fundamental definitions related to dynamical systems, along with the various types of existing solutions, the stability analysis of these solutions is presented in the next section.

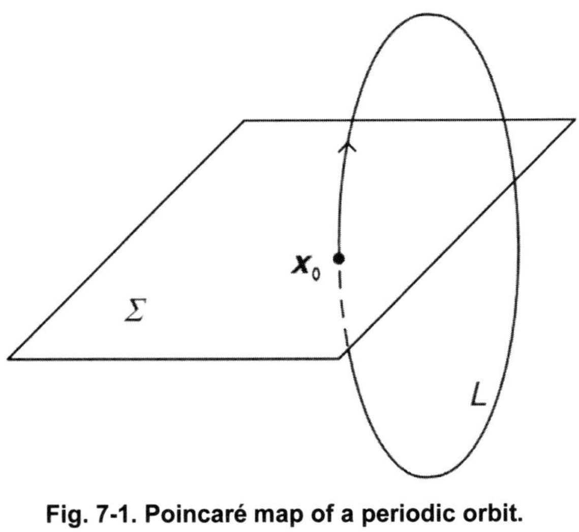

Fig. 7-1. Poincaré map of a periodic orbit.

7.1.1 Introduction to Stability Analysis of Nonlinear Dynamical Systems

There exist different types of stability. For the precise mathematical definitions and types of stability, the reader is referred to the literature [94, 101]. For the purposes of this work, a rather qualitative definition is provided. A steady state $\boldsymbol{x_0}$ is (Lyapunov) stable if and only if there exists a neighborhood V of $\boldsymbol{x_0}$ such that every trajectory with initial condition $\boldsymbol{x} \in V$ remains within V at all times $t > 0$. Furthermore, a steady state is asymptotically stable if and only if there exists a neighborhood V of $\boldsymbol{x_0}$ such that every trajectory with initial condition $\boldsymbol{x} \in V$ reaches arbitrarily close to $\boldsymbol{x_0}$ given enough time $t > 0$. In other words, the ω-limit set of any initial condition within V is $\boldsymbol{x_0}$. Conversely, a steady state $\boldsymbol{x_0}$ is unstable if there exists a neighborhood $V$of $\boldsymbol{x_0}$ such that $\boldsymbol{x_0}$ is the α-limit set of all initial conditions in V. Finally, a steady state $\boldsymbol{x_0}$ is called non-stable if for every neighborhood V, there exists at least one point whose ω-limit set is $\boldsymbol{x_0}$ and one point whose α-limit set is $\boldsymbol{x_0}$.

7.1.2 Equilibrium Point

The stability of an equilibrium point $\boldsymbol{x_0}$ is examined by considering the linear perturbation of the vector field $\boldsymbol{f}(\boldsymbol{x})$ at $\boldsymbol{x_0}$. The eigenvalues of the Jacobian $\boldsymbol{J}(\boldsymbol{x_0})$ of the vector field determine the stability of the solution.

$$\boldsymbol{\delta\dot{x}} = \frac{\boldsymbol{df}}{\boldsymbol{dx_0}}\boldsymbol{\delta x} = \boldsymbol{J}(\boldsymbol{x_0})\boldsymbol{\delta x} \qquad (7.1\text{-}6)$$

with $\boldsymbol{\delta x}(\boldsymbol{0}) = \boldsymbol{\delta x_0}$ representing an initial perturbation from $\boldsymbol{x_0}$. The solution of the linear differential equation Eq. (7.1-6) generally takes the form

$$\boldsymbol{\delta x}(\boldsymbol{t}) = \sum_{i=1}^{N} c_i e^{\lambda_i t} \boldsymbol{a_i} \qquad (7.1\text{-}7)$$

where λ_i and $\boldsymbol{a_i}$ are the eigenvalues and eigenvectors of $\boldsymbol{J}(\boldsymbol{x_0})$, respectively. The constants c_i are determined by the initial condition $\boldsymbol{\delta x_0}$.

For a given N-dimensional vector field $\boldsymbol{f}(\boldsymbol{x})$ the $N \times N$ Jacobian matrix $\boldsymbol{J}$ has N eigenvalues. An equilibrium point whose eigenvalues do not have a real part equal to zero is called hyperbolic. If all eigenvalues have negative real parts, the point $\boldsymbol{x_0}$ is asymptotically stable. Correspondingly, if there exists one eigenvalue with a positive real part, $\boldsymbol{x_0}$ is unstable. Finally, if there exists one eigenvalue with a real part equal to zero, the equilibrium point is non-hyperbolic, a condition that is equivalent to the determinant of $\boldsymbol{J}$ being equal to zero, and the eigenvalues of $\boldsymbol{J}$ are not sufficient to determine its stability.

In the case of an ($N = 2$)-dimensional system, the Jacobian matrix has two eigenvalues λ, which satisfy the following characteristic equation [102]

$$\lambda^2 - \sigma\lambda + \Delta = 0 \qquad (7.1\text{-}8)$$

where σ is the sum and Δ the product of the two eigenvalues. The classification of the hyperbolic equilibrium points and their stability for different values of λ is shown in Fig. 7-2 [102].

7.1.3 Periodic Steady State

In order to determine the stability of a periodic solution $\boldsymbol{x_0}(t)$ the linear perturbation of Eq. (7.1-1), also called a linear variational equation, with respect to the time varying $\boldsymbol{x_0}(t)$ is formed, leading to a system of linear differential equations with periodic coefficients.

$$\boldsymbol{\delta\dot{x}} = \frac{\boldsymbol{df(t)}}{\boldsymbol{dx_0}}\boldsymbol{\delta x} = \boldsymbol{J}[\boldsymbol{x_0}(t)]\boldsymbol{\delta x} \qquad (7.1\text{-}9)$$

with $\boldsymbol{\delta x}(0) = \boldsymbol{\delta x_0}$ and for $\boldsymbol{J}[\boldsymbol{x_0}(t+T)] = \boldsymbol{J}[\boldsymbol{x_0}(t)]$.

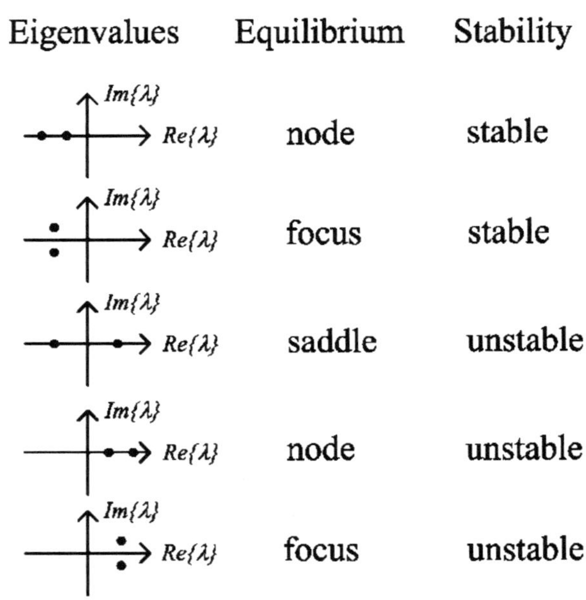

Fig. 7-2. Hyperbolic equilibria of a two-dimensional system.

The solution of Eq. (7.1-6) is derived using Floquet theory [94]

$$\boldsymbol{\delta x}(t) = \sum_{i=1}^{N} c_i\, m_i^{\frac{t}{T}} \boldsymbol{p_i}(t) = \sum_{i=1}^{N} c_i e^{\lambda_i t} \boldsymbol{p_i}(t) \qquad (7.1\text{-}10)$$

where m_i are the Floquet multipliers and $\boldsymbol{p_i}(t)$ are periodic vector functions. The Floquet exponents λ_i are related to the multipliers by

$$m_i = e^{\lambda_i T} \qquad (7.1\text{-}11)$$

It is seen from Eq. (7.1-11) that there is not a unique mapping between multipliers and exponents, as adding to any exponent a complex factor $jk\, 2\pi/T$ with k an arbitrary integer results in the same multiplier.

The stability of $\boldsymbol{x_0}(t)$ is determined by the Floquet multipliers m_i. They can be calculated by direct integration of Eq. (7.1-9) for one period T with initial condition $\boldsymbol{\delta x}(0) = \boldsymbol{I_N}$, where $\boldsymbol{I_N}$ is the identity diagonal square matrix of dimension N. The result of the integration is the monodromy matrix C whose eigenvalues are the desired Floquet mutlipliers m_i [94].

A periodic solution $\boldsymbol{x_0}(t)$ of an autonomous system has at least one Floquet multiplier with magnitude equal to 1, or equivalently a Floquet exponent equal to zero. Furthermore, a periodic solution $\boldsymbol{x_0}(t)$ is stable if the remaining multipliers have a magnitude less than one ($|m_i|<1$). Correspondingly, if one multiplier with magnitude larger than 1 exists, the solution is unstable.

7.1.4 Lyapunov Exponents

The Lyapunov exponents are defined as

$$\mu_i = \lim_{t\to\infty} \frac{1}{t} \ln\left|e^{\lambda_i t}\right| \qquad (7.1\text{-}12)$$

and can be considered a generalization of both the characteristic eigenvalues of the equilibrium point and the Floquet multipliers of the periodic steady state [101]. In fact, the Lyapunov exponents can be used to determine the stability of quasi-periodic and chaotic steady-state solutions.

One can easily see from Eq. (7.1-7) that the Lyapunov exponents of the equilibrium point correspond to the real part of the characteristic eigenvalues.

$$\mu_{i,eq} = Re\{\lambda_i\} \qquad (7.1\text{-}13)$$

Correspondingly, the Lyapunov exponents of the periodic steady state are equal to the natural logarithm of the magnitude of the Floquet multipliers divided by the period of the solution, which is equal to the real part of the Floquet exponents.

$$\mu_{i,pss} = \frac{\ln|m_i|}{T} = Re\{\lambda_i\} \tag{7.1-14}$$

7.2 Bifurcations of Nonlinear Dynamical Systems

In practice, a dynamical system described by Eq. (7.1-1) depends on a set of parameters $\boldsymbol{\xi}$ of dimension k ($\boldsymbol{\xi} \in \mathbb{R}^k$), which enter the definition of the vector field:

$$\dot{\boldsymbol{x}} = \boldsymbol{f}(\boldsymbol{x}, \boldsymbol{\xi}) \tag{7.2-1}$$

A parameter corresponds to some circuit control voltage or bias voltage/current or any other physical parameter such as the dimension of a transmission line. As the parameter vector varies, the solutions of Eq. (7.2-1) change. The change of stability of a specific steady-state solution, the creation of new steady-state solutions or elimination of existing ones, as one or more parameters of a nonlinear system vary is called a *bifurcation* [100,102]. The corresponding parameter values for which a bifurcation occurs are called *bifurcation values*. A *bifurcation diagram* is a plot of a selected state variable(s) corresponding to a limit set versus the system parameter(s). An example of a bifurcation diagram is the plot of the DC voltage at a selected circuit node or the oscillation amplitude versus the external bias voltage of the oscillator. Bifurcations are classified into *local* and *global*. Local bifurcations are detected by studying the vector field f in a neighborhood of a limit set. In contrast, local information is not sufficient to detect global bifurcations. Typically in this book we study systems where one parameter is varied ($k = 1$).

7.2.1 Bifurcations of Equilibrium Points

Let us consider such a continuous time system with one parameter that has a hyperbolic equilibrium point. As the parameter varies, there are two ways that a hyperbolic point can become nonhyperbolic. In the first one, a simple real eigenvalue becomes zero ($\lambda_1 = 0$). In this case the system is going through a bifurcation known as *fold* bifurcation. Fold bifurcation is also known as *turning-point* or *saddle-node* bifurcation. In a fold bifurcation the equilibrium point curve presents an infinite slope at the parameter value $\xi = \xi_0$ where one real eigenvalue becomes zero. This is seen in the bifurcation diagram of a one-dimensional system shown in Fig. 7-3.

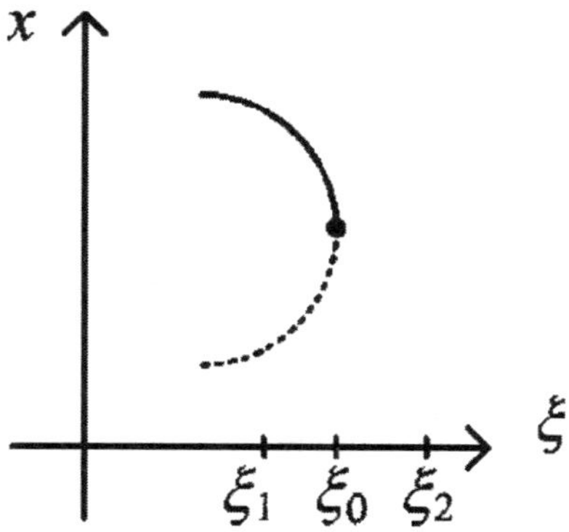

Fig. 7-3. Fold bifurcation.

The infinite slope at ξ_0 corresponding to a real zero eigenvalue leads to a folding, a turning point of the solution curve. For larger values of the parameter $\xi > \xi_0$ no solutions exist, whereas for $\xi < \xi_0$ two solutions exist. In fact, one solution branch contains stable nodes indicated by a solid line, whereas the other unstable (saddle) solutions, and this is indicated by a dotted line [102]. In the case of a one-dimensional system, the unstable solution is a node, whereas in the general case it is a saddle. At the critical value $\xi = \xi_0$, the node and saddle collide, hence the name saddle-node bifurcation.

In the second case, a pair of simple complex eigenvalues fall on the imaginary axis ($\lambda_0 = \pm j\omega_0$ with $\omega_0 > 0$). In this case, the system undergoes a *Hopf* bifurcation, and a limit cycle is born or is extinguished. It is straightforward to see that a Hopf bifurcation requires that the system be at least second order ($n \geq 2$).

An example of a Hopf bifurcation in a two-dimensional system is shown in Fig. 7-4. In a supercritical Hopf bifurcation (Fig. 7-4a), a stable limit cycle is born as the parameter goes through the bifurcation value ξ_0. At the same time the stable equilibrium solution becomes unstable. In subcritical Hopf bifurcation (Fig. 7-4b), an unstable limit cycle is created while an unstable equilibrium point becomes stable.

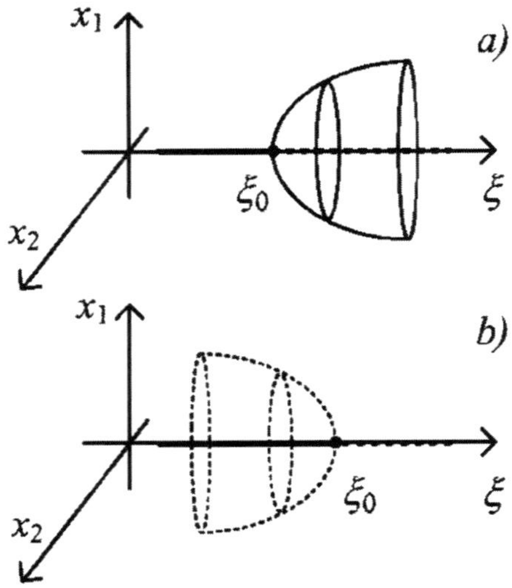

Fig. 7-4. Hopf bifurcation: (a) supercritical, (b) subcritical.

7.2.2 Bifurcations of Periodic Orbits

Correspondingly, a hyperbolic limit cycle is a limit cycle that does not have any Floquet multipliers with magnitude equal to one [102]. A periodic steady state of an autonomous system has one Floquet multiplier equal to one, and therefore, it is hyperbolic if the remaining multipliers do not have magnitude equal to one.

Given a periodic steady state with frequency $\omega = 2\pi/T$, there exist three types of bifurcations for one-parameter systems [102], corresponding to three distinct possibilities that a multiplier crosses the unit cycle as the parameter is varied, shown in Fig. 7-5.

In the *fold* bifurcation (Fig. 7-5a), a real multiplier takes the value $m = 1$. In this case the frequency of the limit cycle remains the same, something that can be inferred from Fig. 7-5a) and Eq. (7.1-10), which describes the linear perturbation of the dynamical system around the periodic steady state. A Floquet multiplier equal to 1 leads to a perturbation that evolves as

$$\boldsymbol{\delta x}(t) = ce^{j\omega t}\boldsymbol{p}(t) \tag{7.2-2}$$

with n integer and therefore it does not perturb the oscillation frequency of the system. A fold bifurcation leads to a change in the stability of the periodic steady-state solution, much as the fold bifurcation of an equilibrium point does for the equilibrium point.

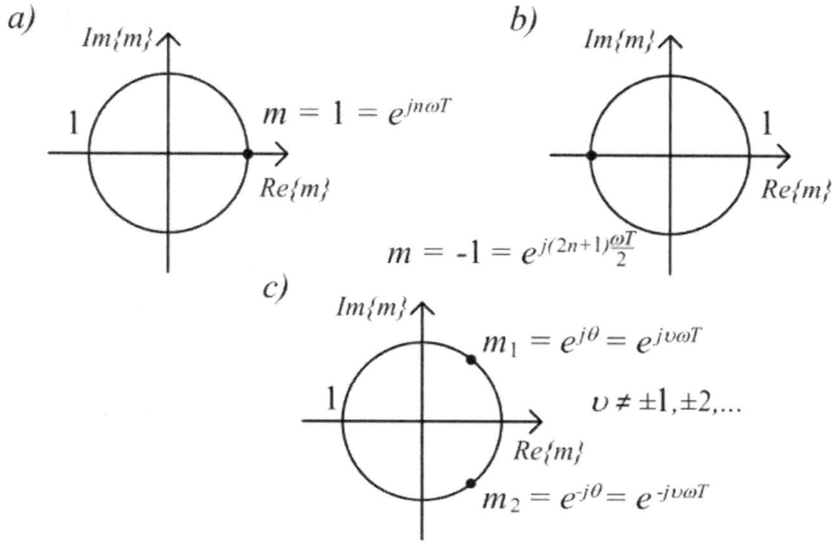

Fig. 7-5. Bifurcations of periodic orbits: (a) fold, (b) flip, and (c) Neimark–Sacker bifurcation.

A *flip* bifurcation occurs when the multiplier becomes $m = -1$ (Fig. 7-5b). In contrast to the fold bifurcation, a flip bifurcation leads to the existence of a new limit cycle whose oscillation frequency is half of the original one. Due to this fact, a flip bifurcation is also known as a *period-doubling* bifurcation. The contribution to the linear variational equation of the $m = -1$ multiplier is

$$\boldsymbol{\delta x}(t) = ce^{j\frac{\omega}{2}t}\boldsymbol{p}(t) \tag{7.2-3}$$

where one can observe the appearance of a term with a frequency half of the original one.

Finally, in a *Neimark–Sacker* or *torus* bifurcation, a pair of complex conjugate multipliers appear on the unit cycle (Fig. 7-5c). As a result, a new frequency, which is not harmonically related to the orininal one, appears in the system, leading to the onset of a quasi-periodic solution. The contribution to the linear variational equation of this Floquet multiplier is expressed as

$$\boldsymbol{\delta x}(t) = ce^{j\upsilon\omega t}\boldsymbol{p}(t) \tag{7.2-4}$$

where υ is not an integer.

7.3 The Averaging Method and Multiple Time Scales

The averaging method is typically used to analyze the periodic steady-state solutions of weakly nonlinear systems

$$\dot{\boldsymbol{x}} = \varepsilon \boldsymbol{f}(\boldsymbol{x}, \varepsilon) \tag{7.3-1}$$

and perturbations of the linear oscillator systems [100, 103], with $\varepsilon \ll 1$. It was originally developed by Krylov and Bogoliubov [104]. The method is particularly suitable to analyze the perturbed linear oscillator problem described by

$$\ddot{x} + \omega^2 x = \varepsilon f(x) \tag{7.3-2}$$

where $\omega, x, f \in \mathbb{R}$. The van der Pol differential equation belongs to this class of systems with $f(x) = (1 - x^2)\dot{x}$. In the case of weakly coupled oscillators, an equation of the form Eq. (7.3-2) is used to describe each oscillator, and $f(x)$ contains the nonlinear term of the free-running (uncoupled) oscillator as well as contributions from external coupled signals from other oscillators, which can be linear or nonlinear. The averaging theorem [100] Uc585947 states that there exists a change of coordinates $\boldsymbol{x} = \boldsymbol{y} + \varepsilon \boldsymbol{w}(\boldsymbol{y}, \varepsilon)$, which transforms Eq. (7.3-1) to the averaged system

$$\dot{\mathbf{y}} = \varepsilon \bar{\boldsymbol{f}}(\mathbf{y}) \tag{7.3-3}$$

where

$$\bar{\boldsymbol{f}}(\mathbf{y}) = \frac{1}{2\pi} \int_0^{2\pi} \boldsymbol{f}(\mathbf{y}(z), 0) dz \tag{7.3-4}$$

The system given by Eq. (7.3-3) is an autonomous system, whereas Eq. (7.3-1) can be nonautonomous. The essential property of the averaged system that is extensively applied in the study of coupled oscillator systems is that a hyperbolic periodic steady state of Eq. (7.3-3) corresponds to a hyperbolic equilibrium point of Eq. (7.3-1) and that both steady states have the same stability [100]. This essentially means that the eigenvalues of the linearized system of an equilibrium point of Eq. (7.3-3) determine its stability. This is quite useful as obtaining the Floquet multipliers of a microwave oscillator may not be a trivial task. Furthermore, for the cases that are considered in this book, the bifurcations of the averaged system are the same as those of the original system [100].

In order to transform the perturbed linear oscillator problem in the standard form Eq. (7.3-1), the following transformation (known as the van der Pol transformation [100, 103]) is commonly applied to Eq. (7.3-2) before averaging:

$$x = A\cos(\omega t + \phi)$$
$$\dot{x} = -\omega A \sin(\omega t + \phi) \tag{7.3-5}$$

In this case, Eq. (7.3-2) becomes

$$\dot{A} \approx -\frac{\varepsilon}{\omega}\sin(\omega t + \phi) f\,(A, \phi, \varepsilon) = \varepsilon g_1(A, \phi, \varepsilon)$$
$$\dot{\phi} \approx -\frac{\varepsilon}{A\omega}\cos(\omega t + \phi) f(A, \phi, \varepsilon) = \varepsilon g_2(A, \phi, \varepsilon) \tag{7.3-6}$$

Applying Eq. (7.3-4), the averaged solution is obtained:

$$\dot{\bar{A}} = \varepsilon \bar{g}_1(\bar{A}, \bar{\phi})$$
$$\dot{\bar{\phi}} = \varepsilon \bar{g}_2(\bar{A}, \bar{\phi}) \tag{7.3-7}$$

It should be noted that the transformation given by Eqs. (7.3-5) and subsequent application of the perturbation method limits the analysis of the system given by Eq. (7.3-2) locally near the oscillation frequency ω in the frequency domain. The system can be studied near a different harmonic by modifying appropriately the transformation, that is, setting $n\omega$ in place of ω where n is the desired harmonic order. In practice, considering the oscillator behavior near the fundamental frequency is sufficient for the study of high Q oscillators because higher harmonics are small and therefore can be ignored in the analysis. Specifically, the averaged van der Pol differential equation for which $f(x) = (1 - x^2)\dot{x}$ becomes [103]

$$\dot{\bar{A}} = \frac{\varepsilon}{2}\bar{A}\left(1 - \frac{\bar{A}^2}{4}\right)$$
$$\dot{\bar{\phi}} = 0 \tag{7.3-8}$$

The above system leads to a nontrivial steady-state oscillation with amplitude $\bar{A} = 2$ obtained by requiring that $\dot{\bar{A}} = 0$.

7.4 Averaging Theory in Coupled Oscillator Systems

Kurokawa considered the oscillator equivalent of a series resistance, inductance, capacitance (RLC) resonator connected in series with a negative resistance and applied the averaging theory to study the properties of noisy oscillators and injection-locked oscillators [105].

The theory of Kurokawa applied to the study of coupled oscillator arrays was introduced to the antennas and microwaves communities by Stephan [1] with an aim toward antenna-array applications such as power-combining arrays and phased arrays. The work of Stephan focused on taking advantage of the dynamical properties of coupled oscillator array topologies in order to generate constant phase shift distributions among the array elements in a continuously variable manner. A parallel RLC resonator in parallel with a negative resistance was used to model each oscillator, leading to a dual form of the one used by Kurokawa.

It should be noted that there is significant theoretical work in the literature regarding coupled oscillator systems, also called distributed and ladder oscillators, considering the various operating modes and stability of one- and two-dimensional arrays. Notable references are [106, 107, 108, and 109]. The latter work by Endo and Mori [109] presented an elegant way to obtain a formulation equivalent to a perturbed van der Pol equation in vector form for an array of coupled oscillators modeled as a parallel RLC resonator with a negative resistance, and it will be given in the next section. An efficient analysis of coupled oscillator arrays for quasi-optical power combining and the stability of the various existing operating modes was proposed by York and Compton [110] utilizing only the phase dynamics of the array, or in other words the second equation of Eqs. (7.3-7).

We may distinguish among power-combining applications where the stability of the various operating modes of coupled oscillator systems is with an aim to secure excitation of only the in-phase mode, and applications where an arbitrary phase distribution among the oscillator elements is required (such as beamforming and phased arrays). The latter may be viewed as a generalization of the former.

Following their initial work, York produced a general formulation for coupled-oscillator arrays based on the fundamental harmonic approximation and the averaging method that is essentially used to date in most approximate analysis methods for such systems [95, 111]. Furthermore, York introduced an elegant way to achieve constant progressive phase shifts among the array oscillator elements in a continuous fashion by only modifying the oscillation frequency of the end elements of the array. In 2004, Heath presented an elegant and unifying formulation of the application of the method of averaging (specifically the Lindstedt method was used to derive the slow time differential equations [94]) in coupled-oscillator arrays along with a detailed stability analysis of various different coupling network topologies [112]. The latter formulation is given here:

$$\dot{A}_m = \mu(A_{0m}^2 - A_m^2)A_m + \sum_{i=1}^{N} K_{mi}A_i \cos(\phi_i - \phi_m + \Phi_{im}) \tag{7.4-1}$$

$$A_m\dot{\phi}_m = \Delta\omega_m A_m + \sum_{i=1}^{N} K_{mi}A_i \sin(\phi_i - \phi_m + \Phi_{im})$$

where an array of N oscillators is assumed. The variable A_m represents the slowly varying averaged amplitude of oscillator m, as given in Eqs. (7.3-7) with the bar suppressed for simplicity. Correspondingly, the phase of oscillator m is given by $\theta_m = \omega_0 t + \phi_m$ where ϕ_m is the averaged time-varying component of the oscillator phase corresponding to Eqs. (7.3-7) (with the bar suppressed). When uncoupled to the rest of the array elements, each oscillator m has a periodic steady state with amplitude A_{0m} and frequency $\omega_m = \omega_0 + \Delta\omega_m$. Furthermore, each individual oscillator satisfies a van der Pol differential equation of nonlinearity constant ε, which appears in Eqs. (7.4-1) through $\mu = \varepsilon\,\omega_m/(2Q)$ where Q is the external quality factor of the resonator of each oscillator element calculated using a reference load admittance G_L. Coupling among the oscillator elements is included in the form of a square complex matrix $T = [t_{mi}]$ of dimension N, with $t_{mi} = T_{mi}e^{j\Phi_{mi}}$. Note that T is a transfer function (unitless). If, for example, an admittance matrix is used to express the coupling among oscillator elements, then T is the admittance matrix normalized to the reference load admittance G_L. In Eqs. (7.4-1) the coupling coefficients also appear in normalized form setting:

$$\kappa = [\kappa_{mi}] = \left[K_{mi}e^{j\Phi_{mi}}\right] = [t_{mi}\omega_m/2Q] \tag{7.4-2}$$

Finally, Eqs. (7.4-1) can be written in a complex-valued compact format letting $a_m = A_m e^{j\phi_m}$:

$$\dot{a}_m = j\Delta\omega_m a_m + \mu(A_{0m}^2 - |a_m|^2)a_m + \sum_{i=1}^{N} \kappa_{mi}a_i \tag{7.4-3}$$

Under weak coupling conditions, the phase dynamics alone are sufficient to analyze the behavior of the coupled oscillator system. We may then consider only the second equation of Eqs. (7.4-1) and assume that the oscillator amplitudes are approximately equal to their uncoupled values $A_m = A_0$. The system of equations pertaining to the phase dynamics provide significant insight and a very computationally efficient method to analyze arrays with a large number of elements. In fact, the analysis results of Part I of this book have focused on the phase dynamics of coupled oscillator arrays. The system of equations limited to the phase dynamics was introduced as the "generalized phase model" by Heath [112]:

$$\dot{\phi}_m = \Delta\omega_m + \sum_{i=1}^{N} K_{mi} \sin(\phi_i - \phi_m + \Phi_{mi}) \tag{7.4-4}$$

When no coupling phase Φ_{mi} is considered, the model is the well-known Kuramoto model [113]. In the special case where a bidirectional symmetrical coupling matrix with $\kappa_{im} = \kappa_{mi}$ is considered, the generalized phase model coincides with the phase model introduced by York [111].

A fixed point of Eq. (7.4-1) corresponds to a periodic steady-state solution, defined for $\dot{A}_m = 0$ and $\dot{\phi}_m = c$ with c an arbitrary constant. Letting c take nonzero values still corresponds to synchronized solutions of the array but for a different frequency than ω_0.

$$\begin{aligned} \mu(A_{0m}^2 - A_m^2)A_m + \sum_{i=1}^{N} K_{mi}A_i \cos(\phi_i - \phi_m + \Phi_{im}) = 0 \\ (\Delta\omega_m - c)A_m + \sum_{i=1}^{N} K_{mi}A_i \sin(\phi_i - \phi_m + \Phi_{im}) = 0 \end{aligned} \tag{7.4-5}$$

Every set (A_m, ϕ_m) that satisfies the above conditions corresponds to an oscillating mode of the array. In principle, there exist up to 2^{N-1} modes [111]. It should be emphasized that, due to the autonomous nature of the coupled oscillator system, it is possible to translate all oscillator phases ϕ_m by the same arbitrarily large value and still obtain the same steady-state solution. This is evidenced by the fact that only phase differences appear in Eqs. (7.4-5). In other words, the steady state is defined by the oscillator phase differences and not their absolute phase.

The stability of the oscillating modes is examined by considering the linear perturbation $(A_m + \delta A_m, \phi_m + \delta\phi_m)$ of Eqs. (7.4-1), which leads to a system of linear differential equations:

$$\delta\dot{A}_m = \mu(A_{om}^2 - 3A_m^2)\delta A_m + \sum_{i=1}^{N} K_{mi}\cos(\phi_i - \phi_m + \Phi_{mi})\,\delta A_i$$
$$-\sum_{i=1}^{N} K_{mi}A_i\sin(\phi_i - \phi_m + \Phi_{mi})\,(\delta\phi_i - \delta\phi_m)$$

(7.4-6)

$$A_m\delta\dot{\phi}_m = -\dot{\phi}_m\delta A_m + \sum_{i=1}^{N} K_{mi}\sin(\phi_i - \phi_m + \Phi_{mi})\,\delta A_i$$
$$+\sum_{i=1}^{N} K_{mi}A_i\cos(\phi_i - \phi_m + \Phi_{mi})\,(\delta\phi_i - \delta\phi_m)$$

In the case of the generalized phase model, one has

$$\delta\dot{\phi}_m = \sum_{i=1}^{N} K_{mi}\cos(\phi_i - \phi_m + \Phi_{mi})\,(\delta\phi_i - \delta\phi_m) \qquad (7.4\text{-}7)$$

Because the steady state is defined by phase differences and not absolute phase values, the perturbation phase values $\delta\phi_m$ of the steady state may not be small. Their differences, however, are assumed to be small, and this allows one to take the linear approximation of the cosine and sine terms in Eqs. (7.4-1) and obtain Eqs. (7.4-6).

7.5 Obtaining the Parameters of the van der Pol Oscillator Model

A useful analytical method is presented, that allows one to obtain the van der Pol differential equation from a parallel resonator with a nonlinear voltage dependent current source. The procedure follows the development presented by Endo and Mori [108], and it represents a time-domain formulation of van der Pol's model described in Section 1.2. This model has very low complexity, and it can be easily incorporated into analysis of large arrays or proof of concept for various topologies of coupled oscillators. In addition, it can be easily introduced into circuit simulators.

Given a nonlinear voltage-dependent current source $i_n = f(v)$ with time derivative $di/dt = g_v(v)\dot{v}$, where $g_v(v) = df(v)/dv$ is a nonlinear conductance, and applying Kirchhoff's current law in the circuit of Fig. 7-6, one has

$$C\dot{v} + g_L v + \frac{1}{L} i_1 + i_n + i_{inj} = 0 \tag{7.5-1}$$

which, after differentiating, becomes

$$C\ddot{v} + (g_L + g_v)\dot{v} + \frac{v}{L} + i_{inj} = 0 \tag{7.5-2}$$

Defining the natural frequency of the tank $\omega_0^2 = 1/LC$, setting $\varepsilon = g_L/(C\omega_0)$, and scaling time $t = \omega_o \tau$ the differential equation takes the form

$$v'' + \varepsilon\left(1 + \frac{g_v}{g_L}\right) v' + v + \frac{\varepsilon}{g_L} i'_{inj} = 0 \tag{7.5-3}$$

where $(\ )'$ indicates the scaled time derivative $d/d\tau$.

If we consider a nonlinear current source modeled by a third-order polynomial, the equation corresponding to the free-running oscillator ($i_{inj} = 0$) can be transformed to the van der Pol equation. Let a voltage-dependent current source be of the form

$$i_n = f(v) = -g_1 v + g_3 v^3 \tag{7.5-4}$$

so that

$$g_v(v) = -g_1 + 3g_3 v^2 \tag{7.5-5}$$

Setting

$$\lambda = \varepsilon \frac{g_1 - g_L}{g_L} = \frac{g_1 - g_L}{C\omega_0} = \frac{1}{Q} \tag{7.5-6}$$

one obtains

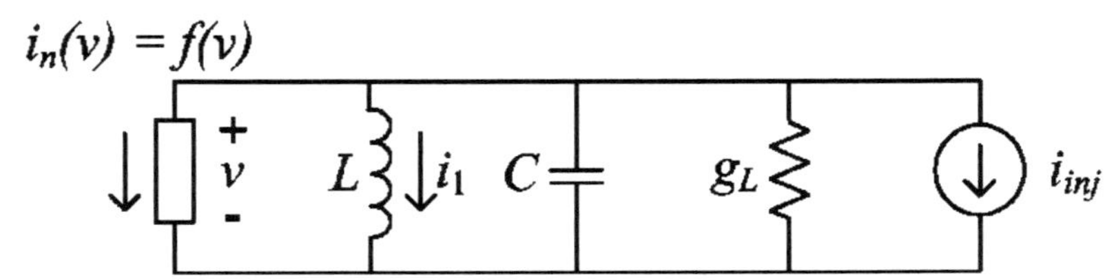

Fig. 7-6. Oscillator model consisting of a parallel RLC resonator and a nonlinear voltage-dependent current source.

$$v'' - \lambda\left(1 - \frac{3g_3}{g_1 - g_L}v^2\right)v' + v + \frac{\lambda}{g_1 - g_L}i'_{inj} = 0 \qquad (7.5\text{-}7)$$

The parameter λ is equal to the inverse of the loaded quality factor, Q, of the oscillator circuit of Fig. 7-6.

Finally, scaling the voltage as

$$x = \sqrt{\beta}v = \sqrt{\frac{3g_3}{g_1 - g_L}}v \qquad (7.5\text{-}8)$$

the differential equation takes the desired form:

$$x'' - \lambda(1 - x^2)x' + x + \lambda\sqrt{\frac{3g_3}{(g_1 - g_L)^3}}i'_{inj} = 0 \qquad (7.5\text{-}9)$$

When no injection signal is present, $i_{inj} = 0$, and this equation becomes the well known van der Pol equation.

$$x'' - \lambda(1 - x^2)x' + x = 0 \qquad (7.5\text{-}10)$$

The approximate solution to the van der Pol oscillation is [94]

$$x = 2\cos\tau \Rightarrow v = \frac{2}{\sqrt{\beta}}\sin\omega_0 t \qquad (7.5\text{-}11)$$

which is identical to the solution provided by the averaging method in Section 7.3.

The approximate parallel model for the oscillator can be extracted using a nonlinear simulator and calculating the admittance at a selected circuit node [114]. Several authors have proposed experimental techniques to evaluate the model parameters [114,115]. Measurement of the oscillator amplitude can be used to obtain the scaling parameter β. Injection locking the oscillator to an external signal and measuring the locking bandwidth can be used to estimate the Q and subsequently the second parameter λ of the van der Pol model [114]. Alternatively, a low-frequency sinusoidal modulating signal can be introduced in the bias circuitry of the oscillator, resulting in a phase-modulated oscillator output. The parameter λ can then be obtained by measuring the relative amplitude of the modulation sidebands [115].

7.6 An Alternative Perturbation Model for Coupled-Oscillator Systems

It is possible to formulate an alternative practical model for analyzing coupled oscillator arrays by considering that the periodic steady state of the coupled system is a perturbation of the free running steady state of the individual oscillator elements when they are uncoupled to each other. This assumption holds when the coupling is weak, as is typically the case when designing such systems. The model described in this section was proposed in Ref. 116.

An advantage of this formulation is that there is no underlying assumption about the oscillator nonlinearity model, such as for example a third-order nonlinearity used in the van der Pol model, and each individual oscillator can be designed using any numerical technique. The uncoupled free-running steady state is expressed in the slow time (at the fundamental frequency component) as

$$Y_0(V,\omega,\mu)V_0 e^{j\phi_0} = 0 \tag{7.6-1}$$

where Y_0 is the admittance looking into a properly selected node of the circuit and V_0 is the oscillation amplitude at that node (Fig. 7-7).

This is merely application of Kirchhoff's current law at the node under consideration. Y_0 typically contains both linear and nonlinear terms and depends on the oscillation frequency $\omega = \omega_0$ and amplitude $V = V_0$. Generally, one may assume that Y_0 depends on a number of additional circuit parameters, such as bias voltages. In Eq. (7.6-1) a single parameter $\mu = \mu_0$ is considered which corresponds to some DC voltage that allows for frequency tuning. Assuming a nonzero periodic steady-state amplitude V_o, Eq. (7.6-1) is satisfied for $Y_0 = 0$. A free-running oscillator is an autonomous system characterized by an arbitrary time reference, which translates in an arbitrary phase reference in the frequency domain. As a result, $\phi_0 = 0$ may be set without loss of generality.

A coupled oscillator array of N elements is then described by

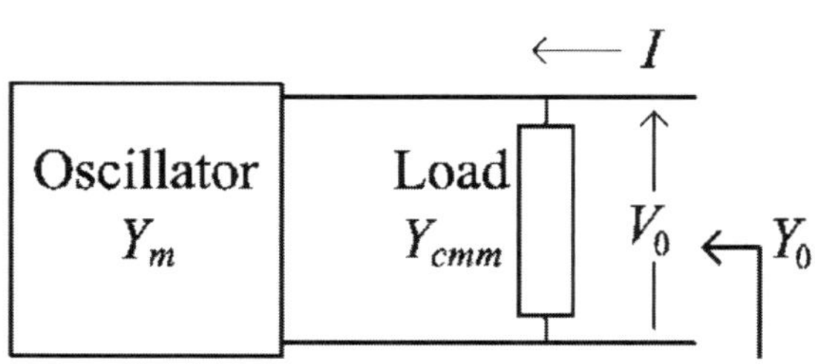

Fig. 7-7. Oscillator 1-port equivalent circuit.

$$Y_m V_m e^{j\phi_m} + \sum_{i=1}^{N} Y_{cmi} V_i e^{j\phi_i} = 0 \tag{7.6-2}$$

where the coupling is represented by the admittance matrix $Y_c(\omega) = [Y_{cmi}(\omega)]$, which typically is frequency dependent. The coupling results in a steady state that can be expressed as a perturbation of the individual oscillator free-running steady state as follows:

$$V_m = V_0 + \Delta V_m(t)$$

$$\phi_m = \phi_{0m} + \Delta\phi_m(t)$$

$$\mu_m = \mu_{0m} + \Delta\mu_m$$

$$\omega = \omega_0 + \dot{\phi}_m(t) - \frac{j\dot{V}_m}{V_m} \tag{7.6-3}$$

$$Y_{cmi}(\omega) = Y_{cmi}(\omega_0) + \frac{\partial Y_{cmi}}{\partial \omega_0}(\omega - \omega_0)$$

$$Y_m(V, \omega, \mu) = \frac{\partial Y_m}{\partial V_0}(V_m - V_0) + \frac{\partial Y_m}{\partial \omega_0}(\omega - \omega_0) + \frac{\partial Y_m}{\partial \mu_0}(\mu_m - \mu_0)$$

The perturbation assumption has been used in the first place by York, Liao, and Lynch [33], and it is described in Section 1.3 dealing with the injection-locked oscillator. However, in their analysis they proceed to assume a specific nonlinear dependence of the adminttance on the amplitude V_m whereas here no such assumption is made. Furthermore, it should be noted that the frequency expansion has been done using the well-known Kurokawa transformation [105] introduced in Section 1.3. The commonly used coupling networks have a broadband frequency response relative to the oscillator locking bandwidth, which allows us to consider a constant coupling term $Y_{cmi}(\omega) = Y_{cmi}(\omega_0)$. Narrowband coupling networks were studied by Lynch and York [117]. After some straightforward manipulation, one obtains

$$\frac{\partial Y_m}{\partial \omega_0}\left[-j\dot{V}_m + V_m\dot{\phi}_{\rm m}\right] + \frac{\partial Y_m}{\partial \mu_0}\Delta\mu_m V_m + \frac{\partial Y_m}{\partial V_0}(V_m - V_0)V_m$$
$$+\sum_{i=1}^{N} Y_{cmi}(\omega_0)V_i e^{j(\phi_i-\phi_m)} = 0 \qquad (7.6\text{-}4)$$

This system of differential equations represents the basis for an alternative model formulation for a system of coupled oscillators. This formulation has been essentially introduced in Ref. 116 and refined in Ref. 118 as well as in subsequent works as a basis to study several properties of coupled oscillator arrays.

In order to appreciate the similarities and differences with the original model of Section 7.4, the amplitude and phase equations are decoupled by first dividing with $\partial Y_m/\partial\omega_0$ and then considering real and imaginary parts. Let for simplicity

$$C_V = j\frac{\partial Y}{\partial \omega_0}^{-1}\frac{\partial Y}{\partial V_0} = C_V^R + jC_V^I$$

$$C_\mu = j\frac{\partial Y}{\partial \omega_0}^{-1}\frac{\partial Y}{\partial \mu_0} = C_\mu^R + jC_\mu^I \qquad (7.6\text{-}5)$$

$$C_c = [C_{cmi}] = \left[j\frac{\partial Y}{\partial \omega_0}^{-1} Y_{mi}(\omega_0)\right] = \left[|C_{cmi}|e^{j\Psi_{mi}}\right]$$
$$= \left[C_{cmi}^R + jC_{cmi}^I\right]$$

Using the above, Eq. (7.6-4) becomes

$$\dot{V}_m + C_V^R(V_m - V_0)V_m + C_\mu^R\Delta\mu_m V_m$$
$$+\sum_{i=1}^{N}|C_{cmi}|V_i\cos(\phi_{0i} - \phi_{0m} + \Psi_{mi}) = 0$$

$$(7.6\text{-}6)$$

$$V_m\dot{\phi}_m + C_V^I(V_m - V_0)V_m + C_\mu^I\Delta\mu_m V_m$$
$$+\sum_{i=1}^{N}|C_{cmi}|V_i\sin(\phi_{0i} - \phi_{0m} + \Psi_{mi}) = 0$$

Furthermore, letting $v_m = V_m e^{j\phi_m}$ it is possible to express Eq. (7.6-6) in a compact complex form:

$$\dot{v}_m + C_V(|v_m| - V_0)v_m + C_\mu \Delta\mu_m v_m + \sum_{i=1}^{N} C_{cmi} v_i = 0 \qquad (7.6\text{-}7)$$

The corresponding generalized phase model associated with this formulation is obtained considering only the phase dynamics, which results in

$$\dot{\phi}_m + C_\mu^I \Delta\mu_m + \sum_{i=1}^{N} |C_{cmi}| \sin(\phi_{0i} - \phi_{0m} + \Psi_{mi}) = 0 \qquad (7.6\text{-}8)$$

The periodic steady-state solution is obtained by setting $\dot{V}_m = 0$ and $\dot{\phi}_m = \Delta\omega$, leading to

$$\begin{aligned} &C_V^R(V_m - V_0)V_m + C_\mu^R \Delta\mu_m V_m \\ &\qquad + \sum_{i=1}^{N} |C_{cmi}| V_i \cos(\phi_{0i} - \phi_{0m} + \Psi_{mi}) = 0 \\ &V_m \Delta\omega + C_V^I(V_m - V_0)V_m + C_\mu^I \Delta\mu_m V_m \\ &\qquad + \sum_{i=1}^{N} |C_{cmi}| V_i \sin(\phi_{0i} - \phi_{0m} + \Psi_{mi}) = 0 \end{aligned} \qquad (7.6\text{-}9)$$

or in complex notation

$$\begin{aligned} &C_V(V_m - V_0)V_m + jV_m \Delta\omega + C_\mu \Delta\mu_m V_m \\ &\qquad + \sum_{i=1}^{N} C_{cmi} V_i e^{j(\phi_{0i} - \phi_{0m})} = 0 \end{aligned} \qquad (7.6\text{-}10)$$

As was the case with the model represented by Eq. (7.4-1) the steady state of the array is defined by phase differences and not absolute phases. In order to study the stability of the steady-state solution, we form the linear perturbation of Eq. (7.6-4) using $(V_m + \delta V_m, \phi_m + \delta\phi_m)$, where small-amplitude perturbations δV_m and small-phase perturbation differences $(\delta\phi_m - \delta\phi_i)$ are considered leading to

$$\delta\dot{V}_m + \left[C_V^R(2V_m - V_0) + C_\mu^R(\mu_m - \mu_0)\right]\delta V_m + \sum_{i=1}^{N} |C_{cmi}| \cos(\phi_{0i} - \phi_{0m} + \Psi_{mi})\, \delta V_i - \sum_{i=1}^{N} |C_{cmi}| V_i \sin(\phi_{0i} - \phi_{0m} + \Psi_{mi})\, (\delta\phi_i - \delta\phi_m) = 0$$

(7.6-11)

$$V_m\delta\dot{\phi}_m + \left[\Delta\omega + C_V^R(2V_m - V_0) + C_\mu^I(\mu_m - \mu_0)\right]\delta V_m + \sum_{i=1}^{N} |C_{cmi}| \sin(\phi_{0i} - \phi_{0m} + \Psi_{mi})\, \delta V_i + \sum_{i=1}^{N} |C_{cmi}| V_i \cos(\phi_{0i} - \phi_{0m} + \Psi_{mi})\, (\delta\phi_i - \delta\phi_m) = 0$$

Correspondingly, the generalized phase model stability is then determined by

$$\delta\dot{\phi}_m + \sum_{i=1}^{N} |C_{cmi}| \cos(\phi_{0i} - \phi_{0m} + \Psi_{mi})\, (\delta\phi_i - \delta\phi_m) = 0 \qquad (7.6\text{-}12)$$

The eigenvalues of the linear variational equation determine the stability of the steady-state solutions. In practice, it is more computationally efficient to formulate and process the array equations as matrix equations, and this is the topic of the next section.

7.7 Matrix Equations for the Steady State and Stability Analysis

From a computational point of view, it is easier to express the various systems of equations of the previous sections in matrix form. In order to do so, the following notation and properties are used. Bold letterface indicates a column vector or a matrix. The dimension of the vector or square matrix is N unless noted otherwise; $\mathbf{0}$ and $\mathbf{1}$ indicate a vector of zeros and a vector of ones, respectively. The notation $[a_m]$ and $[a_{mi}]$ define a vector and a matrix $\boldsymbol{\alpha}$, respectively. The function dg( ) converts a vector to a square diagonal matrix of size N. It is straightforward to show that for any two vectors $\boldsymbol{a}$ and $\boldsymbol{b}$,

$\mathrm{dg}(\boldsymbol{a})\boldsymbol{b} = \mathrm{dg}(\boldsymbol{b})\boldsymbol{a}$. The superscript $(\,)^H$ indicates the conjugate transpose of a matrix or vector, whereas superscripts $(\,)^R$ and $(\,)^I$ indicate real and imaginary part. One can then rewrite Eq. (7.4-1) in matrix form

$$\dot{\boldsymbol{A}} + j\mathrm{dg}(\boldsymbol{A})\dot{\boldsymbol{\phi}} = \mathrm{dg}(\boldsymbol{p}_1(\boldsymbol{A}) + j\boldsymbol{\Delta\omega})\boldsymbol{A} + \boldsymbol{\Phi}^H\boldsymbol{\kappa\Phi A} \tag{7.7-1}$$

where $\boldsymbol{\Phi} = \mathrm{dg}\left[e^{j\phi_m}\right]$. The system given by Eq. (7.7-1) can be integrated numerically after separating real and imaginary parts,

$$\begin{bmatrix} \boldsymbol{I} & \boldsymbol{0} \\ \boldsymbol{0} & \mathrm{dg}(\boldsymbol{A}) \end{bmatrix} \begin{bmatrix} \dot{\boldsymbol{A}} \\ \dot{\boldsymbol{\phi}} \end{bmatrix} = \begin{bmatrix} \mathrm{dg}\left(\boldsymbol{p}_1(\boldsymbol{A})\right)\boldsymbol{A} + \left(\boldsymbol{\Phi}^H\boldsymbol{\kappa\Phi}\right)^R\boldsymbol{A} \\ \mathrm{dg}(\boldsymbol{\Delta\omega})\boldsymbol{A} + \left(\boldsymbol{\Phi}^H\boldsymbol{\kappa\Phi}\right)^I\boldsymbol{A} \end{bmatrix} \tag{7.7-2}$$

where the vector function $\boldsymbol{p}_1(\boldsymbol{A})$ is defined as $\boldsymbol{p}_1(\boldsymbol{A}) = [\mu(A_{om}^2 - A_m^2)]$. The generalized phase model (7.4-4) is then given by

$$\dot{\boldsymbol{\phi}} = \boldsymbol{\Delta\omega} + \left(\boldsymbol{\Phi}^H\boldsymbol{\kappa\Phi}\right)^I\boldsymbol{1} \tag{7.7-3}$$

The steady state of Eq. (7.7-2) is computed by setting $\dot{\boldsymbol{A}} = 0$ and $\dot{\boldsymbol{\phi}} = c\boldsymbol{1}$, which, when substituted in (7.7-2), result in the trivial solution $\boldsymbol{A} = \boldsymbol{0}$ or

$$\begin{bmatrix} \mathrm{dg}\left(\boldsymbol{p}_1(\boldsymbol{A})\right) + \left(\boldsymbol{\Phi}^H\boldsymbol{\kappa\Phi}\right)^R \\ \mathrm{dg}(\boldsymbol{\Delta\omega} - c\boldsymbol{1}) + \left(\boldsymbol{\Phi}^H\boldsymbol{\kappa\Phi}\right)^I \end{bmatrix} = \begin{bmatrix} \boldsymbol{0} \\ \boldsymbol{0} \end{bmatrix} \tag{7.7-4}$$

The linear variational equation of Eq. (7.7-1) is also written as a matrix linear differential equation as follows:

$$\boldsymbol{\delta}\dot{\boldsymbol{A}} + j\mathrm{dg}(\boldsymbol{A})\boldsymbol{\delta}\dot{\boldsymbol{\phi}} = \boldsymbol{D}_A\boldsymbol{\delta A} + \boldsymbol{D}_\phi\,\boldsymbol{\delta\phi} \tag{7.7-5}$$

with

$$\begin{aligned} \boldsymbol{D}_A &= \mathrm{dg}\left(\boldsymbol{g}_1(\boldsymbol{A}) + j(\boldsymbol{\Delta\omega} - c\boldsymbol{1})\right) + \boldsymbol{\Phi}^H\boldsymbol{\kappa\Phi} \\ \boldsymbol{D}_\phi &= j[\boldsymbol{\Phi}^H\boldsymbol{\kappa\Phi}\,\mathrm{dg}(\boldsymbol{A}) - \mathrm{dg}(\boldsymbol{\Phi}^H\boldsymbol{\kappa\Phi}\,\boldsymbol{A})] \end{aligned} \tag{7.7-6}$$

where $\boldsymbol{g}_1(\boldsymbol{A}) = [\mu(A_{0m}^2 - 3A_m^2)]$. The complex system of Eq. (7.7-5) is separated into real and imaginary parts as

$$\begin{bmatrix} \boldsymbol{I} & \boldsymbol{0} \\ \boldsymbol{0} & \mathrm{dg}(\boldsymbol{A}) \end{bmatrix} \begin{bmatrix} \boldsymbol{\delta}\dot{\boldsymbol{A}} \\ \boldsymbol{\delta}\dot{\boldsymbol{\phi}} \end{bmatrix} = \begin{bmatrix} \boldsymbol{D}_A^R & \boldsymbol{D}_\phi^R \\ \boldsymbol{D}_A^I & \boldsymbol{D}_\phi^I \end{bmatrix} \begin{bmatrix} \boldsymbol{\delta A} \\ \boldsymbol{\delta\phi} \end{bmatrix} \tag{7.7-7}$$

or

$$\begin{bmatrix}\delta\dot{A}\\ \delta\dot{\boldsymbol{\phi}}\end{bmatrix} = \boldsymbol{D}\begin{bmatrix}\delta A\\ \delta\boldsymbol{\phi}\end{bmatrix} = \begin{bmatrix}\boldsymbol{D}_A^R & \boldsymbol{D}_\phi^R\\ \mathrm{dg}(\boldsymbol{A})^{-1}\boldsymbol{D}_A^I & \mathrm{dg}(\boldsymbol{A})^{-1}\boldsymbol{D}_\phi^I\end{bmatrix}\begin{bmatrix}\delta A\\ \delta\boldsymbol{\phi}\end{bmatrix} \tag{7.7-8}$$

where $\mathrm{dg}(\boldsymbol{A})^{-1}$ is a diagonal matrix with the inverse of the steady-state oscillator amplitudes in its diagonal. The inversion operation is guaranteed to exist under the assumption that the steady-state solution corresponds to nonzero amplitudes for all oscillators.

Correspondingly, the linear variational equation of the generalized phase model is

$$\delta\dot{\boldsymbol{\phi}} = \boldsymbol{D}_G\delta\boldsymbol{\phi} = [\boldsymbol{\Phi}^H\boldsymbol{\kappa}\,\boldsymbol{\Phi} - \mathbf{dg}(\boldsymbol{\Phi}^H\boldsymbol{\kappa}\,\boldsymbol{\Phi}\mathbf{1})]^R\delta\boldsymbol{\phi} \tag{7.7-9}$$

The matrix differential equation pertaining to the coupled-oscillator dynamics according to the alternative model Eq. (7.6-4) becomes

$$\dot{\boldsymbol{V}} + j\mathrm{dg}(\boldsymbol{V})\dot{\boldsymbol{\phi}} + \boldsymbol{C}_V\boldsymbol{p}_V(\boldsymbol{V}) + \boldsymbol{C}_\mu\mathrm{dg}(\boldsymbol{V})\Delta\boldsymbol{\mu} + \boldsymbol{\Phi}^H\boldsymbol{C}_c\boldsymbol{\Phi}\boldsymbol{V} = \mathbf{0} \tag{7.7-10}$$

where $\boldsymbol{p}_V(\boldsymbol{V}) = (\mathrm{dg}(\boldsymbol{V}) - V_o\boldsymbol{I})\boldsymbol{V}$. After separating into real and imaginary parts, one obtains

$$\begin{bmatrix}\boldsymbol{I} & \mathbf{0}\\ \mathbf{0} & \mathrm{dg}(\boldsymbol{V})\end{bmatrix}\begin{bmatrix}\dot{\boldsymbol{V}}\\ \dot{\boldsymbol{\phi}}\end{bmatrix} = \begin{bmatrix}\boldsymbol{C}_V^R\boldsymbol{p}_V(\boldsymbol{V}) + \boldsymbol{C}_\mu^R\mathrm{dg}(\boldsymbol{V})\Delta\boldsymbol{\mu} + \left(\boldsymbol{\Phi}^H\boldsymbol{C}_c\boldsymbol{\Phi}\right)^R\boldsymbol{V}\\ \boldsymbol{C}_V^I\boldsymbol{p}_V(\boldsymbol{V}) + \boldsymbol{C}_\mu^I\mathrm{dg}(\boldsymbol{V})\Delta\boldsymbol{\mu} + \left(\boldsymbol{\Phi}^H\boldsymbol{C}_c\boldsymbol{\Phi}\right)^I\boldsymbol{V}\end{bmatrix} \tag{7.7-11}$$

The steady-state solution is then given by the nonlinear system of algebraic equations

$$\begin{bmatrix}\boldsymbol{C}_V^R\boldsymbol{p}_V(\boldsymbol{V}) + \boldsymbol{C}_\mu^R\mathrm{dg}(\boldsymbol{V})\Delta\boldsymbol{\mu} + \left(\boldsymbol{\Phi}^H\boldsymbol{C}_c\boldsymbol{\Phi}\right)^R\boldsymbol{V}\\ \mathrm{dg}(\boldsymbol{V})\Delta\omega\mathbf{1} + \boldsymbol{C}_V^I\boldsymbol{p}_V(\boldsymbol{V}) + \boldsymbol{C}_\mu^I\mathrm{dg}(\boldsymbol{V})\Delta\boldsymbol{\mu} + \left(\boldsymbol{\Phi}^H\boldsymbol{C}_c\boldsymbol{\Phi}\right)^I\boldsymbol{V}\end{bmatrix} = \begin{bmatrix}\mathbf{0}\\ \mathbf{0}\end{bmatrix} \tag{7.7-12}$$

Due to the perturbation assumption, one may consider a linear approximation of the steady state as follows:

$$\begin{bmatrix}\boldsymbol{C}_V^R\boldsymbol{I} + \frac{1}{V_0}\left(\boldsymbol{\Phi}^H\boldsymbol{C}_c\boldsymbol{\Phi}\right)^R & \boldsymbol{C}_\mu^R\boldsymbol{I}\\ \boldsymbol{C}_V^I\boldsymbol{I} + \frac{1}{V_0}\left(\boldsymbol{\Phi}^H\boldsymbol{C}_c\boldsymbol{\Phi}\right)^I & \boldsymbol{C}_\mu^I\boldsymbol{I}\end{bmatrix}\begin{bmatrix}\Delta\boldsymbol{V}\\ \Delta\boldsymbol{\mu}\end{bmatrix} + \begin{bmatrix}\left(\boldsymbol{\Phi}^H\boldsymbol{C}_c\boldsymbol{\Phi}\right)^R\mathbf{1}\\ \Delta\omega\mathbf{1} + \left(\boldsymbol{\Phi}^H\boldsymbol{C}_c\boldsymbol{\Phi}\right)^I\mathbf{1}\end{bmatrix} = \begin{bmatrix}\mathbf{0}\\ \mathbf{0}\end{bmatrix} \tag{7.7-13}$$

For a given frequency offset $\Delta\omega$ and phase distribution along the array elements contained in Φ, one may solve the above linear system of $2N$ equations for the N steady-state oscillator amplitudes and N control perturbations. Alternatively, one may fix the control parameter of one arbitrarily selected oscillator and solve the steady-state system for the N steady-state oscillator amplitudes, the $N-1$ remaining control perturbations, and the frequency offset $\Delta\omega$.

The stability of the steady-state solution is obtained taking the linear variational equation of Eq. (7.7-10), leading to the following linearized system of differential equations:

$$\delta\dot{\boldsymbol{V}} + j\mathrm{dg}(\boldsymbol{V})\delta\dot{\boldsymbol{\phi}} = \boldsymbol{D}_V\delta\boldsymbol{V} + \boldsymbol{D}_\phi\,\delta\boldsymbol{\phi} \tag{7.7-14}$$

where

$$\begin{aligned}\boldsymbol{D}_V &= -j\Delta\omega\boldsymbol{1} - \mathrm{dg}\big(\boldsymbol{C}_V\boldsymbol{g}_V(\boldsymbol{V}) + \boldsymbol{C}_\mu^R\Delta\boldsymbol{\mu}\big) - \boldsymbol{\Phi}^H\boldsymbol{C}_c\boldsymbol{\Phi} \\ \boldsymbol{D}_\phi &= -j[\boldsymbol{\Phi}^H\boldsymbol{C}_c\boldsymbol{\Phi}\mathrm{dg}(\boldsymbol{V}) - \mathrm{dg}(\boldsymbol{\Phi}^H\boldsymbol{C}_c\boldsymbol{\Phi}\boldsymbol{V})]\end{aligned} \tag{7.7-15}$$

where $\boldsymbol{g}_V(\boldsymbol{V}) = 2\boldsymbol{V} - V_0\boldsymbol{I}$. One then separates real from imaginary parts to obtain the desired system of linear differential equations:

$$\begin{bmatrix}\boldsymbol{I} & \boldsymbol{0}\\ \boldsymbol{0} & \mathrm{dg}(\boldsymbol{V})\end{bmatrix}\begin{bmatrix}\delta\dot{\boldsymbol{V}}\\ \delta\dot{\boldsymbol{\phi}}\end{bmatrix} = \begin{bmatrix}\boldsymbol{D}_V^R & \boldsymbol{D}_\phi^R\\ \boldsymbol{D}_V^I & \boldsymbol{D}_\phi^I\end{bmatrix}\begin{bmatrix}\delta\boldsymbol{V}\\ \delta\boldsymbol{\phi}\end{bmatrix} \tag{7.7-16}$$

or

$$\begin{bmatrix}\delta\dot{\boldsymbol{V}}\\ \delta\dot{\boldsymbol{\phi}}\end{bmatrix} = \boldsymbol{K}\begin{bmatrix}\delta\boldsymbol{V}\\ \delta\boldsymbol{\phi}\end{bmatrix} = \begin{bmatrix}\boldsymbol{D}_V^R & \boldsymbol{D}_\phi^R\\ \mathrm{dg}(\boldsymbol{V})^{-1}\boldsymbol{D}_V^I & \mathrm{dg}(\boldsymbol{V})^{-1}\boldsymbol{D}_\phi^I\end{bmatrix}\begin{bmatrix}\delta\boldsymbol{V}\\ \delta\boldsymbol{\phi}\end{bmatrix} \tag{7.7-17}$$

The $2N$ eigenvalues of the square matrix $\boldsymbol{K}$ determine the stability of the solution. It should be noted that due to the autonomous nature of the coupled-oscillator array, one eigenvalue of $\boldsymbol{K}$ is always zero. The solution is stable if all remaining eigenvalues of $\boldsymbol{K}$ have negative real parts.

One can easily verify that $\boldsymbol{K}$ is unchanged to phase shifts that are common to all oscillators. This is due to the fact that matrix $\boldsymbol{D}_\phi = \boldsymbol{D}_\phi^R + j\,\boldsymbol{D}_\phi^I$ contains only phase differences between the various elements. It is then possible to reduce the system by one, thus eliminating the zero eigenvalue. Selecting an arbitrary element (for example, element j) as a reference, the N phase equations of Eq. (7.7-17) can be reduced by one by subtracting row $(N+j)$, the equation which corresponds to the phase of oscillator j, from every other equation. The equation that corresponds to the phase of oscillator j can then be eliminated. Furthermore, the elements of column $(N+j)$ from row $(N+1)$ to $(2N)$ are multiplied by zero. In addition, in the amplitude equations, due to the fact that

D^R_{ϕ} contains only phase differences, it is possible to subtract $\delta\phi_j$ from all phases forming $\boldsymbol{D^R_{\phi}\delta\phi} \leftarrow \boldsymbol{D^R_{\phi}}(\boldsymbol{\delta\phi} - \delta\phi_j\mathbf{1})$. As a result, column $(N + j)$ can also be eliminated because it is being multiplied by zero. The remaining square matrix $\widetilde{\boldsymbol{K}}$ of dimension $N - 1$ has the same eigenvalues with $\boldsymbol{K}$ minus the zero eigenvalue. Matrix $\widetilde{\boldsymbol{K}}$ corresponds to the system of $2N - 1$ linear differential equations

$$\begin{bmatrix} \boldsymbol{\delta\dot{V}} \\ \boldsymbol{\delta\dot{\widetilde{\phi}}} \end{bmatrix} = \widetilde{\boldsymbol{K}} \begin{bmatrix} \boldsymbol{\delta V} \\ \boldsymbol{\delta\widetilde{\phi}} \end{bmatrix} \tag{7.7-18}$$

where the vector $\boldsymbol{\delta\widetilde{\phi}}$ of dimension $N - 1$ contains phase difference terms relative to oscillator j. The spectral abscissa of a square matrix is the maximum real part of its eigenvalues [119]. Therefore a steady-state solution is stable if the spectral abscissa of $\widetilde{\boldsymbol{K}}$ is negative.

7.8 A Comparison between the Two Perturbation Models for Coupled Oscillator Systems

The similarity of the two models is made obvious by comparing the two expressions corresponding to the generalized phase model, Eq. (7.4-4) and Eq. (7.6-8). The first model is defined for a parallel RLC tank with a nonlinear voltage-dependent current source that exhibits a third-order nonlinearity similar to the one described in Section 7.5 and shown in Fig. 7-6. In this case, the admittance looking at the output node of the circuit is given by

$$Y(V, \omega) = (Y_N(V) + G_L) + \frac{1}{jL\omega} + jC\omega \tag{7.8-1}$$

where $Y_N(V)$ contains the nonlinear admittance of the current source at the fundamental frequency component. The total admittance contains a real nonlinear admittance term that is amplitude-dependent, plus the load admittance, and an imaginary term which is frequency-dependent. As a result, a real derivative versus the amplitude and an imaginary admittance derivative versus the frequency are obtained:

$$\begin{aligned} \frac{\partial Y}{\partial V} &= \frac{\partial Y_N}{\partial V} \\ \frac{\partial Y}{\partial \omega} &= j2C \end{aligned} \tag{7.8-2}$$

Furthermore, if we consider that for a parallel resonant circuit the external quality factor is given by

$$Q = \frac{\omega_0 C}{G_L} \tag{7.8-3}$$

it is straightforward to verify that the Eqs. (7.4-4) and (7.6-8) are identical.

In general, active devices in microwave frequencies exhibit nonlinear susceptance as well as admittance, in addition to a nonlinear admittance that is frequency-dependent. In other words, a more general expression for the admittance at an oscillator circuit node is

$$Y(V, \omega) = G_N(V, \omega) + jB_N(V, \omega) \tag{7.8-4}$$

As a result, it is possible to view the alternative model (7.6-4) using complex admittance derivatives as a generalized version of (7.4-1).

7.9 Externally Injection-Locked COAs

The coupled-oscillator array is an autonomous system that behaves like a single distributed oscillator. However, there are several applications that require the array to be injection-locked to an external signal. The reason can be to control the phase distribution among the array elements [1], to reduce the array phase noise [97], to fix the array frequency [120], or to introduce modulation to the array [121, 122].

An external injection signal introduces an additive forcing term in the time-domain expression of the perturbed oscillator equation. In Section 7.5, it was demonstrated that the topology of a parallel RLC tank with a nonlinear voltage-dependent current source and an external-injection current term leads to the forced van der Pol equation.

The coupled-oscillator system of differential equations is derived by applying Kirchhoff's current or voltage law at a selected circuit node or loop of each oscillator circuit in the array. Specifically, in the parallel-tank topology, the corresponding equations are obtained by applying Kirchhoff's current law at the output nodes of each oscillator. In addition, these nodes correspond to the nodes where the coupling network is connected to each oscillator. However, this does not always have to be the case, and the coupling network may be connected to a different oscillator node.

The external-injection signal typically may not be applied at the circuit node where the coupled oscillator system is derived. In this case it is necessary to derive analytically, or using a circuit simulator, a transfer function, which relates the applied injection signal to an induced current or voltage at the node or loop where the system equation is applied. Alternatively, it is possible to consider that the nonlinear-oscillator admittance is a function of the injection signal. It should be noted that in the general case there may be more than one injection signal applied and that the injection signal may be coupled to the coupled-oscillator array using different topologies, such as direct injection or radiation coupling, as shown in Fig.7-8 [123].

Following the formulation of Chang et al. [123] where the authors assume a parallel resonance model for the oscillator elements, and they consider the external injection signal in the form of an additional current source in parallel with the oscillator tank, one has

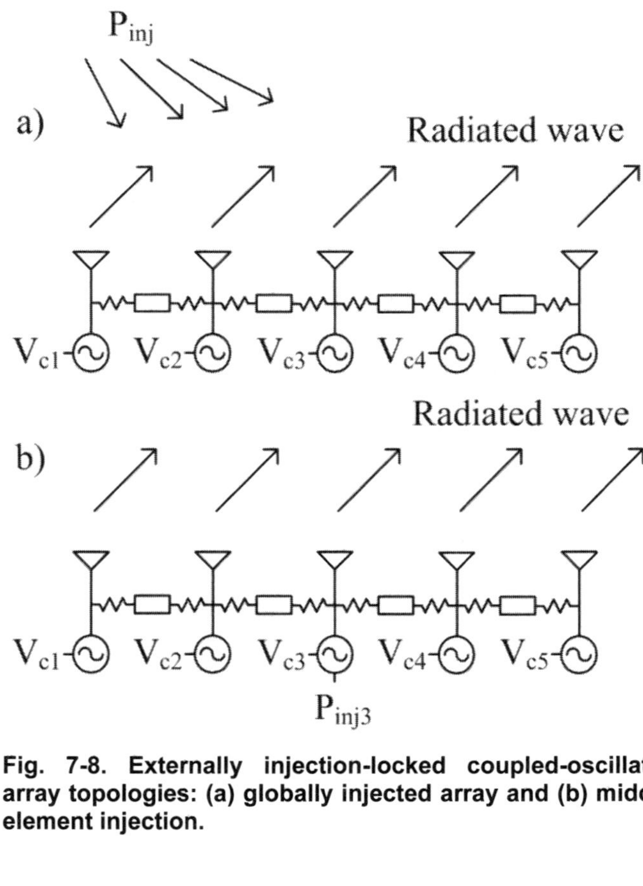

Fig. 7-8. Externally injection-locked coupled-oscillator array topologies: (a) globally injected array and (b) middle element injection.

$$\dot{A}_m = \mu(A_{0m}^2 - A_m^2)A_m + \sum_{i=1}^{N} K_{mi}A_i\cos(\phi_i - \phi_m + \Phi_{mi}) + K_{mp}A_p\cos(\phi_p - \phi_m + \Phi_{mp})$$

$$A_m\dot{\phi}_m = \Delta\omega_m A_m + \sum_{i=1}^{N} K_{mi}A_i\sin(\phi_i - \phi_m + \Phi_{mi}) + K_{mp}A_{mp}\sin(\phi_{mp} - \phi_m + \Phi_{mp}) \quad (7.9\text{-}1)$$

where it is assumed that oscillator m is being injected by an external source mp. The transfer function $\kappa_{mp} = t_{mp}\omega_m/2Q = K_{mp}e^{j\Phi_{mp}}$ consists of a complex normalized term t_{mp} multiplied with a scaling factor $\omega_m/2Q$ as is done for the coupling terms from the other oscillator elements in the array. For the case of an injection-current term in parallel with the oscillator tank, $t_{mp} = 1$. Furthermore, when a single oscillator is considered, Eq. (7.9-1) reduces to Adler's equation.

Alternatively, it is possible to assume that the nonlinear oscillator admittance $Y_m(v_m, \omega, \mu, a_{mp})$ at the node under consideration additionally depends on the injection signal $a_{mp} = A_{mp}e^{j\phi_{mp}} = a_{mp}^R + ja_{mp}^I$ present at an arbitrary node of the oscillator circuit. Assuming a low amplitude-injection signal relative to the oscillator amplitude $\rho_{mp} = A_{mp}/A_m \ll 1$, a Taylor expansion of the oscillator admittance around the free running steady state gives, to first order [124],

$$Y_m(v_m, \omega, \mu, a_{mp}) = Y_m(V_m, \omega, \mu) + Y_{mp}(\phi_m, a_{mp}) \quad (7.9\text{-}2)$$

with

$$Y_{mp}(\phi_m, a_{mp}) = \frac{\partial Y_{mp}}{\partial a_{mp}^R}A_{mp}\cos(\phi_{mp} - \phi_m) + \frac{\partial Y_{mp}}{\partial a_{mp}^I}A_{mp}\sin(\phi_{mp} - \phi_m) \quad (7.9\text{-}3)$$

The first term $Y_m(V_m, \omega, \mu)$ is the one considered in Eq. (7.6-3), where no external injection signal is present. The second term $Y_{mp}(\phi_m, a_{mp})$ is a linear perturbation term due to the external injection signal, which depends on the relative phase between the oscillator and the injection signal. The admittance expression is then introduced in the model presented in Section 7.6 and is repeated here for convenience in order to derive the desired system of equations:

$$Y_m V_m e^{j\phi_m} + \sum_{i=1}^{N} Y_{cmi} V_i e^{j\phi_i} = 0 \tag{7.9-4}$$

As the injection power increases, additional terms in the Taylor expansion can be included in order to improve the accuracy of the approximation [124].

7.10 Phase Noise

Perturbation theory is applied in noise analysis of oscillators as typically noise is modeled as a stochastic forcing term in the oscillator differential equation. The stochastic nature of noise and the nonlinear nature of the oscillator circuits make noise analysis a challenging problem. Applying the averaging theory, Kurokawa [105] presented an elegant analysis of phase noise of free-running and externally injection locked oscillators. A fundamental assumption in his formulation is that noise, described by a time-domain stochastic process $n_m(t)$ can be expanded in a Fourier series around the arbitrarily chosen fundamental frequency ω_0 as

$$n_m(t) = \sum_{n=-\infty}^{+\infty} N_{mn}(t) e^{j\omega_0 t} \tag{7.10-1}$$

with $N_{mn}(t) = G_{mn}(t) + jB_{mn}(t)$ a complex noise process. In the following it is assumed that $n_m(t)$ is a zero-mean white Gaussian process, which results in $G_{mn}(t)$ and $B_{mn}(t)$ being uncorrelated white zero-mean Gaussian processes as well [105].

Extending the work of Kurokawa, Chang et al. [97] studied phase noise in mutually injection-locked coupled-oscillator arrays. Application of the noise expansion and averaging allows us to include the effect of noise in the oscillator formulation Eq. (7.4-1) in terms of $N_{m1}(t) = G_{m1}(t) + jB_{m1}(t)$. In the following, the subscript 1 is dropped for simplicity.

$$\begin{aligned} \dot{A}_m = \mu(A_{0m}^2 - A_m^2)A_m + \sum_{i=1}^{N} K_{mi} A_i \cos(\phi_i - \phi_m + \Phi_{mi}) \\ - \frac{\omega_0}{2QG_L} G_m(t) \end{aligned} \tag{7.10-2}$$

$$A_m \dot{\phi}_m = \Delta\omega_m A_m + \sum_{i=1}^{N} K_{mi} A_i \sin(\phi_i - \phi_m + \Phi_{mi}) - \frac{\omega_0}{2QG_L} B_m(t)$$

The solution of Eq. (7.10-2) is found in the form of a perturbation $(V_m + \delta V_m, \phi_m + \delta\phi_m)$ where (V_m, ϕ_m) is the solution to the noise-free system of Eq. (7.4-1), leading to a forced variational system, which is the same as was considered in the study of the stability of the steady state with the addition of a noise-forcing term. As before, small amplitude-noise perturbations δV_m and small phase-noise-perturbation differences $(\delta\phi_m - \delta\phi_i)$ result in the forced linear system of differential equations:

$$\delta \dot{A}_m = \mu(A_{0m}^2 - 3A_m^2)\delta A_m + \sum_{i=1}^{N} K_{mi} \cos(\phi_i - \phi_m + \Phi_{mi})\, \delta A_i - \sum_{i=1}^{N} K_{mi} A_i \sin(\phi_i - \phi_m + \Phi_{mi})\,(\delta\phi_i - \delta\phi_m) - \frac{\omega_0}{2QG_L} G_m(t)$$

$$A_m \delta\dot{\phi}_m = (\Delta\omega_m - c)\delta A_m + \sum_{i=1}^{N} K_{mi} \sin(\phi_i - \phi_m + \Phi_{mi})\, \delta A_i + \sum_{i=1}^{N} K_{mi} A_i \cos(\phi_i - \phi_m + \Phi_{mi})\,(\delta\phi_i - \delta\phi_m) - \frac{\omega_0}{2QG_L} B_m(t) \tag{7.10-3}$$

Correspondingly, the alternative model in the presence of noise is modified by including an additive complex noise term $N_m(t) = G_m(t) + jB_m(t)$ in Eq. (7.6-4), which leads to forcing terms $G_m(t)$ and $B_m(t)$ in the left hand side of the first and second equations of Eq. (7.6-11), respectively. For compactness, the formulation is not repeated here.

Following Chang [97], we proceed to solve Eq. (7.10-3) by first applying a Fourier transform

$$\begin{aligned} j\Omega\delta\breve{A}_m &= \mu(A_{om}^2 - 3A_m^2)\delta\breve{A}_m \\ &\quad + \sum_{i=1}^{N} K_{mi}\cos(\phi_i - \phi_m + \Phi_{im})\,\delta\breve{A}_i \\ &\quad - \sum_{i=1}^{N} K_{mi}A_i\sin(\phi_i - \phi_m + \Phi_{mi})\left(\delta\breve{\phi}_i\right. \\ &\quad \left. - \delta\breve{\phi}_m\right) - \frac{\omega_0}{2QG_L}\breve{G}_m \end{aligned}$$

$$\begin{aligned} j\Omega A_m\delta\breve{\phi}_m &= (\Delta\omega_m - c)\delta\breve{A}_m \\ &\quad + \sum_{i=1}^{N} K_{mi}\sin(\phi_i - \phi_m + \Phi_{mi})\,\delta\breve{A}_i \\ &\quad + \sum_{i=1}^{N} K_{mi}A_i\cos(\phi_i - \phi_m + \Phi_{mi})\left(\delta\breve{\phi}_i\right. \\ &\quad \left. - \delta\breve{\phi}_m\right) - \frac{\omega_0}{2QG_L}\breve{B}_m \end{aligned} \tag{7.10-4}$$

The frequency Ω indicates offset from the fundamental ω_o, and the hat indicates a Fourier transformed variable. The linear system of Eq. (7.10-4) is processed easier in matrix form. Using the formulation of Section 7.7, it is possible to write Eq. (7.10-4) in the form

$$\{j\Omega\boldsymbol{I} - \boldsymbol{D}\}\begin{bmatrix}\boldsymbol{\delta\breve{A}} \\ \boldsymbol{\delta\breve{\phi}}\end{bmatrix} = \boldsymbol{N}\begin{bmatrix}\boldsymbol{\delta\breve{A}} \\ \boldsymbol{\delta\breve{\phi}}\end{bmatrix} = \begin{bmatrix}\boldsymbol{\breve{G}_n} \\ \boldsymbol{\breve{B}_n}\end{bmatrix} \tag{7.10-5}$$

or

$$\begin{bmatrix}\boldsymbol{\delta\breve{A}} \\ \boldsymbol{\delta\breve{\phi}}\end{bmatrix} = \boldsymbol{P}\begin{bmatrix}\boldsymbol{\breve{G}_n} \\ \boldsymbol{\breve{B}_n}\end{bmatrix} \tag{7.10-6}$$

with $\boldsymbol{P} = \boldsymbol{N}^{-1} = [j\Omega\boldsymbol{I} - \boldsymbol{D}]^{-1}$ where the noise terms have been normalized as $\boldsymbol{\breve{G}_n} = -\frac{\omega_0}{2QG_L}\left[\breve{G}_m\right]$ and $\boldsymbol{\breve{B}_n} = -\frac{\omega_0}{2QG_L}\left[\breve{B}_m\right]$ for compactness. It should be clarified that the identity matrix in Eq. (7.10-5) is of dimension $2N$. Correspondingly, the formulation pertaining to the generalized phase model is

$$\{j\Omega\boldsymbol{I} - \boldsymbol{D_G}\}\boldsymbol{\delta\breve{\phi}} = \boldsymbol{N_G}\boldsymbol{\delta\breve{\phi}} = \boldsymbol{\breve{B}_n} \tag{7.10-7}$$

or

$$\delta\breve{\boldsymbol{\phi}} = \boldsymbol{P}_G \breve{\boldsymbol{B}}_n \tag{7.10-8}$$

with $\boldsymbol{P}_G = \boldsymbol{N}_G^{-1} = [j\Omega\boldsymbol{I} - \boldsymbol{D}_G]^{-1}$.

The noise correlation matrix $\boldsymbol{S}(\Omega)$ of the oscillator array is given by

$$\boldsymbol{S}(\Omega) = \left\langle \begin{bmatrix} \delta\breve{\boldsymbol{A}} \\ \delta\breve{\boldsymbol{\phi}} \end{bmatrix} [\delta\breve{\boldsymbol{A}}^H \quad \delta\breve{\boldsymbol{\phi}}^H] \right\rangle = \begin{bmatrix} \boldsymbol{S}_{AA} & \boldsymbol{S}_{A\phi} \\ \boldsymbol{S}_{\phi A} & \boldsymbol{S}_{\phi\phi} \end{bmatrix} \tag{7.10-9}$$

where the superscript H denotes the conjugate transpose operation. The various noise contributions AM-AM AM-PM, PM-AM, and PM-PM are easily identified.[1] The operator $\langle \quad \rangle$ denotes ensemble average, and following Ref. 97 for white Gaussian processes, one has

$$\begin{aligned} \langle \breve{\boldsymbol{G}}_n \breve{\boldsymbol{G}}_n^H \rangle = \langle \breve{\boldsymbol{B}}_n \breve{\boldsymbol{B}}_n^H \rangle &= \left(\frac{\omega_o \sigma}{2QG_L}\right)^2 \boldsymbol{I} \\ \langle \breve{\boldsymbol{G}}_n \breve{\boldsymbol{B}}_n^H \rangle = \langle \breve{\boldsymbol{B}}_n \breve{\boldsymbol{G}}_n^H \rangle &= \boldsymbol{0} \end{aligned} \tag{7.10-10}$$

with σ^2 as the noise variance. Identical oscillators have been assumed and identical noise sources have been applied at each oscillator for simplicity. The spectral density of the oscillator array is given by the diagonal of $\boldsymbol{S}(\Omega)$.

The noise correlation matrix is then given by

$$\boldsymbol{S}(\Omega) = \left(\frac{\omega_0 \sigma}{2QG_L}\right)^2 \boldsymbol{P}\boldsymbol{P}^H \tag{7.10-11}$$

The generalized phase model expression can be used to obtain an approximate, more simplified expression for the correlation matrix $\boldsymbol{S}_{\boldsymbol{G}}(\Omega)$, without considering amplitude noise:

$$\boldsymbol{S}_{\boldsymbol{G}}(\Omega) = \left(\frac{\omega_0 \sigma}{2QG_L}\right)^2 \boldsymbol{P}_G \boldsymbol{P}_G^H \tag{7.10-12}$$

Note that $\boldsymbol{S}_{\boldsymbol{G}}(\Omega)$ is a square matrix of dimension N containing all correlation terms among the noise quantities of the individual oscillators. The phase noise spectra $\boldsymbol{S}_{\boldsymbol{G}\boldsymbol{\phi}}(\Omega)$ of the individual oscillators in the array are given by the diagonal elements of $\boldsymbol{S}_{\boldsymbol{G}}(\Omega)$, or

$$\boldsymbol{S}_{\boldsymbol{G}\boldsymbol{\phi}}(\Omega) = \left(\frac{\omega_0 \sigma}{2QG_L}\right)^2 \mathrm{dg}\{\boldsymbol{P}_G \boldsymbol{P}_G^H\}\boldsymbol{1} \tag{7.10-13}$$

[1] AM is amplitude modulation, and PM is phase modulation.

The phase noise spectrum $S_{1\phi}(\Omega)$ of a single oscillator element uncoupled to the rest of the array corresponds to $\boldsymbol{D_G} = \mathbf{0}$ and is given by

$$S_{1\phi}(\Omega) = \left(\frac{\omega_0 \sigma}{2QG_L}\right)^2 \frac{1}{\Omega^2} \tag{7.10-14}$$

This expression is in agreement with the one given by Kurokawa in Ref. 105 and demonstrates the dependence of $\Omega^{-2}$of the phase-noise spectrum for the case of white Gaussian noise sources.

The expression for the phase noise spectrum vector $\boldsymbol{S_{G\phi}}(\Omega)$ of each oscillator element in the array finally can be written

$$\boldsymbol{S_{G\phi}}(\Omega) = S_1(\Omega)\Omega^2 \mathrm{dg}\{\boldsymbol{P_G P_G^H}\}\mathbf{1} \tag{7.10-15}$$

In addition to the phase noise of the individual coupled oscillator elements $\delta\phi_m$, in quasi-optical power combining applications, one is also interested in the phase noise of the combined output of the oscillators $\delta\phi_T$. Assuming small perturbations, one may write the combined far-field amplitude $V(t)$ as [97]

$$\begin{aligned} V(t) &= \sum_{m=1}^{N} A_m \cos(\omega_0 t + \phi_m) \\ &\approx NA \sum_{m=1}^{N} \cos(\omega_0 t + \phi_T) \end{aligned} \tag{7.10-16}$$

with

$$\delta\phi_T = \frac{1}{N}\sum_{m=1}^{N} \delta\phi_m = \frac{1}{N}\mathbf{1}^T \boldsymbol{\delta\phi} \tag{7.10-17}$$

The phase noise spectrum $S_{GT}(\Omega)$ of the combined output is given by

$$S_{GT}(\Omega) = \langle \delta\breve{\varphi}_T \delta\breve{\varphi}_T^H \rangle = \frac{1}{N^2}\langle \mathbf{1}^T \boldsymbol{\delta\breve{\varphi}\delta\breve{\varphi}}^H \mathbf{1} \rangle \tag{7.10-18}$$

which, with the help of Eq. (7.10-15), becomes

$$S_{GT}(\Omega) = \frac{S_1(\Omega)\Omega^2}{N^2} \mathbf{1}^T \boldsymbol{P_G P_G^H} \mathbf{1} \tag{7.10-19}$$

Evaluation of the individual oscillator phase noise and the combined output phase noise is generally possible only by numerically evaluating Eqs. (7.10-15) and (7.10-19), respectively. Nonetheless, Chang et al. [97] were able to analytically study several cases commonly found in the literature.

One of the results obtained by Chang et al. [97] corresponds to the case where $\boldsymbol{D_G} = [\boldsymbol{\Phi}^H\boldsymbol{\kappa}\boldsymbol{\Phi} - \mathrm{dg}(\boldsymbol{\Phi}^{\mathrm{H}}\boldsymbol{\kappa}\boldsymbol{\Phi}\mathbf{1})]^R$, repeated here for convenience, is a symmetric matrix ($\boldsymbol{D}_G^T = \boldsymbol{D}_G$). As one can see, $\boldsymbol{D_G}$ depends both on the coupling network, through $\boldsymbol{\kappa}$, and on the steady-state phase distribution of the various oscillator array elements, through $\boldsymbol{\Phi}$. It can be easily verified that $\boldsymbol{D_G}\mathbf{1} = \mathbf{0}$, which reflects the fact that the steady state is unchanged to within a common constant phase term added to all oscillator elements, or, in other words, the fact that the steady state is defined by the phase differences of the various elements. Using the above two properties, Chang et al. [97] have shown by analytically evaluating $\boldsymbol{P}_G\boldsymbol{P}_G^H$ that $\mathbf{1}^T\boldsymbol{P}_G\boldsymbol{P}_G^H\mathbf{1} = \Omega^{-2}N$, which results in

$$S_{GT}(\Omega) = \frac{S_1(\Omega)}{N} \qquad (7.10\text{-}20)$$

This is an important result indicating that the phase noise of the combined array output is reduced by a factor N compared to the individual free-running oscillator phase noise (as indicated in Section 6.4). It remains to identify under which conditions $\boldsymbol{D_G}$ is symmetric. One characteristic example is when an in-phase steady-state solution is assumed ($\boldsymbol{\Phi} = \boldsymbol{I}$) and a reciprocal coupling network matrix with zero coupling phase is $\boldsymbol{\kappa}^T = \boldsymbol{\kappa} = \boldsymbol{\kappa}^R$.

In the case of a reciprocal coupling network of near-neighbor bilateral coupling with zero coupling phase, $\boldsymbol{D_G}$ is symmetric for any constant phase distribution among the oscillator elements. It was also shown that in this case the individual oscillator phase noise is also reduced by a factor N when the oscillators are in-phase. The oscillator phase noise for steady states with phase distributions with nonzero progressive phase $\Delta\phi_p$ degrades with increasing $\Delta\phi_p$ up to the point where the array loses stability and the phase noise becomes equal to the free-running oscillator phase-noise value.

Finally, it was shown by Chang et al. [97] that there is no phase noise improvement in the case of unilaterally coupled oscillators, for both the individual elements and the combined-array output (as indicated in Section 6.4).

The phase noise of externally injection-locked oscillators has been investigated by Kurokawa [105], where it was shown that the injected oscillator phase-noise spectrum follows the phase-noise profile of the injection-locking signal for small frequency offsets near the carrier, and it converges to the free-running oscillator phase-noise spectrum for large frequency offsets. The formulation of Kurokawa [105] was extended to externally injection locked coupled oscillator arrays by Chang et al. [123]. It is straightforward to obtain the formulation pertaining to the externally injection-locked coupled-oscillator arrays by properly including in Eq. (7.10-2) terms due to injection sources as shown in (7.9-1). Chang et al. [123] investigated several topologies including (a) a globally injected linear array and (b) arrays where a different single elements within the array are injected. In summary, the results showed that for small offsets the array phase-noise profile follows the injection-locking source phase-noise profile. However, for large offsets from the carrier the globally illuminated case showed a different behavior than the single-element illumination topology. In the former the phase noise improves with increasing number of array elements, whereas in the latter the phase noise degrades with increasing number of elements. Furthermore, the array phase-noise performance of the single-element injection case improves as one injects an element closer to the array center.

7.11 Modulation

Several authors have considered the use of coupled-oscillator arrays in communication system applications. It is possible to distinguish among (a) architectures where the coupled-oscillator array signal is modulated and (b) architectures employing a coupled-oscillator array as the local oscillator in a multi-antenna up-converting or down-converting transceiver. The first topology has been studied by Kykkotis et al. [99]. Due to the limiting properties of oscillators, modulation formats that lead to large variations in the signal envelope are not recommended as the oscillator dynamics will tend to smooth these variations and introduce distortion. However, constant envelope modulation formats (such as constant phase modulation (CPM) and Gaussian minimum shift keying (GMSK)) represent excellent candidates to be employed in such systems. In Ref. 99, the modulation is applied in the coupled-oscillator array through an external injection signal. Additionally, it is possible to introduce modulation through the frequency-tuning bias voltage of the individual oscillators, as was proposed by Pogorzelski [63].

A formulation based on Eq. (7.9-3) where the effect of the external-injection signal is included in the oscillator admittance was used by Collado and Georgiadis [124] to analyze the performance of such systems as the modulation bandwidth increases. The effect of the modulation on the maximum stable progressive constant phase shift among the oscillator elements was investigated, and it was shown that the presence of modulation leads to a reduction of the maximum achievable scanning range. In Fig. 7-9, the effect of sinusoidal phase modulation in the maximum scanning range of a two-element coupled oscillator array is shown. The maximum stable phase difference between the first harmonics of the two oscillators is obtained using the aforementioned model (denoted by RoM in Fig. 7-9), in good agreement with measurements as well as simulation results obtained using a commercial envelope transient circuit simulator. (The principles of nonlinear-circuit simulation methods, such as the envelope transient, are described in Chapter 8.) In addition, measurements of the maximum phase difference of the second harmonics of the oscillators are presented, and they are compared with the results using the envelope transient simulator. As can be seen from Fig. 7-9, extended scanning range can be obtained by considering the phase variation of an oscillator harmonic frequency rather than the fundamental frequency. Such architectures used to provide extended phase-scanning range are described in Section 8.6.

On the other hand, the use of coupled-oscillator arrays to provide the local oscillator signal in multi-element communication transceivers has been studied by Pogorzelski and Chiha [74] and Pogorzelski [125]. The coupled-oscillator array is used to provide a local-oscillator signal to a mixer with a desired phase distribution in order to appropriately steer the array beam without the need for phase shifters and a complex local-oscillator distributed feed network. In addition, more compact front ends can be implemented employing an array of coupled self-oscillating mixers as was proposed by Ver Hoeye et al. [79].

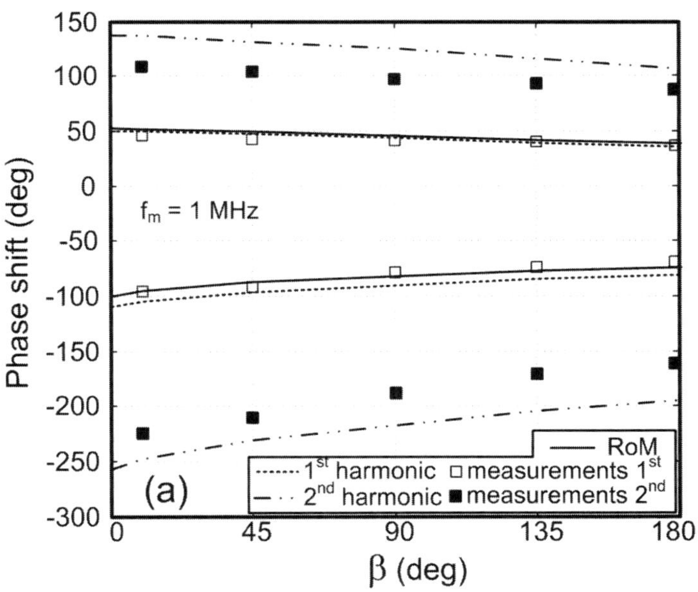

Fig. 7-9. Two-element coupled-oscillator array. Effect of phase modulation index β on stable phase shift among the oscillator elements. Sinusoidal phase modulation of 1-MHz frequency is applied by external injection locking to the one oscillator of the array. The results of the model developed based on Eq. (7.9-3) are denoted by RoM. (Reprinted with permission from Ref. 124, ©2001 IEEE.)

7.12 Coupled Phase-Locked Loops

A phase-locked loop (PLL) is typically used in frequency-generation applications, as well as in phase recovery and phase/frequency modulation/demodulation applications where one oscillator is required to track the phase of a signal present at its input. Therefore, it presents an excellent candidate for generating phase distributions among oscillator elements, which are required in electronic beam-steering applications. Martinez and Compton [126] first proposed the use of a coupled phase-locked loop for phased arrays. Subsequently, Buckwalter et. al. [127] extended their work to study the synchronization properties of such loops, and Chang presented a phase noise analysis [128].

The topology of a coupled PLL system is shown in Fig. 7-10, where a linear array of oscillators is considered. An error signal e is formed by a mixing operation where the outputs of adjacent oscillators are multiplied together. The mixers are used as phase detectors; however, other more sophisticated topologies can also be used where the oscillator outputs are first passed through a frequency divider and are subsequently fed to a digital phase detector, as is typically done in PLL architectures. Finally, the loop is closed by feeding the error signal to each oscillator-control input after it has passed through a loop filter. The relative phases between the oscillator elements are controlled by introducing additional external signals in the error signal path such as x_1 and x_N, shown in Fig. 7-10.

In the following, an introduction to the equations describing the dynamics of a two-element coupled PLL system is presented, following the formulation by Buckwalter et al. [127] and based on the topology indicated in Fig. 7-10. Identical oscillators are assumed where, for simplicity, a linear voltage-to-frequency model relation is considered

$$\dot{\phi}_i = \omega_i + K_v y_i \tag{7.12-1}$$

The index $i = 1,2$ runs through the set of two oscillators. Furthermore, a first-order loop filter is assumed with gain a, one zero τ_z, and one pole τ_p, having a transfer function given by

$$H(\omega) = \alpha \frac{1 + j\omega\tau_z}{1 + j\omega\tau_p} \tag{7.12-2}$$

Identical loop filters are considered for both oscillators.

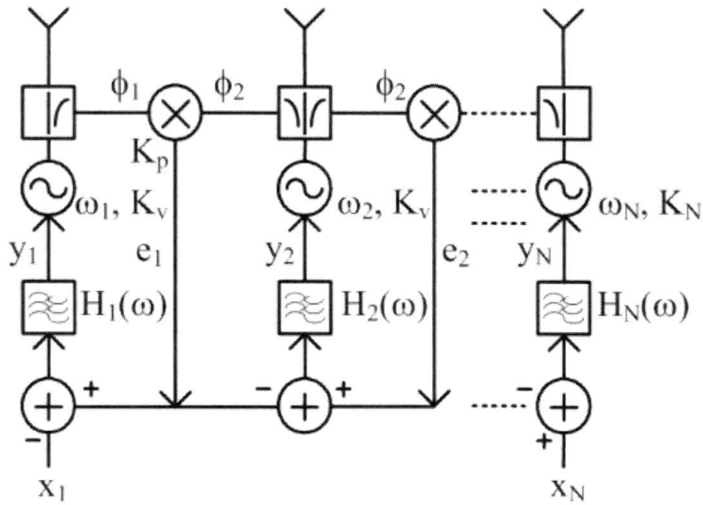

Fig. 7-10. Coupled phase-locked loop architecture.

Based on Fig. 7-10 and considering two oscillators, the following system of equations is derived

$$\tau_p \dot{y}_1 + y_1 = a[\tau_z(\dot{e}_1 - \dot{x}_1) + (e_1 - x_1)] \quad (7.12\text{-}3)$$

$$\tau_p \dot{y}_2 + y_2 = a[\tau_z(\dot{x}_2 - \dot{e}_1) + (x_2 - e_1)] \quad (7.12\text{-}4)$$

$$\dot{\phi}_1 = \omega_1 + K_v y_1 \quad (7.12\text{-}5)$$

$$\dot{\phi}_2 = \omega_2 + K_v y_2 \quad (7.12\text{-}6)$$

$$e_1 = K_p \sin\phi_1 \sin\phi_2 \approx \frac{1}{2} K_p \cos(\phi_2 - \phi_1) \quad (7.12\text{-}7)$$

It should be noted that for the sake of simplicity the higher-frequency mixing product is not considered in the error signal e_1 as it is assumed that it will be greatly attenuated by the loop filter.

Using the above equations, it is possible to derive one equation governing the dynamics of the phase difference $\Delta\phi = \phi_2 - \phi_1$. The external signals x_1 and x_2 can be used to introduce modulation to the loop or simply set some desired phase difference by introducing some offset to the equilibrium point of the loop. Setting external signals equal to zero ($x_1 = x_2 = 0$), it is straightforward to derive the differential equation so that the phase difference between the oscillators satisfies

$$\tau_p \Delta\ddot{\phi} + [1 - \tau_z G \sin\Delta\phi]\Delta\dot{\phi} + G\cos\Delta\phi - \Delta\omega = 0 \qquad (7.12\text{-}8)$$

where $\Delta\omega = \omega_2 - \omega_1$ and $G = \alpha K_v K_p$. The equilibrium points of the coupled PLL correspond to $\Delta\ddot{\phi} = \Delta\dot{\phi} = 0$ and are derived by solving

$$G\cos\Delta\phi - \Delta\omega = 0 \qquad (7.12\text{-}9)$$

It is easy to verify that two solutions exist within the phase interval $[0, 2\pi)$, and perturbation analysis of Eq. (7.12-8) can be used to show that only the one of the two that falls in the interval $[0, \pi)$ is stable [127]. This fact implies that the phase difference of the two oscillators for the topology under consideration can be tuned in the range $[0, \pi)$, by varying the relative frequencies of the two oscillators $\Delta\omega$.

The hold-in range Ω_h of the coupled PLL is the range of the frequency difference among the oscillator elements for which the system remains in a stable equilibrium. The pull-in range, on the other hand, is the range of the frequency difference for which the system will eventually evolve to a stable equilibrium. The hold-in range presents an upper bound to the pull-in range. Based on the above analysis and the stability analysis of the equilibrium points, it was determined by Buckwalter et al. [127] that the hold-in range is equal to

$$\Omega_h = 2G \qquad (7.12\text{-}10)$$

Furthermore, they calculated an approximate value for the pull-in range as given by Eq. 7.12-11 [127]:

$$\Omega_p \approx 2\sqrt{\frac{\sqrt{1 + 4\tau_p^2 G^2} - 1}{2\tau_p^2}} \qquad (7.12\text{-}11)$$

Finally, Buckwalter et al. [127] studied the effect of circuit delay on the hold-in and pull-in range of the system. Such delays are present in the system due to the filter characteristics of the circuit, and they result in complex dynamic behavior and instabilities. We remark that such filter characteristics may be fruitfully interpreted as time delays if the delay is small. Unlike Chapter 5 of this book, the analysis of Buckwalter et al. does include the nonlinear behavior. Recall that in Chapter 5 we introduced coupling delay in oscillator arrays via an exponential of the Laplace transform variable. There the analysis was done in the linear approximation, and thus the solutions did not exhibit any of the complex dynamical behavior arising from nonlinearity. The delay introduced was a true time delay due to propagation through the coupling lines and was not constrained to be small. However, the late time behavior in that situation corresponds to solution at time equal to many delay times, a condition which may be satisfied either by large t or small delay or both.

The model described in this section can be made progressively more complex, by taking into account the high-frequency mixing product at the output of the phase detector in the formulation or by using a higher-order loop filter and digital phase detectors.

7.13 Conclusion

In this chapter we revisited the analysis of coupled-oscillator arrays and presented two approximate models that describe the amplitude and phase dynamics at the fundamental frequency of oscillation of the coupled oscillator arrays. We presented a compact matrix formulation of the models, which can be used to efficiently analyze the transient behavior of the arrays, determine the various steady-state solutions, and examine their stability. In addition, we provided a formulation that enables one to consider external injection-locking signals to the array, which can be used to introduce modulation into the array. These models were used to provide an overview of the phase-noise analysis of coupled-oscillator arrays. Such approximate models can be used to simulate large coupled-oscillator arrays in a computationally efficient manner. Finally, it was pointed out that PLLs can be substituted for VCOs in coupled systems, resulting in behavior quite similar to that of the arrays discussed previously. In the next chapter we describe nonlinear simulation methods that can be used to accurately simulate and design oscillator circuits and coupled-oscillator arrays.

Chapter 8

Numerical Methods for Simulating Coupled-Oscillator Arrays

Coupled-oscillator arrays present a challenge to the designer due to difficulties both in the accurate simulation of oscillator elements and in the requirement for computationally efficient simulation techniques for large arrays. In addition, coupled-oscillator array design is made more difficult by the presence of multiple operating modes and stability considerations. As a result, a number of approximations need to be used to reduce the simulation time. These are describing function models for nonlinear elements [15, 118, 129, 130], along with perturbation models, infinite array approximations and continuum models shown in Chapter 3 [38, 39] and in Chapter 4 [42–44], respectively.

The progress in recent years in nonlinear simulation techniques has led to more accurate analysis and optimization methods for nonlinear circuits such as oscillators and mixers, as well as arrays [120,131]. Furthermore, these nonlinear simulation tools can be combined with electromagnetic simulation in order to analyze radiating structures and nonlinear antennas and arrays.

In this chapter, an introduction to numerical methods for simulating nonlinear circuits is presented [131, 132], focused on the simulation of autonomous circuits such as oscillators, followed by an introduction to convex optimization principles [133]. Nonlinear simulation techniques are demonstrated in order to trace the steady-state solutions of coupled-oscillator arrays and investigate their stability [116].

8.1 Introduction to Numerical Methods

The recent advances in numerical methods for simulating nonlinear microwave circuits permit one to model oscillator and coupled-oscillator array circuits efficiently and accurately. In this section a brief introduction to the principles of commonly used methods will be presented, with an aim toward obtaining the periodic steady state of oscillator circuits. The reader is prompted to the literature for an advanced and detailed description of the various methods, such as for example Refs. 101, 131, 132, and 134. Among the various existing numerical methods, transient simulation, harmonic balance, and envelope-transient simulation are described next.

8.1.1 Transient Simulation

A general nonlinear circuit is considered where a vector x of size N contains the state variables of the circuit, namely node voltages and currents. The circuit is described by a nonautonomous system of differential equations obtained by applying Kirchhoff's current law at the circuit nodes as well as the voltage law at the circuit branches, as introduced in Eq. (7.1-2) and repeated here for convenience:

$$\dot{\boldsymbol{x}} = \boldsymbol{f}(\boldsymbol{x}, \boldsymbol{u}, t) \tag{8.1-1}$$

with an initial condition $\boldsymbol{x}(\boldsymbol{t_0}) = \boldsymbol{x_0}$, where typically $t_0 = 0$. A vector $\boldsymbol{u}$ of size P including external, known, forcing terms has been included for generality.

The system is classified as an initial value problem [132], and the computation of its solution over a given time interval is known as transient simulation. There exist various discrete time numerical integration methods that are used to perform a transient simulation [101]. Assuming an integration time step q, the values of the state variable vector x_k at time $t_k = kq$ are generally computed as follows [101]:

$$\boldsymbol{x_{k+1}} = \sum_{n=0}^{m-1} a_n \boldsymbol{x_{k-n}} + q \sum_{n=-1}^{m-1} b_n \boldsymbol{f}(\boldsymbol{x_{k-n}}, \boldsymbol{u_{k-n}}, t_{k-n}) \tag{8.1-2}$$

The number of evaluations of the state variables and vector field that are required for the evaluation of the next state $k + 1$ are called the steps of the algorithm, and these steps define the order of the algorithm. An algorithm is called explicit if the future state depends only on past values of the state variables and the vector field, which corresponds to $b_{-1} = 0$. If $b_{-1} \neq 0$, the algorithm is called implicit.

A commonly used single-step, explicit integration algorithm is the forward Euler algorithm, which is defined as

$$\boldsymbol{x}_{k+1} - \boldsymbol{x}_k = q\, f(\boldsymbol{x}_k, \boldsymbol{u}_k, t_k) \tag{8.1-3}$$

In contrast, the backward Euler algorithm is a single-step implicit algorithm

$$\boldsymbol{x}_{k+1} - \boldsymbol{x}_k = q\, \boldsymbol{f}(\boldsymbol{x}_{k+1}, \boldsymbol{u}_{k+1}, t_{k+1}) \tag{8.1-4}$$

where the evaluation of the state vector at time $k+1$ requires the computation of the vector field at the same time step. The resulting nonlinear system of algebraic equations maybe solved using some numerical root-finding algorithm. Typically the Newton–Raphson algorithm is used to compute the solution at each time step [132]. Assuming a nonlinear system $\boldsymbol{h}(\boldsymbol{x})$ of algebraic equations with unknown the steady state $\boldsymbol{x} = \boldsymbol{x}_{k+1}$ at time step $t_k = (k+1)q$,

$$\boldsymbol{h}(\boldsymbol{x}) = \boldsymbol{x} - \boldsymbol{x}_k - q\, \boldsymbol{f}(\boldsymbol{x}, \boldsymbol{u}_{k+1}, t_{k+1}) = \boldsymbol{0} \tag{8.1-5}$$

the Newton–Raphson algorithm is an iterative algorithm that requires an initial guess $\boldsymbol{x}_0$ as a starting point, and it proceeds to find the roots of $\boldsymbol{h}(\boldsymbol{x})$ by calculating successive approximations of the unknown steady-state vector as

$$\boldsymbol{x}^{(j+1)} = \boldsymbol{x}^{(j)} - \left[\boldsymbol{jh}\left(\boldsymbol{x}^{(j)}\right)\right]^{-1} \boldsymbol{h}\left(\boldsymbol{x}^{(j)}\right) = \boldsymbol{0} \tag{8.1-6}$$

where j is the iteration index, and $\boldsymbol{jh}\left(\boldsymbol{x}^{(j)}\right)$ is the Jacobian of the nonlinear function $\boldsymbol{h}\left(\boldsymbol{x}^{(j)}\right)$ [132]. The steady-state vector at the previous time step j is a good candidate for an initial guess $\boldsymbol{x}_0 = \boldsymbol{x}_k$. It can be shown that if the initial guess is close enough to a solution given by Eq. (8.1-5), if the nonlinear function $\boldsymbol{h}$ is continuously differentiable, and the Jacobian $\boldsymbol{jh}$ is not singular, the sequence given by Eq. (8.1-6) converges to a root of $\boldsymbol{h}$.

Many different numerical integration algorithms (8.1-2) exist, depending on the choice of the various a_n and b_n coefficients. Selection of the appropriate integration algorithm depends on computational complexity, accuracy, and numerical stability considerations [101]. Moreover, modern integration routines adaptively adjust the integration step and order of the integration algorithm.

In order to obtain the periodic steady state of an oscillator, one needs to integrate Eq. (8.1-1) for a sufficient time interval in order to allow all transient responses to decay. As a result, transient simulation is not an efficient method to analyze the behavior of oscillator and coupled oscillator systems. Conversely, however, transient simulation provides a way to examine the stability of the solutions, as time-domain integration converges only to stable steady-state solutions.

There exist time-domain algorithms such as the shooting methods that minimize the evaluation of the initial transient state in order to efficiently obtain the desired periodic steady state. In this case, one solves the system of differential equations given by Eq. (8.1-1) subject to a periodic boundary condition $\boldsymbol{x}(t) = \boldsymbol{x}(t+T)$, where T is the period of the steady state. The reader is prompted to the literature for a description of these methods [101,132].

8.1.2 Harmonic Balance Simulation

Frequency-domain methods are particularly suited for the analysis of systems where a periodic solution exists. In this case it is possible to represent the steady-state solution by a trigonometric polynomial of degree M. The selected value of M is a trade-off between accuracy and computational efficiency. Specifically, assuming a state vector $\boldsymbol{x}(t)$ of size N, along with a vector of external forcing signals $\boldsymbol{u}(t)$ of size P, we can write

$$\boldsymbol{x}(t) = \sum_{k=-M}^{M} \boldsymbol{X_k} e^{jk\omega t} \tag{8.1-7}$$

$$\boldsymbol{u}(t) = \sum_{k=-M}^{M} \boldsymbol{U_k} e^{jk\omega t} \tag{8.1-8}$$

where ω is the angular fundamental frequency of the periodic steady state. The frequency-domain state vector and external forcing signal vector are defined by the N by $2M+1$ matrix $\boldsymbol{X} = [\boldsymbol{X_k}]$ and by the P by $2M+1$ matrix $\boldsymbol{U} = [\boldsymbol{U_k}]$, respectively. Similarly, the vector field $\boldsymbol{f}(\boldsymbol{x},\boldsymbol{u},t)$ is a periodic function and can also be expanded in a Fourier series as

$$\boldsymbol{f}(\boldsymbol{x},\boldsymbol{u},t) = \sum_{k=-M}^{M} \boldsymbol{F_k}(\boldsymbol{X},\boldsymbol{U})\, e^{jk\omega t} \tag{8.1-9}$$

where $\boldsymbol{F} = [\boldsymbol{F_k}(\boldsymbol{X},\boldsymbol{U})]$ is the frequency-domain vector field N by $2M+1$ matrix, and it depends both on $\boldsymbol{X}$ and $\boldsymbol{U}$.

In a typical piecewise harmonic balance algorithm implementation [134], the circuit is divided into a linear subcircuit and a nonlinear subcircuit, and Kirchhoff's laws are applied in the nodes that connect the two subcircuits. The response of the nonlinear subcircuit is computed in the time domain, and a fast Fourier transform algorithm is used to convert the related data to the frequency domain. As a result, in order to compute the frequency-domain vector field matrix $\boldsymbol{F}$, one first applies the inverse Fourier transform to the state $\boldsymbol{X}$ and external signal $\boldsymbol{U}$ vectors in order to obtain their time-domain expressions $\boldsymbol{x}$ and

$\boldsymbol{u}$, then computes the time-domain vector field $\boldsymbol{f}(\boldsymbol{x},\boldsymbol{u},t)$, and finally applies the Fourier transform to $\boldsymbol{f}(\boldsymbol{x},\boldsymbol{u},t)$ in order to obtain $\boldsymbol{F}$.

By introducing Eqs. (8.1-7), (8.1-8), and (8.1-9) into the original time-domain system of differential equations given by Eq. (8.1-1) and then balancing the coefficients of the exponential terms $e^{jk\omega t}$, a system of algebraic equations is obtained:

$$\boldsymbol{H}(\boldsymbol{X},\boldsymbol{U}) = \boldsymbol{\Omega X} - \boldsymbol{F}(\boldsymbol{X},\boldsymbol{U}) = \boldsymbol{0} \tag{8.1-10}$$

The matrix $\boldsymbol{\Omega}$ contains the angular frequency terms generated by the time derivative operation on the Fourier series expansion in Eq. (8.1-7). The above system of algebraic equations is efficiently solved using root finding algorithms such as for example the Newton–Raphson algorithm [134,132] described in the previous section.

It should be noted that when the steady state is expanded using only a first-order trigonometric polynomial ($M = 1$), the corresponding formulation is known as the describing function [131], and it can be used to obtain insightful analytical expressions. The coupled-oscillator models of the previous chapter are describing function formulations.

Harmonic balance is able to handle quasi-periodic solutions by properly extending the polynomial basis and the time to a frequency-domain transform algorithm [132].

8.1.3 Conversion Matrix

In microwave mixer circuits, a quasi-periodic steady-state solution exists with two or more fundamental frequency components. In the simplest scenario, two fundamental frequencies need to be considered corresponding to the local oscillator signal and the RF input signal to the mixer. Correspondingly, a two-fundamental-frequency harmonic balance algorithm needs to be used in order to evaluate the steady state.

However, in typical mixer operation, the local oscillator signal has significantly larger power than the RF input to the mixer. As a result, it is possible to evaluate the periodic steady state in the absence of the RF input signal, defined by the local oscillator signal and using a harmonic balance algorithm with a single fundamental-frequency component. The effect of the RF input signal is then considered as a linear perturbation of the previously defined steady state, leading to a computationally efficient algorithm known as the conversion matrix method.

Assuming a linear perturbation of the steady-state solution $\boldsymbol{x}(t) = \boldsymbol{x_0}(t) + \boldsymbol{\delta x}(t)$, along with an external RF signal $\boldsymbol{u}(t)$, the initial system of differential equations becomes

$$\dot{\boldsymbol{x}}_{\boldsymbol{0}}(\boldsymbol{t}) = \boldsymbol{f}(\boldsymbol{x_0}, \boldsymbol{0}, t)$$
$$\boldsymbol{\delta}\dot{\boldsymbol{x}}(t) = \boldsymbol{d_x f}(\boldsymbol{x_0}, \boldsymbol{0}, t)\boldsymbol{\delta x}(t) + \boldsymbol{d_u f}(\boldsymbol{x_0}, \boldsymbol{0}, t)\boldsymbol{u}(t) \tag{8.1-11}$$

where $\boldsymbol{x_0}(t)$ is the solution that corresponds to the large local-oscillator signal in the absence of the RF input, and $\boldsymbol{d_x f}$ and $\boldsymbol{d_u f}$ are the time-varying Jacobians of the vector field $\boldsymbol{f}$ versus the state vector $\boldsymbol{x}(t)$ and versus the input RF signal vector $\boldsymbol{u}(t)$, respectively, evaluated at $\boldsymbol{x_0}(t)$ and $u(t) = 0$. Both equations of the above system are solved in the frequency domain by applying the harmonic balance algorithm as Eq. (8.1-10). The frequency domain coefficients of the Jacobian matrices involved in the second equation are obtained at no additional computational cost during the Newton–Raphson harmonic balance computation of the large signal steady state corresponding to the first equation of Eq. (8.1-11) [131, 134].

8.1.4 Envelope Transient Simulation

The envelope transient simulation is a combination of the transient and harmonic balance simulation methods proposed by Sharrit [135] and by Ngoya and Larcheveque [136]. In effect, one represents the state variables, external forcing terms, and vector field by Fourier-series expansions of time-varying phasors:

$$\boldsymbol{x}(t) = \sum_{k=-M}^{M} \boldsymbol{X_k}(t)e^{jk\omega t} \tag{8.1-12}$$

$$\boldsymbol{u}(t) = \sum_{k=-M}^{M} \boldsymbol{U_k}(t)e^{jk\omega t} \tag{8.1-13}$$

$$\boldsymbol{f}(\boldsymbol{x}, \boldsymbol{u}, t) = \sum_{k=-M}^{M} \boldsymbol{F_k}(\boldsymbol{X}(t), \boldsymbol{U}(t))\, e^{jk\omega t} \tag{8.1-14}$$

Consequently, a transformed system of differential equations is obtained that has the form

$$\dot{\boldsymbol{X}} = -\boldsymbol{\Omega X} + \boldsymbol{F}(\boldsymbol{X}, \boldsymbol{U}) = -\boldsymbol{H}(\boldsymbol{X}, \boldsymbol{U}) \tag{8.1-15}$$

The above system is solved using time-domain integration. The advantage of envelope transient simulation over the traditional transient simulation is that the time-varying phasors $\boldsymbol{X}(t)$ are slowly varying, allowing one to use a much larger time step in the simulation. Being a time-domain simulation, envelope transient simulation may also be used to verify the stability of a steady-state solution as it converges only to stable solutions.

8.1.5 Continuation Methods

Once a steady-state solution is obtained, continuation methods can be used to obtain the families of steady-state solutions that occur as one or more parameters of the circuit under consideration are varied. Continuation techniques provide an initial condition that is close to the required steady-state solution, so that the application of the Newton–Raphson or any other root finding algorithm that is being used converges quickly and efficiently.

Assuming a parameter $p = p_0$ for which the steady-state solution $\boldsymbol{x} = \boldsymbol{x_0}$ has been evaluated, it is then possible to obtain the steady-state solution $\boldsymbol{x_e}$ corresponding to the parameter value $p = p_e$ by considering a sequence of values $p_0 < p_1 < p_2 \dots < p_e$ and progressively evaluating the steady state corresponding to each parameter value by using the solution at each step as the initial condition for the evaluation of the next step [132].

In order to reduce the steps of the continuation method, the already obtained steady-state values are extrapolated. Assuming that the steady-state solution at step k has been obtained by solving the harmonic balance system [Eq. (8.1-10)]

$$\boldsymbol{H_k}(\boldsymbol{X_k}, p_k) = 0 \qquad (8.1\text{-}16)$$

the solution corresponding to p_{k+1} is approximated by linear extrapolation as

$$\boldsymbol{J_X H_k}(\boldsymbol{X_{k+1}} - \boldsymbol{X_k}) + \frac{\partial \boldsymbol{H_k}}{\partial p}(p_{k+1} - p_k) = 0 \qquad (8.1\text{-}17)$$

where $\boldsymbol{J_X H_k}$ is the Jacobian matrix of the harmonic balance system. The above matrix equation can be solved in order to obtain an initial condition for the state vector X_{k+1}:

$$\boldsymbol{X_{k+1}} = \boldsymbol{X_k} - [\boldsymbol{J_X H_k}]^{-1} \frac{\partial \boldsymbol{H_k}}{\partial p}(p_{k+1} - p_k) \qquad (8.1\text{-}18)$$

Continuation methods based on Eq. (8.1-18) may fail due to singularities in the Jacobian matrix $\boldsymbol{JH_k} = [\boldsymbol{J_X H_k} \quad \partial \boldsymbol{H_k}/\partial p]$, which result from the existence of multiple solutions versus the parameter under consideration. In this case, tracing of the steady-state solutions can be accomplished by parameter switching [137]. Parameter switching corresponds to tracing the steady-state solutions versus another, different circuit parameter or steady-state variable, for which the corresponding Jacobian matrix is not singular.

8.2 Obtaining Periodic Steady-State Solutions of Autonomous Circuits in Harmonic-Balance Simulators

Autonomous circuits, such as free-running oscillators, present an additional difficulty in harmonic-balance simulators due to the fact that the frequency basis of the trigonometric polynomial expansion is unknown. The autonomous nature of the oscillators is expressed in the time domain by the lack of a time reference, which translates in the frequency domain to an arbitrary phase of one of the harmonic components of its state variables. This fact is explored by Rizzoli and Neri [134], where it is proposed that the harmonic-balance system of algebraic equations can be extended by one more equation defined by arbitrarily setting the phase of one of the harmonics of a circuit state variable to a specific value. As an example, the phase of the fundamental-harmonic component may be set to zero leading to

$$\mathrm{Im}\{X_1\} = 0 \tag{8.2-1}$$

Augmenting the harmonic-balance system by one equation allows one to additionally augment the number of unknowns by the fundamental frequency ω. Nonetheless, the Newton–Raphson algorithm may still converge to a DC (nonoscillating) solution due to the difficulty in selecting a suitable initial condition that is sufficiently close to the desired oscillating steady state.

Ch.-R Chang proposed an alternative method [138] in which an oscillator circuit is represented by a one-port equivalent circuit by looking into the terminals of the oscillator load, as shown in Fig. 8-1. The steady-state oscillation condition in the frequency domain is expressed by the total resistance or admittance at the load being equal to zero, known as the Kurokawa condition [129]. This condition, expressed at each harmonic k, is written as

$$Y_k = \frac{I_k}{V_k} = Y_{k,osc} + Y_{k,load} = 0 \tag{8.2-2}$$

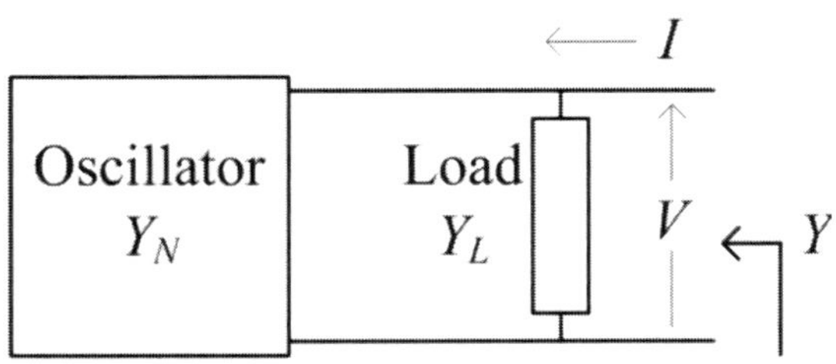

Fig. 8-1. Oscillator 1-port equivalent circuit.

Enforcing the above condition in addition to the harmonic balance system enables the algorithm to avoid converging to solutions corresponding to zero harmonic components V_k and I_k, such as the nonoscillating DC solution. In fact, in order to avoid the DC solution, it is necessary to impose the admittance condition only at the fundamental harmonic component [138]

$$Y_1 = 0 \tag{8.2-3}$$

which leads to two additional real equations in the harmonic balance system

$$\begin{aligned} \mathrm{Re}\{Y_1\} &= 0 \\ \mathrm{Im}\{Y_1\} &= 0 \end{aligned} \tag{8.2-4}$$

As a result, two additional variables can be introduced to the extended harmonic balance system: the unknown frequency ω and the oscillation amplitude V_1 at the load. The additional advantage of this formulation is that the designer may impose in a circuit optimization problem the desired oscillation frequency and amplitude at the load. A dual formulation may also be obtained by considering a series one-port equivalent circuit and enforcing the oscillation condition by setting the total impedance equal to zero.

The condition given by Eq. (8.2-3) was implemented by Quere et al. in commercial simulators [139], allowing for a practical design and optimization methodology for autonomous circuits. According to Ref. 139, one needs to introduce into the simulator an ideal probe circuit such as the one shown in Fig. 8-2. The probe is connected in parallel to a selected circuit node and consists of an ideal sinusoidal source of a given amplitude V_S, phase ϕ_S, and frequency f_S, connected in series with a current meter I_S and an ideal filter. The filter is such that it presents infinite impedance for frequencies other than the ideal source frequency f_S, thus restricting the effect of the probe to f_S.

In free-running oscillator simulation, the phase of the probe is set to an arbitrary but fixed value, for example zero. An optimization loop is run in order to find the nonzero amplitude and frequency of the probe that correspond to zero admittance $Y_S = I_S/V_S$. Each iteration of the loop is a harmonic-balance analysis. The result (V_S, f_S) of the optimization defines the oscillating steady state. Alternatively, in the case of an externally injection-locked oscillator, the frequency f_S is known and corresponds to the frequency of the external source. In this case, the pair (V_S, ϕ_S) represents the unknowns of the optimization loop, as the oscillation phase is not arbitrary any more; rather, it depends on the injection source.

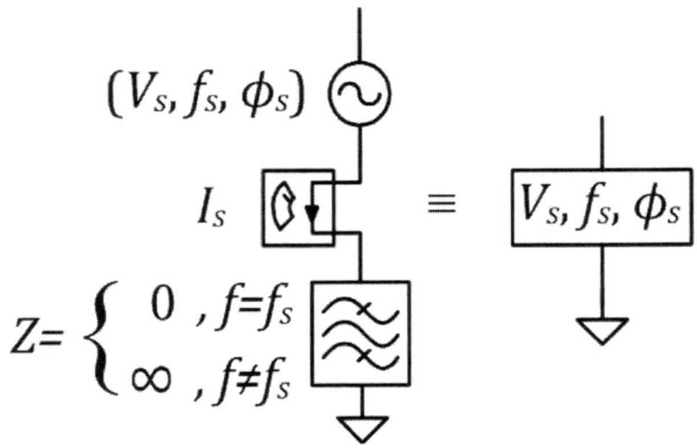

Fig. 8-2. Ideal probe circuit used for oscillator simulation in harmonic balance.

The ideal probe can also be used to initialize an envelope transient simulation to the oscillating steady state [140]. The optimization loop is first run in order to obtain the oscillating steady state (V_S, f_S), and subsequently an envelope-transient analysis is executed with the probe connected to the circuit only for an initial small time interval. The probe is then disconnected from the circuit (for example with the help of a time-dependent switching resistor), and the circuit is left to evolve for the remaining time interval according to its dynamics. This way, the envelope-transient analysis can be used to verify the stability of the steady-state solution. Once the probe is disconnected from the circuit, if the solution is unstable, the circuit will evolve to a different steady state.

8.3 Numerical Analysis of a Voltage-Controlled Oscillator

The simulation tools described in the previous section are now used to design a voltage-controlled oscillator that may serve as the array element in the coupled-oscillator array numerical analysis examples of the following sections. The oscillator circuit is based on the pseudomorphic high electron mobility transistor (pHEMT) device shown in Fig. 8-3.

A series resonator is connected at the gate terminal of the device, and a feedback capacitor is introduced at the source terminal. The feedback capacitance guarantees the presence of a negative resistance at the gate terminal. At the output, a matching network composed of two inductors is formed at the drain terminal. A frequency-tuning varactor is connected at the source terminal. The pHEMT device is self-biased, using a resistor placed at the source terminal of the device. Additionally, a 50-ohm (Ω) termination is used at the gate terminal in order to accommodate a port for an external injection signal, and also bias the gate terminal at 0 V DC, ensuring a negative gate-source (V_{GS}) voltage. The values of the various circuit components and bias are shown in Table 8-1.

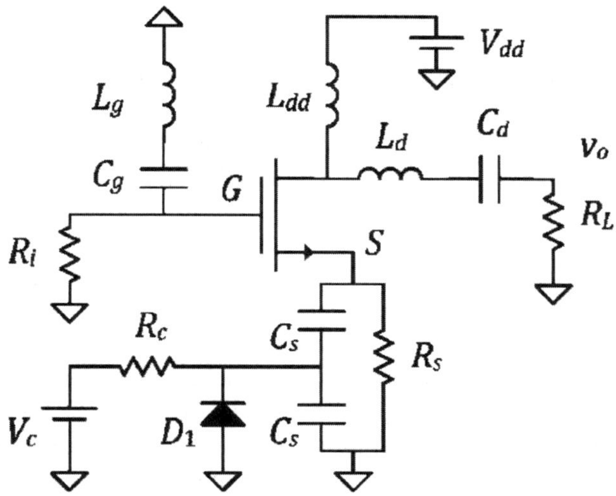

Fig. 8-3. Oscillator circuit schematic.

Table 8-1. Oscillator circuit component values.

Parameter	Value
L_{dd} (nH)	0.5
L_d (nH)	0.2
C_d (pF)	1.0
L_g (nH)	3.3
C_g (pF)	0.5
C_s (pF)	1.5
R_c (k Ω	5.0
R_i (Ω)	50.0
R_L (Ω)	50.0
R_s (Ω)	25.0
V_{dd} (V)	1.5
V_c (V)	0-15
pHEMT	NE3210S01
D_1	MA46H070-1056

nH = nanohenry, pF = picofarad

Harmonic balance optimization using an ideal probe to ensure convergence to the oscillating steady state was used for the design. The probe was connected to the output node v_0; however, other nodes may also be used such as any of the pHEMT terminals. The use of the output node is convenient because one can directly optimize the oscillator output power.

The VCO frequency and output power are shown in Fig. 8-4 and Fig. 8-5, respectively. The oscillator is consuming about 22.5 mW from a 1.5-V supply for all values of the control voltage. Its DC-to-RF conversion efficiency is approximately 9.5 percent at $V_c = 0$ V and reduces to 6.3 percent as V_c reaches 15 V.

The ideal probe is then used to simulate oscillator synchronization properties when an external injection source is applied at the gate terminal. In the first case, an external source with fixed power is connected to the circuit and its frequency is varied in order to obtain the synchronization curves. The control voltage is fixed to 10 V. The result is shown in Fig. 8-6 for two values of available power of the injection source. The synchronization bandwidth is equal to the frequency interval contained between the two edges of the closed curves defined by the infinite slope of the power versus frequency curves. The free-running steady state is represented by a point in the plot corresponding to a frequency of 9.892 GHz and power of 2.9 decibels referenced to milliwatts (dBm). As the injection power increases, the synchronization curves become larger, and they eventually open [141].

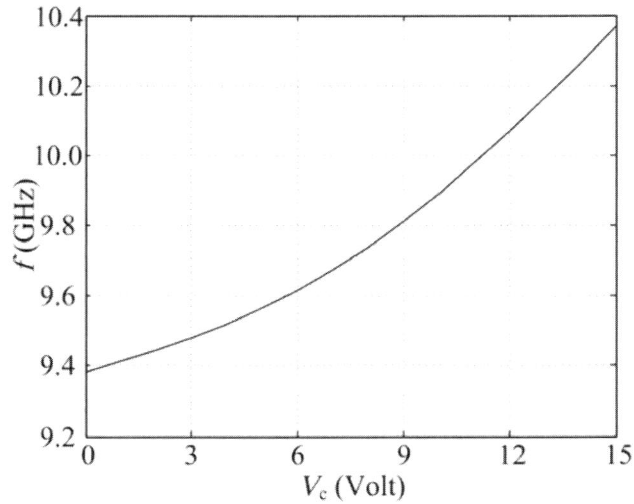

Fig. 8-4. VCO frequency versus the control voltage.

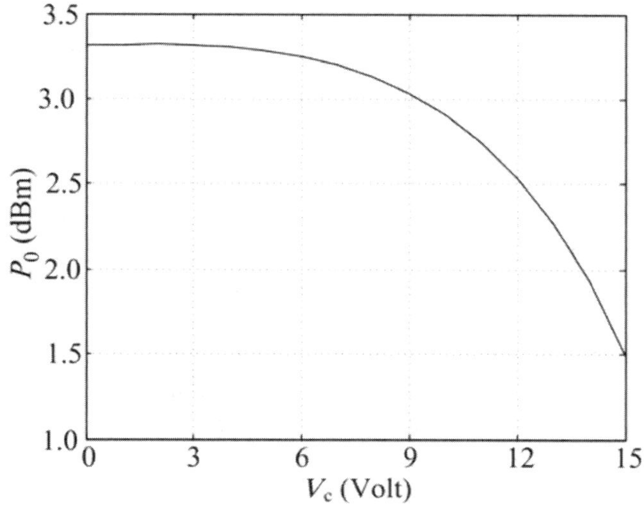

Fig. 8-5. VCO output power versus the control voltage.

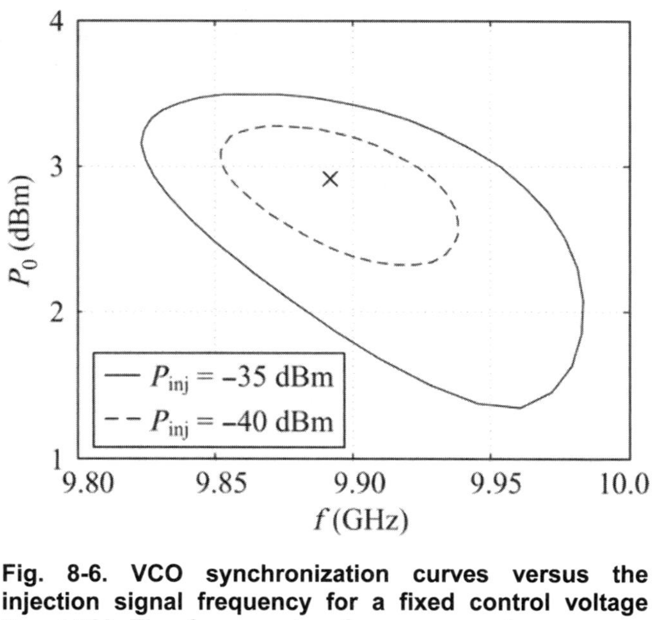

Fig. 8-6. VCO synchronization curves versus the injection signal frequency for a fixed control voltage $V_c = 10$ V. The free-running frequency and power are indicated by a point 'x' in the plot.

Alternatively, one may fix the injection signal frequency and obtain the synchronization curves versus the control voltage V_c, which corresponds to varying the free-running frequency of the oscillator. The corresponding curves obtained for an injection-signal frequency of 9.892 GHz (equal to the free-running oscillator frequency at $V_c = 10$ V) are shown in Fig. 8-7. It should be noted that the curves of Fig. 8-6 and Fig. 8-7 are generated by sweeping the phase difference between the injection signal and the oscillator output by 360 deg.

The points of infinite slope are the turning points of the curve, and due to the fact that the synchronization curves are closed, for any frequency or control voltage between them, there exist two solutions for the oscillator power.

In Fig. 8-7, the free-running frequency and power are indicated by a point '×' in the plot. For every V_c value between the turning points, two steady-state solutions exist (for example points A_1 and A_2 correspond to $V_c = 10.45$ V).

It was shown in Section 7.2 that turning points correspond to a change of the stability of the steady-state solution; as a result, only one of the two solution branches joined by the two turning points is stable. Each solution branch corresponds to a phase shift variation of 180 deg between the injection signal and the oscillator output. Therefore, in the case of an injection signal with frequency close to the fundamental frequency of the oscillator, one can obtain approximately up to 180 deg of (stable) phase shift range between the injection signal and the oscillator signal.

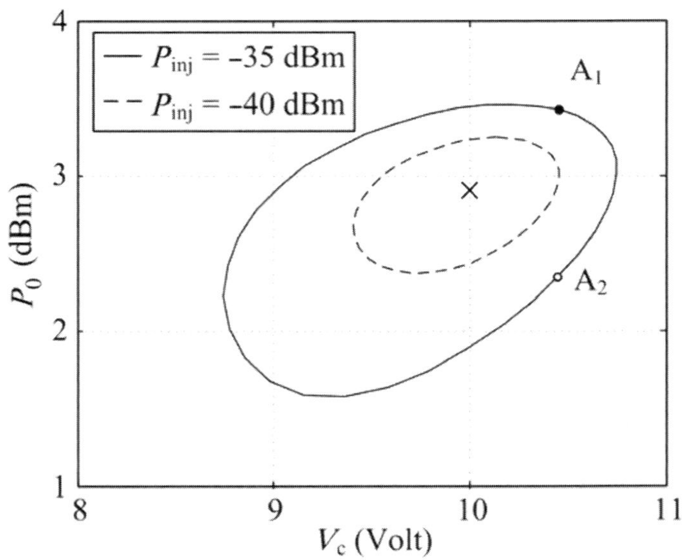

Fig. 8-7. VCO synchronization curves versus the control voltage for a fixed injection-signal frequency $f_{inj} = 9.892$ GHz, equal to the oscillator free-running frequency at $V_c = 10$ V.

It is possible to determine which one of the branches is stable by selecting one point on it and running an envelope transient analysis initialized to the steady-state solution as was described in Section 8.2. For example, for an injection signal power of –35 dBm, the points A_1 and A_2 of output power 2.35 dBm and 3.4 dBm, respectively, correspond to $V_c = 10.45$ V (Fig. 8-7). The stability of A_2 was examined by envelope transient simulation. A simulation time of 100 nanoseconds (ns) and a time step of 5 ns were used, with a Gear time-domain integration algorithm [101]. The result is shown in Fig. 8-8, where one can see that the system evolves to point A_1, demonstrating that the upper branch of Fig. 8-7 is stable. In a similar way, one can also verify that the upper branch of Fig. 8-6 is the stable one.

Finally, it should be noted that the oscillator admittance derivatives of the perturbation model of Section 7.6 that is used to model coupled oscillator arrays, can be easily computed from a harmonic-balance simulation of the single element with an ideal probe placed at the desired oscillator node. The oscillator circuit admittance derivatives are equal to the derivatives of the ideal probe admittance. Once the steady state corresponding to zero admittance looking into the probe has been determined, the probe admittance derivatives can be evaluated using finite differences [116]. As an example, the admittance derivatives for the oscillator of Fig. 8-3, corresponding to the steady state defined by control voltage $V_c = 10$ V, frequency $f_0 = 9.892$ GHz, and amplitude 0.442 V ($P_0 = 2.9$ dBm), are listed in Table 8-2.

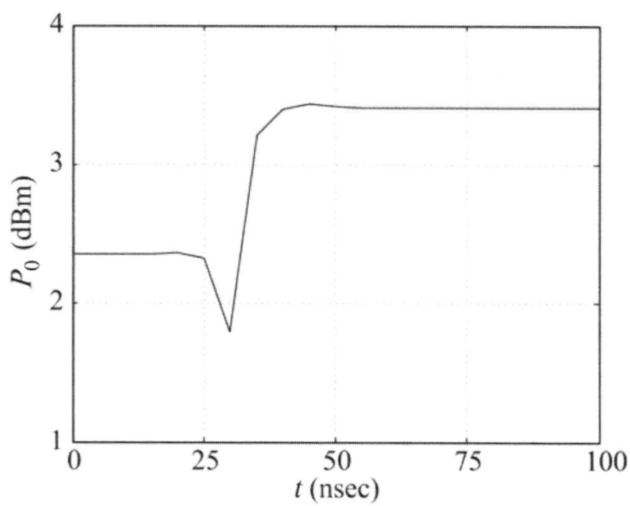

Fig. 8-8. Envelope-transient analysis of the steady state corresponding to A_2 of Fig. 8-7. A_2 is unstable and the system evolves to A_1.

Table 8-2. Oscillator steady-state and corresponding admittance partial derivatives calculated using finite differences

Parameter	Value
P_0 (dBm)	2.9
F_0 (GHz)	9.892
V_c (V)	10.0
$\frac{\partial Y}{\partial V}$ (SV^{-1})	0.0547 + j 0.1957
$\frac{\partial Y}{\partial V_c}$ (SV^{-1})	0.002 – j 0.008
$\frac{\partial Y}{\partial f}$ $(SGHz^{-1})$	-0.015 + j 0.109

8.4 Numerical Analysis of a Five-Element Linear Coupled-Oscillator Array

The VCO of the previous section is used here to create a five-element linear coupled-oscillator array. The array elements are coupled with resistor-loaded transmission-line sections of 50-Ω characteristic impedance and electrical length of 360 deg at a frequency of 9.89 GHz, which corresponds to a control voltage of 10 V in the free-running VCO element. The series resistors in the transmission line coupling sections control the coupling strength among the array elements, as was proposed by Liao and York [142]. The schematic of the array is shown in Fig. 8-9. The coupling network is connected at the oscillator outputs and each oscillator can be externally injected through its gate terminals.

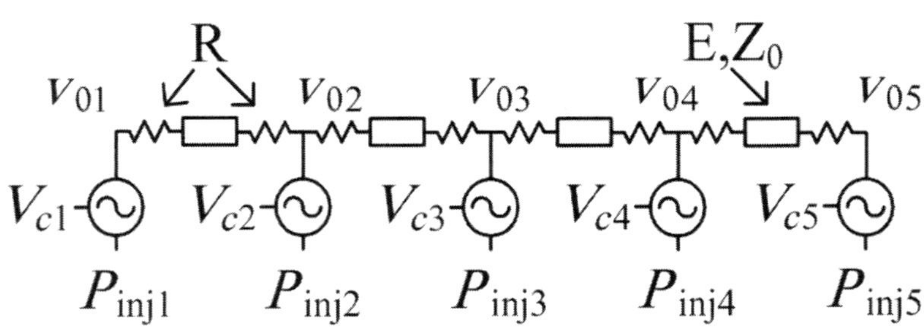

Fig. 8-9. Five-element linear coupled oscillator array.

Harmonic-balance simulation is used to trace the various solutions of the coupled oscillator array, by connecting one oscillator probe at the output node of each oscillator element. That way, it is guaranteed that the simulator will properly converge to the periodic steady state of each oscillator in the array. The five probes extend the harmonic balance system of algebraic equations by ten real equations, thereby allowing the designer to optimize ten additional unknowns. The synchronized solutions (which correspond to a constant phase shift among adjacent oscillator elements) are obtained by sweeping the phase shift while optimizing the five oscillator-output voltages, the common oscillation frequency, and four control voltages, all except the one corresponding to the middle oscillator.

The simulation results are shown in Figs. 8-10 through 8-12, where the output power, the frequency, and the control voltages, respectively, are plotted versus the oscillator phase shift. The coupling-network resistor is set to $R = 270\ \Omega$, and the control voltage of the middle oscillator is fixed at $V_{c3} = 10$ V. The phase shift has been swept from 0 to 180 deg, with the oscillator phases increasing from oscillator 1 to the left and toward oscillator 5 to the right of Fig. 8-9. Due to the symmetry of the array, the solution curves for the remaining phase-shift values (0 to –180 deg) can be obtained by considering the mirror image of the array elements with respect to the central element 3—in other words, replacing element 5 with element 1 and replacing element 4 with element 2.

Figure 8-12 shows the variation of the oscillator-control voltages versus the phase shift. One can see that the edge element-control voltages present a significantly larger variation compared to the inner elements. In fact, the control voltages of elements 2 and 4 remain practically constant for phase shifts up to 90 deg. This represents a numerical verification using a harmonic balance simulation of the proposition of Liao and York [142], where, by only tuning the free-running frequency of the peripheral elements of a coupled-oscillator array, it is possible to generate constant phase-shift distributions among the array elements, thus both minimizing the required number of controls and eliminating the need for phase shifters.

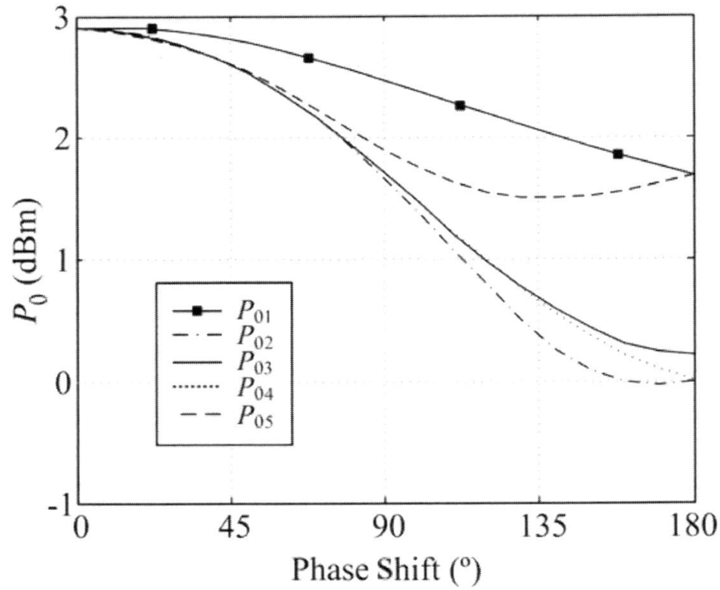

Fig. 8-10. Five-element linear coupled-oscillator array. Output power of each oscillator versus the phase shift among adjacent elements. The coupling resistor is R = 270 Ω.

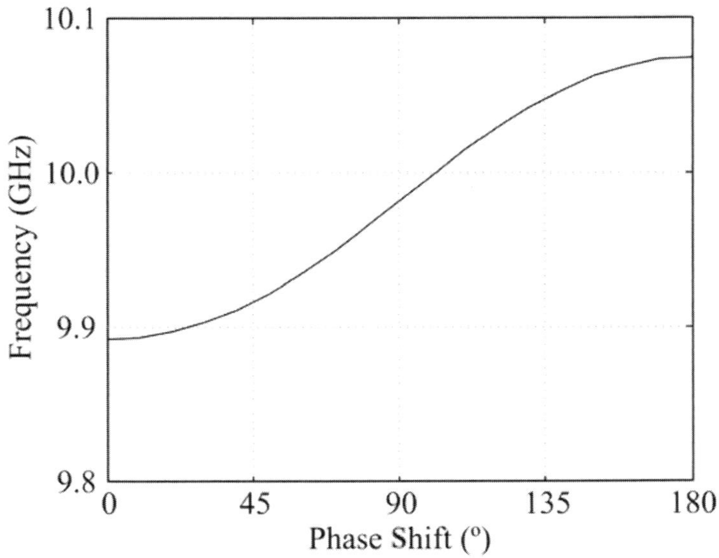

Fig. 8-11. Five-element linear coupled-oscillator array. Array frequency versus the phase shift among adjacent elements. The coupling resistor is R = 270 Ω.

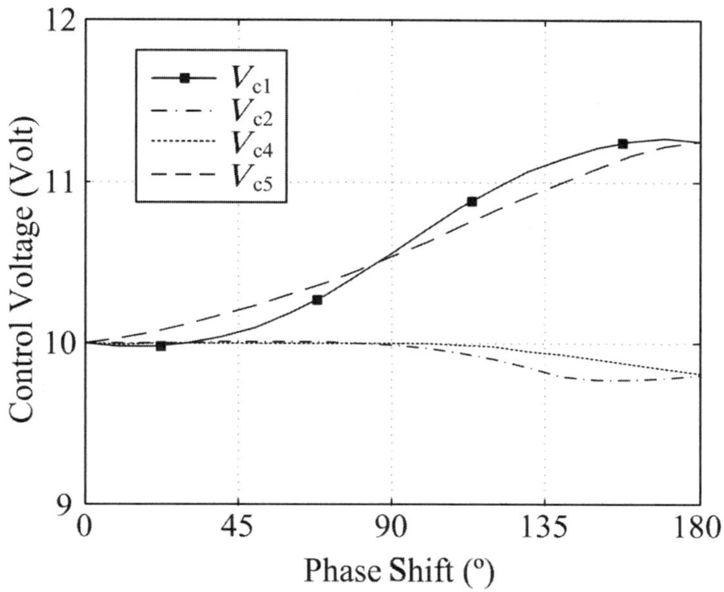

Fig. 8-12. Five-element linear coupled-oscillator array. Control voltage of the oscillator elements versus the phase shift among adjacent elements. The middle oscillator element control voltage is fixed at V_{c3} = 10 V, and the coupling resistor is R = 270 Ω.

However, one can observe from Fig. 8-11 that as the phase shift varies, the array frequency also varies. One way to maintain a constant frequency for every desired phase shift is by allowing the control voltage of one more oscillating element to vary. In this example, one may allow the middle oscillator-element control voltage (V_{c3}) to vary, thus being able to eliminate potentially undesired frequency variations. In Figs. 8-13 and 8-14, the five oscillator amplitudes and control voltages are plotted versus the phase shift for a coupling resistor of $R = 330\,\Omega$, while the array frequency is fixed at 9.892 GHz. The inner-oscillator control voltages take very similar values; however, they need to be varied in order to maintain the frequency of the array constant.

The coupling strength among the oscillator elements is set by the coupling resistor R. In fact the selection of the optimum coupling strength is a trade-off among a number of parameters. As the coupling strength increases, the variation in the output power of the oscillators, the frequency, and the control voltages with the phase shift all increase. As the coupling strength decreases, the oscillators eventually desynchronize due to the presence of noise.

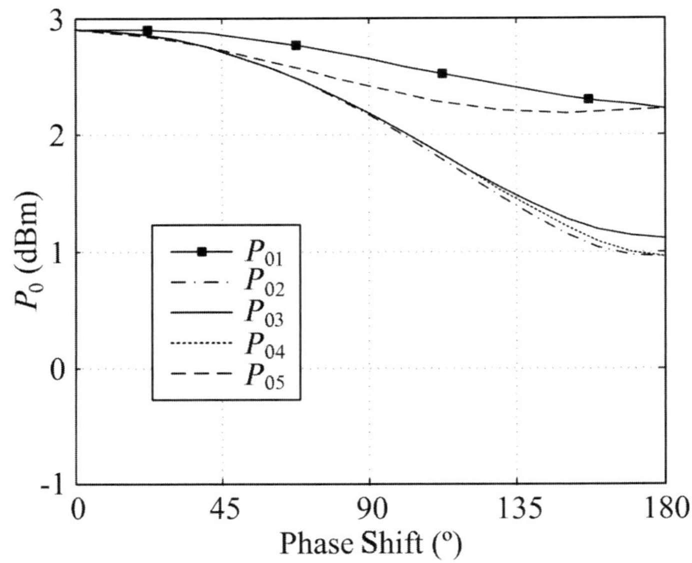

Fig. 8-13. Five-element linear coupled-oscillator array. Output power versus phase shift for a coupling resistor $R = 330\ \Omega$. The array frequency is fixed at $f = 9.892$ GHz, and the control voltages of all elements are allowed to vary.

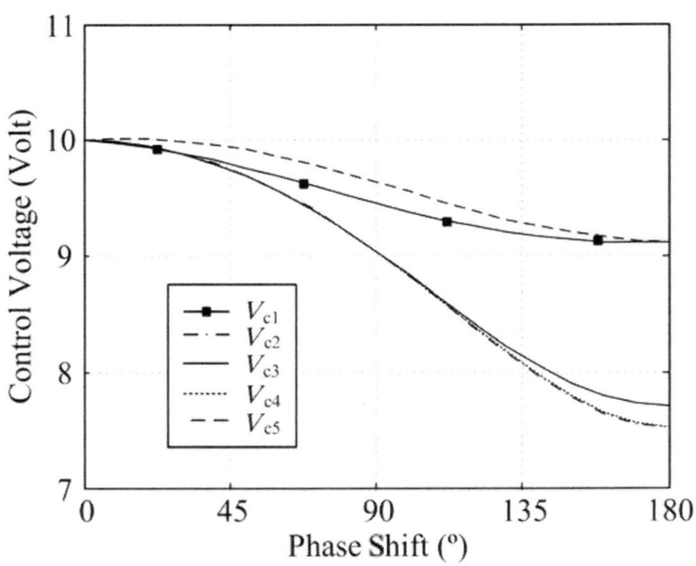

Fig. 8-14. Five-element linear coupled-oscillator array. Oscillator control voltages versus phase shift for a coupling resistor $R = 330\ \Omega$. The array frequency is fixed at $f = 9.892$ GHz.

It is easily verified from Figs. 8-10 to Fig. 8-12 that for the value of the coupling resistor of $R = 270\ \Omega$, a harmonic balance solution for every possible phase shift value exists. As the coupling strength increases, it is not possible to obtain a solution for every phase shift. This is demonstrated in Fig. 8-15, where oscillator output power of the harmonic balance solutions corresponding to a coupling resistor of $R = 220\ \Omega$ is plotted. It is easily seen that solutions exist only up to approximately 120 deg, and they are limited by the presence of a turning point [116]. As the desired phase shift progressively increases, the amplitude of oscillator 2 eventually drops to zero.

The stability of the solutions was also verified using envelope transient analysis, using the method described in Section 8.2. The simulation results showed that the coupled oscillator array with $R = 270\ \Omega$ loses stability for phase shift values larger than approximately 58 deg. Additionally, the perturbation model of Section 7.6 was used to evaluate the constant phase shift steady-state solutions and their stability. The real part of the largest eigenvalue of the linear variational equation of the steady-state solution is shown in Fig. 8-16, where one can see that the perturbation model predicts loss of stability for a phase shift approximately equal to 63 deg, a value that is in relatively good agreement with the result obtained from envelope transient simulation.

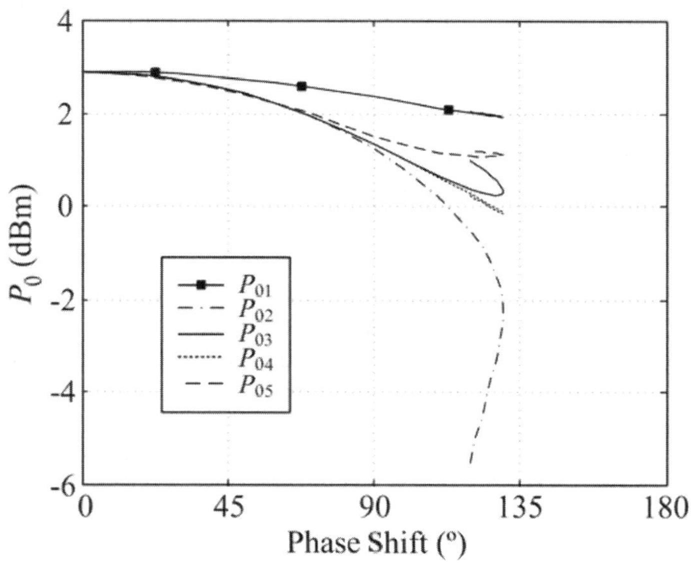

Fig. 8-15. Five-element linear coupled-oscillator array. Output power versus phase shift for a coupling resistor $R = 220\ \Omega$. As the coupling strength increases solutions do not exist for every phase shift value. The middle oscillator element control voltage is fixed at $V_{c3} = 10$ V.

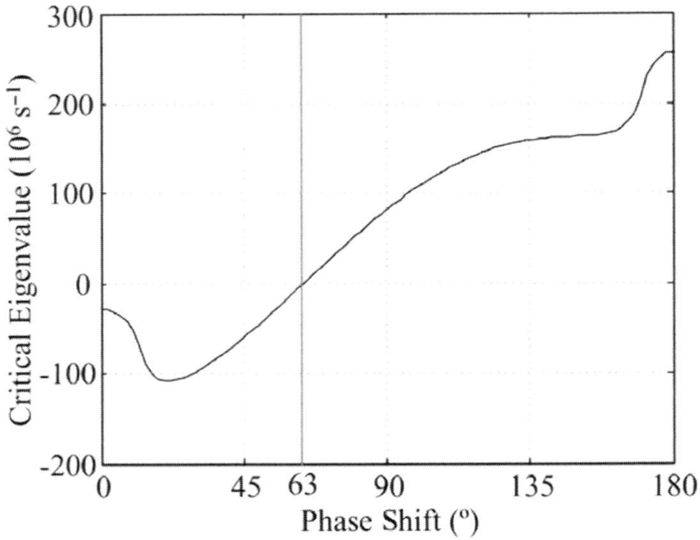

Fig. 8-16. Stability analysis of the steady-state solution using the perturbation model of Section 7.8 showing critical eigenvalue real part versus the phase shift between adjacent elements.

In Fig. 8-16, positive values of the real part correspond to unstable solutions. The coupling resistor is $R = 270\ \Omega$, and the control voltage of the middle oscillator is fixed at $V_{c3} = 10$ V.

Further comparison between the harmonic balance solution and the perturbation model is made in Figs. 8-17 and 8-18, where the amplitude of oscillators 1 and 3 along with the array frequency are plotted versus the phase shift between adjacent elements obtained using both methods. One can see that the agreement becomes worse for large phase offsets where the perturbation is larger. The perturbation model is limited to small perturbations around the free-running steady state, which in this case is near the 0-deg phase shift (in-phase) solution, and, additionally, to oscillators with small harmonic content [116,143]. Nonetheless, the advantage of the perturbation model lies in its computational efficiency which quickly becomes important as the number of array elements increases.

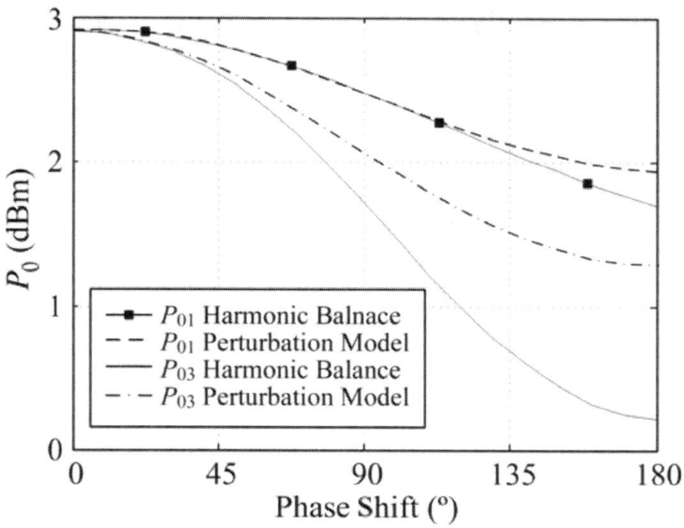

Fig. 8-17. Comparison of harmonic balance simulation and perturbation model of Section 7.8. Output power of oscillators 1 and 3 versus the phase shift between adjacent elements. The coupling resistor is *R* = 270 Ω, and the control voltage of the middle oscillator is fixed at $V_{c3} = 10$ V.

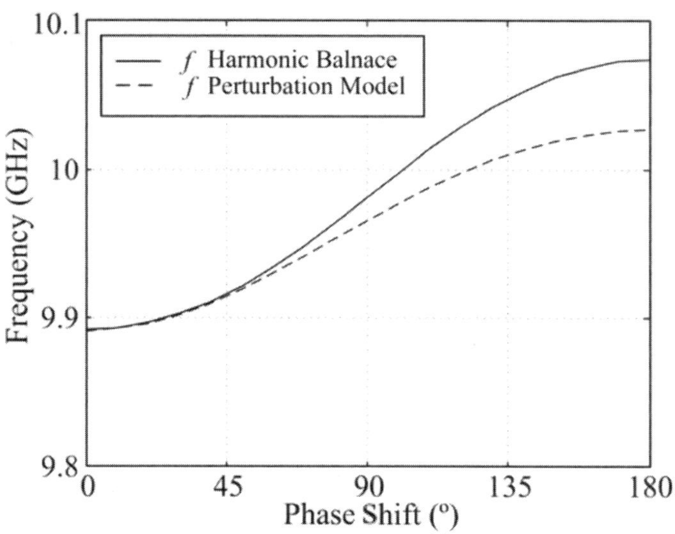

Fig. 8-18. Comparison of harmonic-balance simulation and perturbation model of Section 7.8. Array frequency versus the phase shift between adjacent elements. The coupling resistor is *R* = 270 Ω, and the control voltage of the middle oscillator is fixed at $V_{c3} = 10$ V.

8.5 Numerical Analysis of an Externally Injection-Locked Five-Element Linear Coupled-Oscillator Array

Injection locking the array to an external source signal is desirable in several applications in order to reduce the array phase noise as shown by Chang et al. [123], or introduce modulation to the oscillator signal as considered by Kykkotis et al. [99] and Auckland et al. [122].

The dynamics of the system and the stability of the various solutions depend strongly on the element that is being injected, whether it is located in the center of the array or near the edges [120, 123, 140]. Additionally, the number of elements that are being externally injected strongly influences the number and behavior of the existing solutions. Commonly used topologies are the one proposed by Stephan [1], in which the two end elements of a linear array are injection-locked to an external source, and the topology where the external signal is illuminating all the elements of the coupled oscillator array leading to a globally injection locked array [123], such as the case of a reflectarray or transmit-array antenna.

In the case of an externally injection-locked array, the oscillation frequency is determined by the frequency of the external source. In contrast, the phase difference between the injection source and the element that is being injected must be included in the unknowns of the harmonic-balance system of equations. Similarly with the free-running array case, a probe must be connected to each oscillator element in order to guarantee the convergence of the harmonic balance simulator to the oscillating solution.

The five-element array of Section 8.5 is considered with a coupling resistor of $R = 330\ \Omega$. The middle element (3) is injection-locked to an external signal source through its gate termination. The steady-state solutions corresponding to a constant phase shift among the array elements are traced versus the phase shift among adjacent elements. The additional unknowns in the harmonic balance optimization that can be obtained due to the use of the ideal probes are the five oscillator amplitudes, the four control voltages corresponding to all the elements (except the one being injected), and the phase difference between the injected element and the external source signal. The phase of the injected element is fixed at 0 deg, and the phase of the injection signal ϕ_{inj} is allowed to vary. The control voltage of the injected element is fixed at 10 V. Finally, the frequency of the external signal is 9.892 GHz.

In Fig. 8-19 the phase ϕ_{inj} is plotted versus the phase shift between the oscillator elements for different injection-signal powers. As was the case in Section 8.4, due to the symmetry of the array, solution curves also exist for the phase shift interval between 0 deg and –180 deg, and they can be obtained by taking the mirror image of the array elements with respect to the central element 3. One can see that for a given injection-signal power and oscillator phase shift it is possible to have two solutions corresponding to two different values of ϕ_{inj}.

The output power of the middle oscillator is plotted in Fig. 8-20 versus the inter-oscillator phase shift, for different injection signal-power levels. It can be seen that for lower injection-signal power levels (P_{inj} = –35 dBm), solutions for every inter-element phase shift in the range [0 deg, 180 deg) do not exist. Specifically, a closed solution curve exists for phase shifts up to approximately 100 deg, where a turning point appears. As the injection power increases, the solution curve widens, and the turning point shifts to a larger phase-shift value. Finally for larger injection powers, the curve opens, forming two solution branches. As shown in Fig. 8-20, as many as three solutions may exist for a given phase shift value.

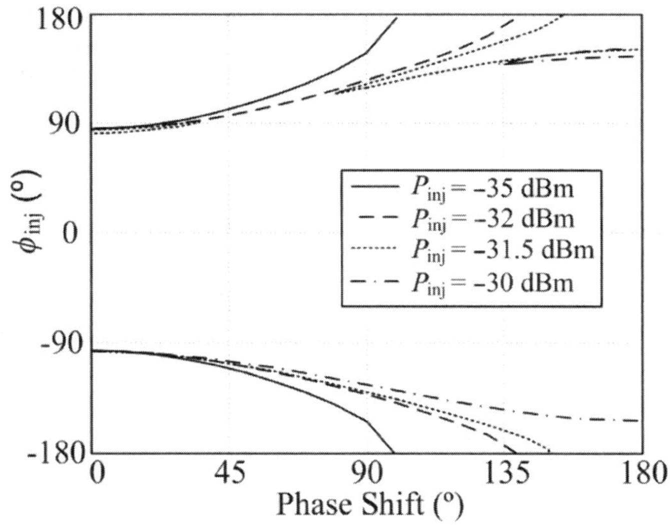

Fig. 8-19. Externally injection-locked five-element linear coupled-oscillator array. Injection-signal phase versus the oscillator phase shift, for different injection-signal power levels. The injection signal frequency is 9.892 GHz, and the coupling resistor $R = 330\ \Omega$. The middle oscillator element-control voltage is fixed at $V_{c3} = 10$ V.

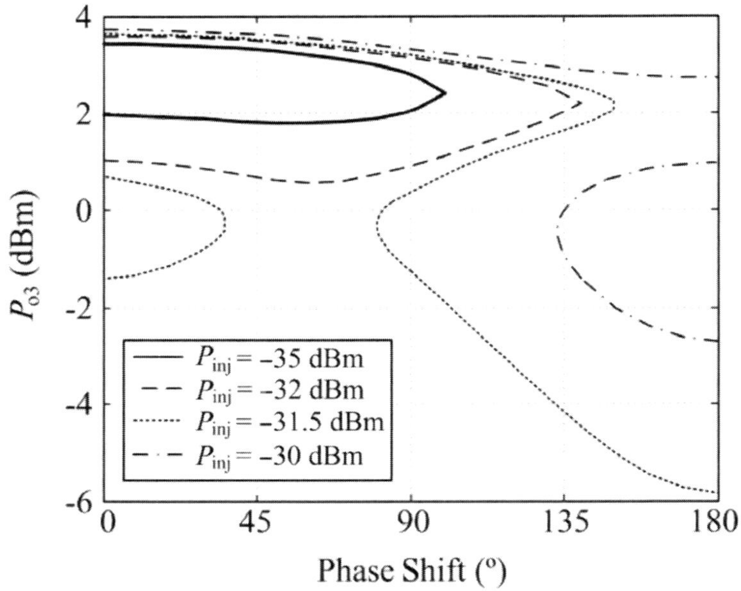

Fig. 8-20. Externally injection-locked five-element linear coupled-oscillator array. Middle oscillator output power versus the phase shift, for different injection signal power levels. The injection signal frequency is 9.892 GHz, and the coupling resistor $R = 330\ \Omega$. The middle oscillator element control voltage is fixed at $V_{c3} = 10$ V.

Subsequently, the stability of the solutions must be examined, in order to determine which of the multiple solutions are stable and will appear in practice. The solution stability may be determined using transient or envelope-transient simulation, or by examining the eigenvalues of the linear variational system of equations corresponding to one of the analytical model formulations of the previous sections. In Ref. 120, Collado and Georgiadis studied the injection-locked solutions of a two-element array, and it was determined that there exists an optimum coupling strength that leads to a maximum stable constant phase-shift range.

8.6 Harmonic Radiation for Extended Scanning Range

When an array of oscillators is coupled at the fundamental frequency, the maximum stable phase-shift range that can be introduced between adjacent array elements is 180 deg. In the case of a coupling network that has a coupling phase of 0 deg, this translates to being able to generate constant phase-shift distributions $\Delta\phi$ in the –90-deg to 90-deg range. Considering a linear array where the radiating elements are placed at a half-wavelength distance, its main beam can be scanned according to $\Delta\phi = kd \sin\theta = \pi \sin\theta$, where θ is measured from broadside, for a maximum of $\theta = 30$ deg.

It is possible to extend the phase-scanning range by considering the fact that the phase variation of the oscillator *N*th harmonic is *N* times the phase variation of its fundamental frequency component, where *N* is the harmonic order. The task of the designer then becomes that of being able to generate sufficient power in the desired harmonic component. Essentially there are two ways of implementing such architectures, either by placing a frequency N-tupler circuit at the output of each oscillator or by properly designing the oscillator elements to have maximum power at the harmonic under consideration.

In Ref. 145, Alexanian et al. proposed a linear array of five coupled oscillators, where each oscillator element is followed by a frequency doubler, as shown in Fig. 8-21. The fundamental frequency of the oscillators is 4 GHz, and their output power is 9 dBm. The prototype array in Ref. 145 used a compact field-effect transistor (FET)-based frequency doubler circuit with 1-dB conversion gain. The theoretical phase-tuning range that can be achieved with this topology is 360 deg.

Based on the same principle, a frequency-tripled two-dimensional coupled-oscillator array operating in X-band was reported by Pogorzelski in Ref. 69. An inter-oscillator phase difference ranging up to 60 deg was tripled to 180 deg. Thus, this array had a demonstrated H-plane scanning range of ±90 deg. The fabricated prototype additionally contained a diagnostic system used to evaluate the phase differences between the various oscillator elements. The array is described in more detail in Section 6.2.

Alternatively, Sanagi et al. [146] proposed a four-element coupled-oscillator array, where the oscillator elements were specifically designed in order to have a high second-harmonic content, thus also obtaining a 360-deg phase-scanning range. The proposed circuit is shown in Fig. 8-22.

The oscillators are coupled using directional couplers. Termination circuits based on the coupler networks are also attached to the edge elements in order to implement a symmetrical coupling network where all oscillators see approximately the same load. Sanagi et al. [146] extended the coupled oscillator model based on the cubic nonlinearity, which was introduced by York [111], in order to study their proposed circuit architecture. Specifically, they considered a nonzero square term in the cubic polynomial describing the current-to-voltage characteristic of the nonlinear device used for the oscillators; additionally, Sanagi et al. [146] introduced in the formulation an additional equation pertaining to the second harmonic. The block diagram of the considered circuit topology is shown in Fig. 8-23, which was used to investigate the effects in the array performance due to coupling both at the fundamental frequency and at the second harmonic. It was shown that as the second harmonic coupling becomes stronger relative to the coupling at the fundamental frequency, the achievable phase tuning range is reduced.

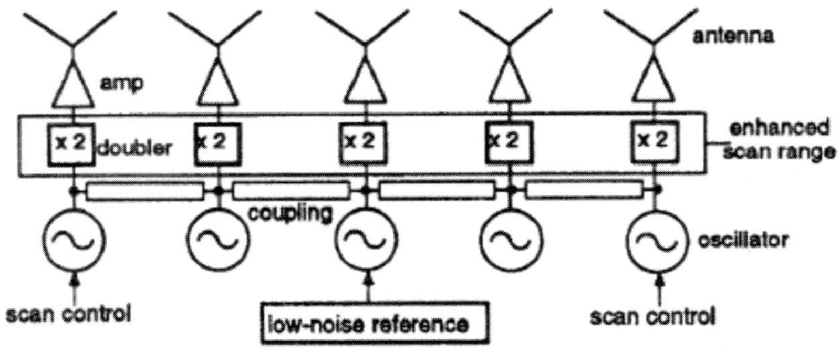

Fig. 8-21. Coupled-oscillator array using frequency doublers for extended scanning range. (Reprinted with permission from Ref. 145, ©1995 IEEE.)

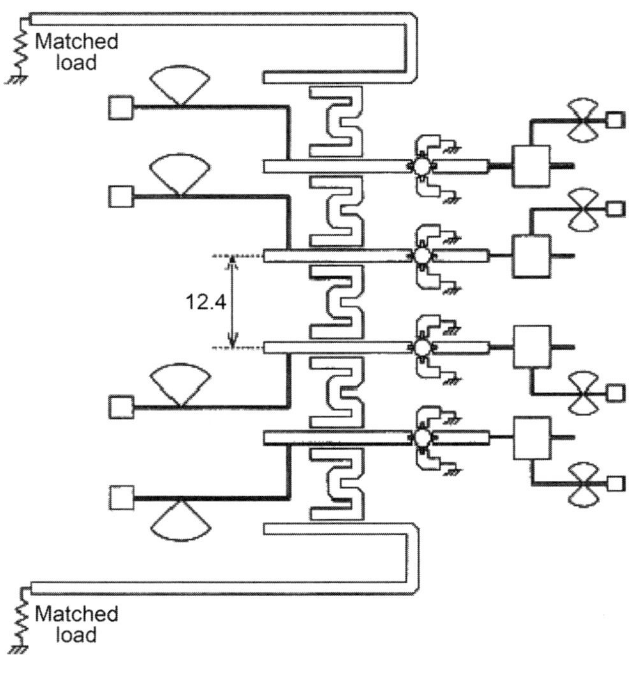

Fig. 8-22. Coupled-oscillator array radiating the second harmonic frequency component (dimensions in millimeters). (Reprinted with permission from Ref. 146. This material is reproduced with permission of John Wiley & Sons, Inc.)

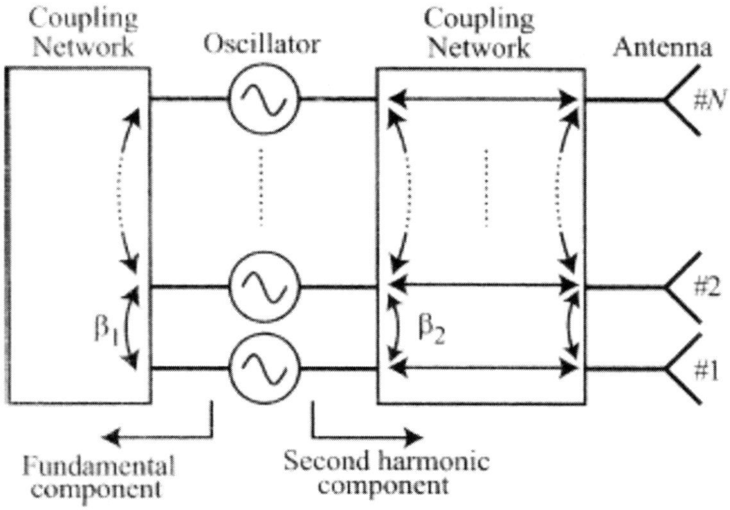

Fig. 8-23. Model of the coupled-oscillator array radiating the second harmonic-frequency component. (Reprinted with permission from Ref. 146. This material is reproduced with permission of John Wiley & Sons, Inc.)

In Ref. 147, Georgiadis proposed a three-element coupled-oscillator array shown in Fig. 8-24, also optimized in order to radiate the second-harmonic output wave. The array operates based on the same principle as the one by Sanagi et al. [146]. In this work, however, the alternative perturbation model for the coupled oscillator array given in Section 7.6 was extended in order to include the formulation for the second-harmonic frequency component.

Furthermore, harmonic-balance analysis was used to trace the steady-state solutions corresponding to constant phase shifts between the array elements at the second-harmonic component. In order to do so, it is necessary to place two ideal probes at each oscillator output node, one at each harmonic [147]. The computational load associated with the optimization of the coupled-oscillator array radiating the second harmonic is increased due to the fact that the number of ideal probes, and therefore optimization goals, required for the simulation is doubled.

8.7 Numerical Analysis of a Self-Oscillating Mixer

Self-oscillating mixers (SOMs) are particularly attractive for low-cost, compact implementations of microwave circuits due to the fact that the same circuit is used to provide a local-oscillator signal as well as for frequency translation.

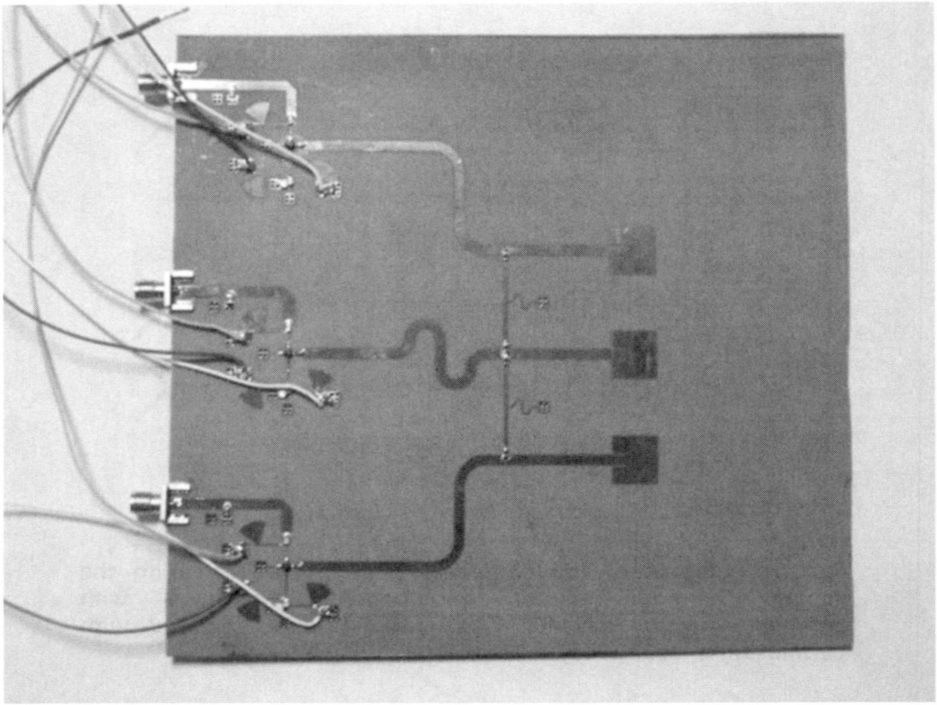

Fig. 8-24. Three-element coupled-oscillator array prototype, designed to radiate the second harmonic frequency component. (Reprinted with permission from Ref. 147, ©2007 IEEE.)

The performance parameters of self-oscillating mixer circuits (such as conversion gain and inter-modulation distortion) can be evaluated using harmonic-balance simulation, provided that an ideal probe is used to enforce the convergence of the simulator to the oscillating steady state. The probe equations are set up in order to make sure that the admittance associated with the probe is equal to zero at the oscillating frequency of the circuit.

The radio-frequency (RF) and intermediate-frequency (IF) signals are treated by introducing a second fundamental frequency component in the harmonic-balance frequency basis, thus using a two-fundamental harmonic balance system of equations. Alternatively, one may consider the effect of the RF and IF signals as a linear perturbation of the oscillating steady state and employ the conversion matrix method to efficiently compute the conversion gain of the self-oscillating mixer. Finally, the RF and IF frequency signals can be efficiently treated using an envelope-transient simulation that has been initialized to the oscillating steady state.

In Ref. 148, Herran et al. optimized the gain associated with a selected mixing product of a self-oscillating mixer by using two ideal probes properly introduced in the circuit and optimizing the reflection coefficients of an ideal multi-harmonic load connected to the circuit input. The circuit schematic that was used is shown in Fig. 8-25.

The first probe, called an auxiliary generator in Fig. 8-25, is used to enforce the oscillation condition at the desired frequency. The admittance looking into this probe is set to zero in order not to perturb the circuit steady state, and the complex admittance or reflection coefficient of the multi-harmonic load at the fundamental frequency that satisfies this condition is found through harmonic balance optimization.

The second ideal generator probe is connected in series with the gate terminal of the FET device, and its frequency corresponds to a desired Nth harmonic that is selected for the mixing process. Mixing products involving the second and third harmonics were considered. The reflection coefficient of the multi-harmonic load at the desired harmonic is set to –1, corresponding to a short circuit. The optimization procedure consists of finding the complex amplitude of the ideal generator which results in a desired mixing gain value. The corresponding admittance looking into the generator must have a positive real part in order for it to correspond to a passive load. In this way, the multi-harmonic load is optimized for a desired mixing gain value and its reflection coefficient at the fundamental frequency and selected harmonic frequency are determined. The final design is obtained by implementing the obtained reflection coefficient values using passive printed or lumped circuit components [148].

A varactor diode may be appropriately placed in the self-oscillating mixer circuit in order to provide a frequency-tuning capability. An externally injection-locked self-oscillating mixer operates as both a mixer and a phase-shifter element, where the phase shift between the input and output of the mixer is varied by changing the free-running frequency of the self-oscillating mixer.

Being a synchronized oscillator, the externally injection locked self-oscillating mixer can be used to provide a continuous phase-shift range of $N \times 180$ deg, where the external injection signal is assumed to have a frequency near the fundamental frequency of oscillation of the self-oscillating mixer, and the Nth oscillator harmonic is used in the mixing operation. Here, the fact that the tuning range of the phase of the oscillator Nth harmonic is N times the tuning range of the phase of its fundamental frequency component being used [145].

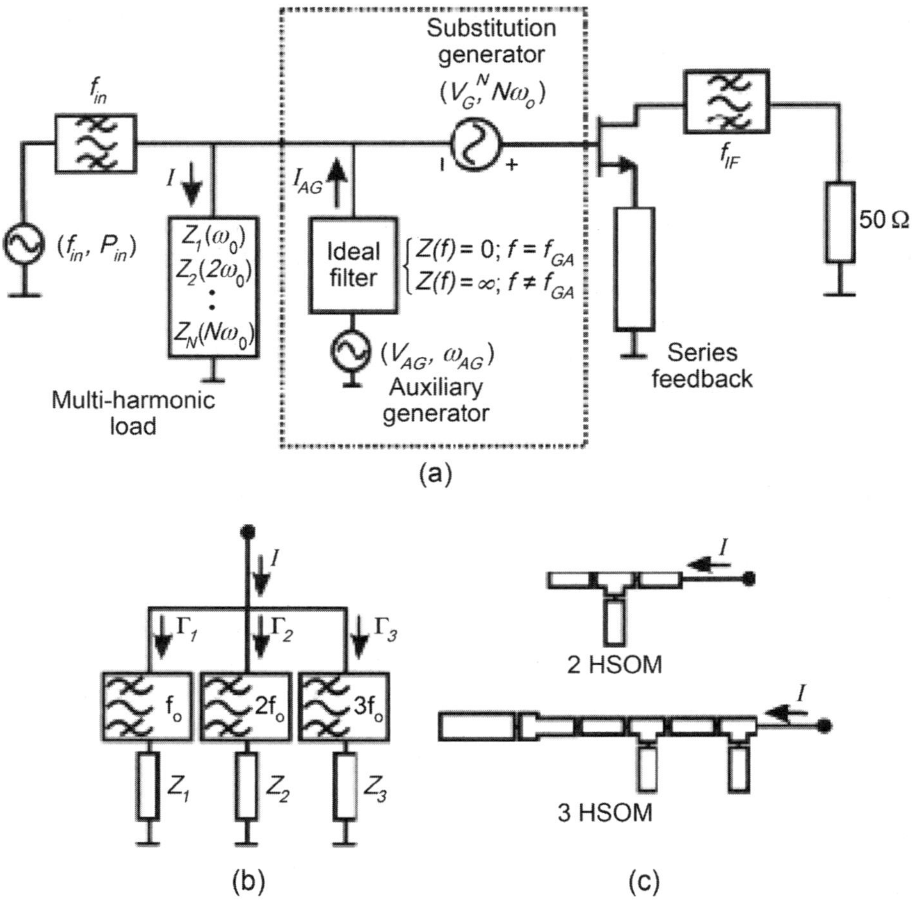

Fig. 8-25. Nonlinear optimization of a self-oscillating mixer. (a) Circuit topology: The input signal of power P_{in} passes through a band-pass filter with center frequency f_{in}, before it is mixed by the active circuit and collected at the output through an intermediate frequency filter of center frequency f_{IF}. The optimization procedure consists of designing a multi-harmonic load with impedance Z_n at harmonic $n\omega_0$. Optimization is performed using an ideal auxiliary generator probe AG with amplitude V_{AG} and frequency ω_{AG} defined as in Fig. 8.2, along with a substitution generator with amplitude V_G^N at the harmonic frequency $N\omega_0$. (b) Definition of the multi-harmonic load using ideal circuit components. The figure indicates the input reflection coefficients Γ_n corresponding to the load impedance Z_n at frequency $n\omega_0$. (c) Implementation of the multi-harmonic load using microstrip components for the cases of a second (2 HSOM) and third (3 HSOM) harmonic self-oscillating mixer, respectively. (Reprinted with permission from Ref. 148, ©2006 IEEE.)

It has been argued in Sections 7.9 and 8.5 dealing with the analysis of externally injection locked oscillator arrays that such architectures can be used to transmit information by introducing phase or frequency modulation in the external injection signal. Furthermore, the effect of modulation in the array scanning range was investigated in Section 7.11. Such topologies are limited to relatively narrowband applications due to the fact that the modulation strongly affects the steady state of the synchronized oscillator signals. Furthermore, specific modulation formats leading to small envelope variations are suitable for such applications due to the fact that the amplitude-limiting properties of the oscillators tend to introduce distortion to the envelope of the modulating signals. Continuous phase modulation (CPM) [149], which is a constant envelope modulation, is a prominent candidate for such systems. A well-known example of CPM is Gaussian minimum-shift keying (GMSK) used in the Global System for Mobile Communications (GSM), second-generation mobile (cellular) communication systems.

However, when modulation is introduced through the RF input signal of the self-oscillating mixer, it does not strongly affect the synchronization state of the mixer due to the fact that the input signal has a low power level and represents only a perturbation of the steady state. As a result, self-oscillating mixers can be used as frequency translation and phase-shifter circuits for input RF signals of arbitrary modulation. Furthermore, proper design of the mixer can allow one to obtain broadband gain, and therefore the self-oscillating mixer is not limited to RF input signals with narrowband modulation.

The use of an injection-locked self-oscillating mixer as a downconverter and phase shifter element was studied by Ver Hoeye [80]. The proposed circuit topology is the same as in Fig. 8-25 with the addition of a varactor diode connected in parallel with the series feedback shorted stub present at the source terminal of the active device in order to provide a frequency tuning capability. The SOM design was performed using the methodology described previously in this section. An oscillation at 3.25 GHz was obtained, and an RF signal of 11.25 GHz was mixed with the third harmonic of the SOM, resulting in an IF output of 1.5 GHz. Phase tuning of as much as 3×180 deg = 540 deg was achieved by utilizing the third harmonic mixing product. The obtained conversion gain was 4.5 dB over a bandwidth of approximately 100 MHz. It is shown in Fig. 8-26 that the conversion gain depends both on the injection power level P_s and on the varactor control voltage V_{cont} or, in other words, the selected phase difference between the input and output SOM terminals. The results have been obtained using a two-fundamental-harmonic balance simulation, and one can observe the closed synchronization curves of the injection locked self-oscillating mixer, which are similar to the ones obtained for the synchronized oscillator in Fig. 8-7. The synchronization curves open as the injection power increases, and there exist two solutions for a given control voltage within the synchronization band limited by the curve edges of infinite slope. Only one of the two solutions is stable and therefore measured experimentally, and in this case it corresponds to the branch with lower conversion gain.

8.8 Conclusion

In this chapter we provided an introduction to nonlinear analysis methods with a special focus on methods of analysis applied to nonlinear circuits such as oscillators, self-oscillating mixers, and coupled oscillator arrays. Such tools can be combined with electromagnetic simulators in order to accurately model the various passive components of the circuits under consideration such as transmission lines, interconnects, resonators, and antennas. Typically, these methods can be used to analyze small arrays consisting of tens of elements or fewer due to their increased computational complexity. Additionally, they can be used to compute the various parameters that are required to formulate the approximate models of the previous chapter such as the nonlinear admittance derivatives, which, in turn, can be used for an efficient less time-consuming simulation and optimization of the arrays.

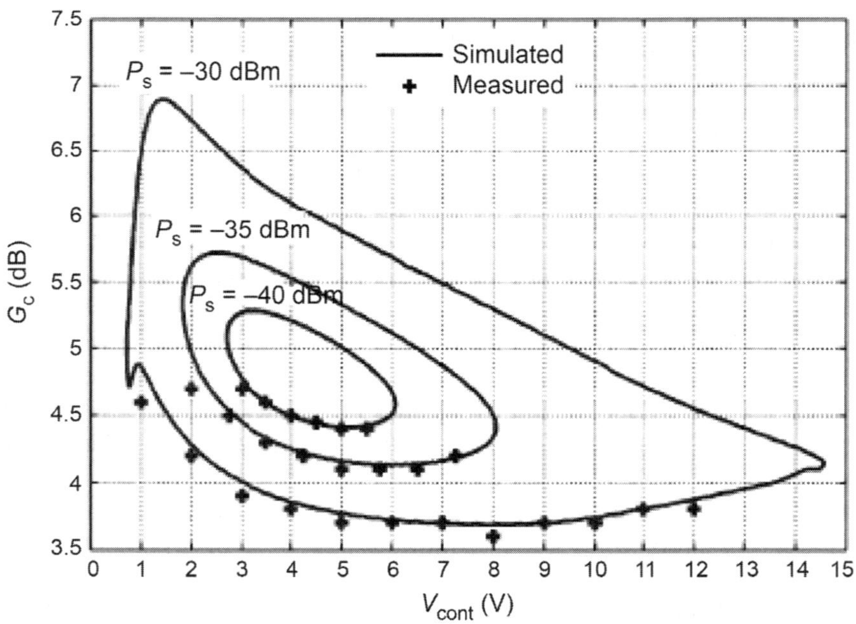

Fig. 8-26. SOM conversion gain versus the varactor control voltage for different injection power levels P_s. (Reprinted with permission from Ref. 80, ©2006 IEEE.)

Chapter 9
Beamforming in Coupled-Oscillator Arrays

In this chapter, convex optimization and other global optimization techniques are used to demonstrate the beamforming capabilities of coupled-oscillator arrays and to optimize the stability of the coupled-oscillator array steady-state solution. An introduction to convex optimization is presented followed by several optimization problems showing the beamforming capabilities of such arrays, such as pattern-nulling, difference-beam generation, and multiple-beam generation [96,118,150,151,152]. A global optimization algorithm is also presented that permits one to optimize the stability of the steady-state solution, and therefore it leads to more robust solutions and maximizes the obtained stable beam-scanning limits [153]. Finally, the operation of a coupled-oscillator array as an adaptive beamforming system is demonstrated [154].

9.1 Preliminary Concepts of Convex Optimization

Convex optimization is a class of optimization problems that has enjoyed an increased scientific interest in the recent years due to the development of very efficient algorithms essentially rendering their solution as easy as the solution of linear programs [133]. As a result, convex optimization problems have found wide application in fields such as control and signal processing and, among these, in the problem of antenna array beam-steering and beamforming. Due to this fact, in this chapter we first present a brief introduction to convex optimization and the mathematical framework required to express the beamforming problem as a convex optimization problem and additionally introduce the coupled-oscillator array solutions presented in the previous chapters as constraints to the problem at hand.

An optimization problem is expressed in the form

$$\begin{aligned} &\text{minimize } f_0(\boldsymbol{x}) \\ &\text{subject to } f_i(\boldsymbol{x}) \le b_i \quad i = 1,\dots,M \end{aligned} \tag{9.1-1}$$

where $\boldsymbol{x}$ is the optimization variable, a vector of dimension N. The real function f_0 is called the objective function of the problem, and real f_i are the M constraints of the problem with limits or bounds b_i [133]. The family of convex optimization problems consists of those optimization problems where both the objective and the constraints satisfy the property of convexity. In convex optimization problems, a local minimum is also a global minimum; and therefore, once a solution is found, it is guaranteed to be optimal. Additionally, there exist many computationally efficient algorithms for solving convex optimization problems, such as the interior point methods [155]. As a result, once an optimization problem is formulated as a convex one, its efficient resolution is guaranteed.

A set C is *convex* if for every two points $\boldsymbol{x}$ and $\boldsymbol{y}$ of dimension N that belong to the set and any real number θ such that $0 \le \theta \le 1$, the point $\boldsymbol{z} = \theta\boldsymbol{x} + (1-\theta)\boldsymbol{y}$ also belongs in C. Geometrically, this means that any point z that lies on the line segment connecting $\boldsymbol{x}$ and $\boldsymbol{y}$ must belong to C. If, instead, θ is allowed to take any real value, then the set C is called *affine*. Correspondingly, an affine set contains every point on the line that is defined by two points $\boldsymbol{x}$ and $\boldsymbol{y}$. These concepts are illustrated in Fig. 9-1.

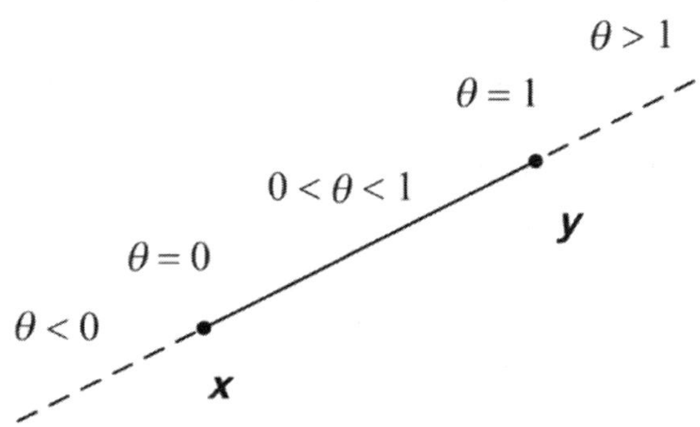

Fig. 9-1. Geometric interpretation of convex and affine sets.

Some well known convex sets are line segments and half-spaces [133]. A half-space is the solution set of a linear inequality of the form

$$\{\boldsymbol{x} \mid \boldsymbol{a}^T\boldsymbol{x} \le \boldsymbol{b}\} \tag{9.1-2}$$

where $\boldsymbol{a} \neq \boldsymbol{0}$. The hyperplane defined by the linear equality

$$\{\boldsymbol{x} | \boldsymbol{a}^T\boldsymbol{x} = \boldsymbol{b}\} \tag{9.1-3}$$

is an affine set ($\boldsymbol{a} \neq \boldsymbol{0}$) that separates the space of N-dimensional real vectors into two convex half-spaces corresponding to the inequalities $\boldsymbol{a}^T\boldsymbol{x} \le \boldsymbol{b}$ and $\boldsymbol{a}^T\boldsymbol{x} \ge \boldsymbol{b}$. A hyperplane is defined by a point $\boldsymbol{x_0}$ and a nonzero vector $\boldsymbol{a}$, and it contains all vectors $\boldsymbol{x}$ such that the difference vector $\boldsymbol{x} - \boldsymbol{x_0}$ is orthogonal to $\boldsymbol{a}$ (Fig. 9-2).

A norm ball with center $\boldsymbol{x_c}$ and radius r is a convex set defined by

$$\{\boldsymbol{x} | \|\boldsymbol{x} - \boldsymbol{x_c}\| \le r\} \tag{9.1-4}$$

where $\|\boldsymbol{x}\|$ is a properly defined norm of $\boldsymbol{x}$, such as for example the Euclidean norm. Furthermore, a norm cone is is a convex set defined as the set of $(\boldsymbol{x}, t)$ pairs such that

$$\{(\boldsymbol{x}, t) | \|\boldsymbol{x}\| \le t\} \tag{9.1-5}$$

If the Euclidean norm $\|\boldsymbol{x}\|_2 = \sqrt{\sum_{i=1}^{N} x_i^2}$ is considered, then the corresponding norm cone is called a second-order cone, or ice-cream cone [133].

A real function f_i is convex if its domain is a convex set and if for any two vectors $\boldsymbol{x}$ and $\boldsymbol{y}$ in its domain, the following inequality holds

$$f_i(a\boldsymbol{x} + b\boldsymbol{y}) \le af_i(\boldsymbol{x}) + bf_i(\boldsymbol{y}) \tag{9.1-6}$$

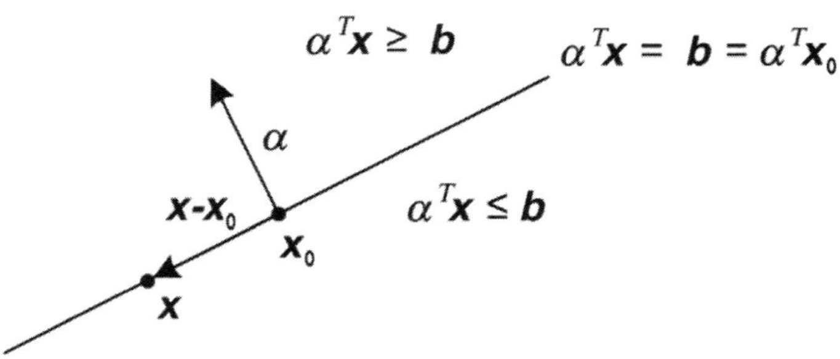

Fig. 9-2. Geometric interpretation of hyperplane and hyperspace.

where a and b are non-negative real numbers such that $a + b = 1$. One can easily verify from Eq. (9.1-6) that a linear function is convex. Another commonly used convex constraint is a linear matrix inequality [156]

$$\boldsymbol{F}(\boldsymbol{x}) = \boldsymbol{F_0} + \sum_{i=1}^{N} x_i \boldsymbol{F_i} > 0 \tag{9.1-7}$$

where $\boldsymbol{x} = [x_n]$ is a vector of dimension N and $\boldsymbol{F_i} = \boldsymbol{F_i^T}$ are real symmetric matrices of dimension M. A real square matrix $\boldsymbol{F}(\boldsymbol{x})$ is positive definite $\boldsymbol{F}(\boldsymbol{x}) > 0$ if for any nonzero vector $\boldsymbol{u}$, $\boldsymbol{u^T}\boldsymbol{F}(\boldsymbol{x})\boldsymbol{u} > 0$. Many convex constraints such as linear inequalities, convex quadratic inequalities, and Lyapunov matrix inequalities can be cast in the form of a linear matrix inequality. According to Lyapunov theory, the system of differential equations

$$\dot{\boldsymbol{x}} = \boldsymbol{A}\boldsymbol{x} \tag{9.1-8}$$

is stable if and only if exists a positive definite matrix $\boldsymbol{P} > \boldsymbol{0}$ such that

$$\boldsymbol{A^T}\boldsymbol{P} + \boldsymbol{P}\boldsymbol{A} < 0 \tag{9.1-9}$$

The above inequality is known as a Lyapunov matrix inequality. The inequality of Eq. (9.1-9) with the matrix $\boldsymbol{P}$ as unknown can be cast in the form of a linear matrix inequality [156].

The minimization of the maximum eigenvalue of a matrix $\boldsymbol{A}$ subject to a linear matrix inequality constraint $\boldsymbol{B}(\boldsymbol{x}) > 0$ is a convex problem defined as [156]

$$\begin{gathered}\text{minimize } \lambda \\ \text{subject to } \lambda \boldsymbol{I_N} - \boldsymbol{A}(\boldsymbol{x}) > \boldsymbol{0}\,, \boldsymbol{B}(\mathbf{x}) > \boldsymbol{0}\end{gathered} \tag{9.1-10}$$

with $\boldsymbol{A}$ and $\boldsymbol{B}$ symmetric matrices that depend affinely on $\boldsymbol{x}$. If one defines an extended unknown vector $\boldsymbol{y} = [\boldsymbol{x} \quad \lambda]^T$ and $\boldsymbol{c} = [\boldsymbol{0_N} \quad 1]^T$, the eigenvalue minimization problem can be written as minimization of a linear function subject to a linear matrix inequality

$$\begin{gathered}\text{minimize } \boldsymbol{c^T}\boldsymbol{y} \\ \text{subject to } \boldsymbol{F}(\boldsymbol{y}) > \boldsymbol{0}\end{gathered} \tag{9.1-11}$$

where $\boldsymbol{B}(\mathbf{x}) > \boldsymbol{0}$ together with $\boldsymbol{\lambda I_N} - \boldsymbol{A}(\boldsymbol{x}) > \boldsymbol{0}$ have been formulated as a single linear matrix inequality $\boldsymbol{F}(\boldsymbol{y}) > \boldsymbol{0}$.

Linear programming and least-squares optimization are two well-known examples of convex optimization problems. In linear programming, both the objective and the constraints are linear functions

$$\begin{gathered}\text{minimize } \boldsymbol{c^T}\boldsymbol{x} \\ \text{subject to } \boldsymbol{a_i^T}\boldsymbol{x} \le \boldsymbol{b_i},\ i = 1, \ldots, M\end{gathered} \tag{9.1-12}$$

In least squares optimization the objective function is a sum of squares which is a convex function and there are no constraints:

$$\text{minimize } \|\boldsymbol{Ax}-\boldsymbol{b}\|_2^2 = \sum_{i=1}^{M}\left(\boldsymbol{a}_i^T\boldsymbol{x}-b_i\right)^2 \qquad (9.1\text{-}13)$$

where $\boldsymbol{A} = [a_{mn}]$ is an M by N matrix, $\boldsymbol{a}_i$ is a vector of dimension N containing the elements of column i of matrix $\mathbf{A}$, and $\boldsymbol{b}$ is a vector of dimension M.

Finally, the minimization of the maximum generalized eigenvalue of a pair of symmetric matrices $\boldsymbol{A}$ and $\boldsymbol{B}$ that depend affinely on $\boldsymbol{x}$, subject to an additional linear matrix inequality constraint $\boldsymbol{C}(\boldsymbol{x}) > 0$, is a quasi-convex optimization problem [156] expressed as

$$\begin{gathered}\text{minimize } \lambda \\ \text{subject to } \lambda\boldsymbol{B}(\boldsymbol{x}) - \boldsymbol{A}(\boldsymbol{x}) > \boldsymbol{0}\,, \boldsymbol{B}(\boldsymbol{x}) > \boldsymbol{0}\,, \boldsymbol{C}(\boldsymbol{x}) > \boldsymbol{0}\end{gathered} \qquad (9.1\text{-}14)$$

A real function f is quasi-convex if and only if its domain is a convex set, and for any two vectors $\boldsymbol{x}$ and $\boldsymbol{y}$ in its domain, along with a real number θ, such that $0 \leq \theta \leq 1$, the following inequality holds [133]:

$$f(\theta\boldsymbol{x} + (1-\theta)\boldsymbol{y}) \leq \max\{f(\boldsymbol{x}), f(\boldsymbol{y})\} \qquad (9.1\text{-}15)$$

Convex functions are also quasi-convex but not vice versa. The standard formulation of a quasi-convex optimization problem has a quasi-convex objective and convex constraints. The generalized eigenvalue minimization problem given by Eq. (9.1-14) can be written in the standard format [133, 156]. Similarly to convex optimization problems, quasi-convex optimization problems can also be solved efficiently.

9.2 Beamforming in COAs

The ability to generate constant phase distributions among the coupled-oscillator array elements by tuning the frequency of only the edge array elements has been one of the most attractive properties of coupled-oscillator arrays as they can be used in beam-scanning applications eliminating the need for phase shifters or a complicated local-oscillator feed network. If, however, one is allowed to tune the frequency of more or all the array elements, then additional features maybe introduced in the radiated pattern such as placement of nulls at desired far-field angular directions.

Once a constant progressive phase shift is established among the array elements, the main beam direction is steered toward a desired direction. In Ref. 157, Steyskal showed that additional nulls maybe formed in the radiation pattern at desired angular directions by introducing small perturbations to the phases of the array elements. This method was used by Heath [96] in conjunction with the generalized phase model to demonstrate beamforming capabilities using coupled-oscillator arrays. Finally, Georgiadis et al. [118] extended Heath's work by including both amplitude and phase perturbations. In the following, a description of this beamforming methodology is provided.

The array factor of a uniform linear antenna array of N elements is given by

$$F(\theta) = \sum_{n=1}^{N} V_n e^{j(nkd \sin\theta + \phi_n)} \tag{9.2-1}$$

where the element distance is d, and the angular direction θ is measured from broadside. The main beam is steered at θ_0 when the excitation amplitudes are equal to $V_n = V_0$ and the element phases are set as $\phi_{0n} = -nkd \sin\theta_0$. The array factor is then written in compact form

$$F(\theta) = V_0 \boldsymbol{u}^{\boldsymbol{H}} \boldsymbol{1}_{\boldsymbol{N}} \tag{9.2-2}$$

where $\boldsymbol{u}(\theta) = \left[e^{-j(nkd \sin\theta + \phi_{on})}\right]$. If one introduces a perturbation in the excitation amplitudes and phases $\boldsymbol{x} = [\boldsymbol{\Delta V^T} \quad \boldsymbol{\Delta\phi^T}]^T$, the array factor is approximated to first order as

$$F(\theta) = V_0 \boldsymbol{u}^{\boldsymbol{H}} \boldsymbol{1}_{\boldsymbol{N}} + u^H [\boldsymbol{I_N} \quad jV_0 \boldsymbol{I_N}] \boldsymbol{x} \tag{9.2-3}$$

A constraint in the array factor at angle θ_1 is introduced by imposing $|F(\theta_1)| \leq f_1$, where f_1 is a desired maximum level at θ_1. Given $M < N$ level constraints, one may form a complex vector $\boldsymbol{U} = \boldsymbol{C} + \boldsymbol{jS} = [F(\theta_1)\ F(\theta_2) \dots F(\theta_M)]^T$ containing all the constraints and a second one containing $\boldsymbol{f} = [f_1\ f_2 \dots f_M]^T$ and combine them in a matrix inequality

$$\begin{bmatrix} -V_0 \boldsymbol{C} \boldsymbol{1}_{\boldsymbol{N}} - \boldsymbol{f} \\ -V_0 \boldsymbol{S} \boldsymbol{1}_{\boldsymbol{N}} \end{bmatrix} \leq \begin{bmatrix} \boldsymbol{C} & -V_0 \boldsymbol{S} \\ \boldsymbol{C} & V_0 \boldsymbol{C} \end{bmatrix} \boldsymbol{x} \leq \begin{bmatrix} -V_0 \boldsymbol{C} \boldsymbol{1}_{\boldsymbol{N}} + \boldsymbol{f} \\ -V_0 \boldsymbol{S} \boldsymbol{1}_{\boldsymbol{N}} \end{bmatrix} \tag{9.2-4}$$

which can be written in compact form:

$$\boldsymbol{f_l} \leq \boldsymbol{Fx} \leq \boldsymbol{f_h} \tag{9.2-5}$$

The beamforming problem can be formulated as a convex optimization problem as follows:

$$\begin{gathered} \min_{\boldsymbol{x}} t \\ \text{subject to } \|\boldsymbol{x}\| \leq t \\ \boldsymbol{f_l} \leq \boldsymbol{Fx} \leq \boldsymbol{f_h} \end{gathered} \tag{9.2-6}$$

where the linear objective is subject to a second-order cone constraint and a linear inequality. Minimizing the norm of $\boldsymbol{x}$ ensures that the perturbation approximation of the array factor is valid.

The problem given by Eq. (9.2-6) was analytically solved by Georgiadis et al. in Ref. 118 for the case where the inequality constraints are null constraints ($\boldsymbol{f_l} = \boldsymbol{f_h} = \mathbf{0}$). In fact, the analytical solution to this problem when considering phase perturbations only was given by [157]. In this case [118, 157],

$$t_{\min} = (\boldsymbol{C}\mathbf{1}_N)^T(\boldsymbol{S}\boldsymbol{S}^T)^{-T}(\boldsymbol{C}\mathbf{1}_N) \qquad (9.2\text{-}7)$$

and

$$\Delta\boldsymbol{\phi}_{\mathbf{min}} = \boldsymbol{S}^T(\boldsymbol{S}\boldsymbol{S}^T)^{-1}(\boldsymbol{C}\mathbf{1}_N) \qquad (9.2\text{-}8)$$

It is interesting to study Eq. (9.2-7), for the simple case of main-beam direction at $\theta_0$with one nulling constraint at angle θ_1. One then evaluates $t_{\min}$ as

$$t_{\min} = \frac{[\sum_1^N \cos(nkd\alpha)]^2}{\sum_1^N \sin^2(nkd\alpha)} \qquad (9.2\text{-}9)$$

where $a = \sin\theta_1 - \sin\theta_0$. This shows that there exist combinations of θ_0 and θ_1 such that the required perturbation magnitude $t_{\min}$ goes to infinity, for which the optimization problem does not have a solution. These solutions correspond to $mkda = q\pi$, where m and q are integers. One such solution is for $a = 0$, which corresponds to $\theta_1 = \theta_0$—in other words, when the desired null is in the direction of the main lobe. A second solution is when

$$\sin\theta_1 - \sin\theta_0 = \frac{\pi}{kd} \qquad (9.2\text{-}10)$$

which corresponds to a desired null direction θ_1 that depends on the main beam angle θ_0. The existence of such points was also verified numerically for the case of a coupled-oscillator array in Refs. 118 and 150.

In order to apply the pattern constraints to the coupled oscillator array, one needs to limit the perturbation vectors $\boldsymbol{x}$ satisfying Eq. (9.2-6) to the set that corresponds to a coupled-oscillator array steady-state solution. Reference 118 introduced the coupled-oscillator array steady-state solution in Eq. (9.2-6) as an additional linear constraint maintaining the convexity of the optimization problem. The steady-state solution (7.7-12) is first reformulated to reflect the nature of perturbation $\boldsymbol{x}$, which contains both amplitude $V_n = V_0 + \Delta V_n$ and phase perturbations $\phi_n = \phi_{on} + \Delta\phi_n = -nkd\sin\theta_0 + \Delta\phi_n$. Due to the autonomous nature of the coupled oscillator array, the steady-state solution is defined by the relative phases of the oscillator elements. In other words, the phase of one oscillator may be set to an arbitrary value, or alternatively the phases of all oscillators can be changed by an equal amount without affecting the steady state. This is verified by the steady-state expression Eq. (7.6-10) where only phase differences are present. Consequently, a perturbation of the steady state is set by considering the terms $\Delta\phi_n$ such that even though individually they may take large values, their relative differences are kept small. This argument was also used in the early works of Kurokawa [105] when modeling the externally injection-locked oscillator. It is, therefore, possible to approximate the phase exponents appearing in Eq. (7.6-10) as

$$e^{j(\phi_n-\phi_m)} \approx e^{j(\phi_{on}-\phi_{om})}[1+j(\Delta\phi_n-\Delta\phi_m)]$$

and obtain the perturbed steady state as

$$\begin{bmatrix} \boldsymbol{C_V I_N} + \boldsymbol{\Phi}^H \boldsymbol{C}_c \boldsymbol{\Phi} & \boldsymbol{C}_\phi \boldsymbol{I}_N \end{bmatrix} \boldsymbol{x} + \boldsymbol{C}_\mu \boldsymbol{\Delta\mu} + j\Delta\omega \boldsymbol{1}_N + \boldsymbol{\Phi}^H \boldsymbol{C}_c \boldsymbol{\Phi} \boldsymbol{1}_N = 0 \quad (9.2\text{-}11)$$

with

$$\boldsymbol{C}_\phi = \boldsymbol{j}[\boldsymbol{\Phi}^H \boldsymbol{C}_c \boldsymbol{\Phi} - \mathrm{dg}(\boldsymbol{\Phi}^H \boldsymbol{C}_c \boldsymbol{\Phi} \boldsymbol{1}_N)] \quad (9.2\text{-}12)$$

The final system of equations is obtained by separating real and imaginary parts

$$\begin{bmatrix} \boldsymbol{C}_V^R \boldsymbol{I}_N + (\boldsymbol{\Phi}^H \boldsymbol{C}_c \boldsymbol{\Phi} \boldsymbol{1}_N)^R & \boldsymbol{C}_\phi^R \boldsymbol{I}_N & \boldsymbol{C}_\mu^R \boldsymbol{I}_N \\ \boldsymbol{C}_V^I \boldsymbol{I}_N + (\boldsymbol{\Phi}^H \boldsymbol{C}_c \boldsymbol{\Phi} \boldsymbol{1}_N)^I & \boldsymbol{C}_\phi^I \boldsymbol{I}_N & \boldsymbol{C}_\mu^I \boldsymbol{I}_N \end{bmatrix} \begin{bmatrix} \boldsymbol{x} \\ \boldsymbol{\Delta\mu} \end{bmatrix} = -\begin{bmatrix} (\boldsymbol{\Phi}^H \boldsymbol{C}_c \boldsymbol{\Phi} \boldsymbol{1}_N)^R \\ \Delta\omega \boldsymbol{1}_N + (\boldsymbol{\Phi}^H \boldsymbol{C}_c \boldsymbol{\Phi} \boldsymbol{1}_N)^I \end{bmatrix} \quad (9.2\text{-}13)$$

which is written in compact form:

$$[\boldsymbol{G} \quad \boldsymbol{G}_{\Delta\mu}] \begin{bmatrix} \boldsymbol{x} \\ \boldsymbol{\Delta\mu} \end{bmatrix} = \boldsymbol{g} \quad (9.2\text{-}14)$$

Using the above linear constraint for the steady state, it is possible to formulate the beamforming optimization problem for coupled oscillator arrays [118] as follows:

$$\begin{aligned} &\min_{\boldsymbol{x},\Delta\mu} t \\ &\text{subject to } \|\boldsymbol{x}\| + \|\boldsymbol{\Delta\mu}\| \le t \\ &\boldsymbol{f}_l \le \boldsymbol{F}\boldsymbol{x} \le \boldsymbol{f}_h \\ &[\boldsymbol{G} \quad \boldsymbol{G}_{\Delta\mu}] \begin{bmatrix} \boldsymbol{x} \\ \boldsymbol{\Delta\mu} \end{bmatrix} = \boldsymbol{g} \end{aligned} \quad (9.2\text{-}15)$$

where the norm of the vector $\boldsymbol{\Delta\mu}$ is also minimized in order to enforce the perturbation condition pertaining to the derivation of the steady-state constraint. The above formulation was extended to planar arrays in Ref. 150.

Once the steady state is obtained, its stability is examined by considering the linear variational system corresponding to the system of differential equations describing the coupled oscillator dynamics and evaluating the eigenvalues of matrix $\boldsymbol{K}$ or $\widetilde{\boldsymbol{K}}$ in Eq. (7.7-17).

We may further explore the arbitrary phase reference of the coupled oscillator array in order to minimize the number of optimization variables in Eq. (9.2-15). $\boldsymbol{G}$ is a square matrix of dimension $2N$. It has one zero eigenvalue due to the fact that the array steady state is unaffected by applying an arbitrary but constant phase term to all elements. It is therefore possible, without loss of generality, to set the phase perturbation of an arbitrarily selected element j to zero $\Delta\phi_j = 0$ and eliminate the column of $\boldsymbol{G}$ that corresponds to $\Delta\phi_j$. Then, a new steady-state vector $\boldsymbol{y} = \left[\boldsymbol{\Delta V}^{\mathrm{T}} \quad \boldsymbol{\Delta}\widetilde{\boldsymbol{\phi}}^{\mathrm{T}} \quad \Delta\mu_j\right]^T$ of dimension $2N$ is constructed, where $\boldsymbol{\Delta}\widetilde{\boldsymbol{\phi}}$ contains all phase perturbations except $\Delta\phi_j$. Using $\boldsymbol{y}$, Eq. (9.2-14) is rearranged in the form

$$\begin{bmatrix}\widetilde{\boldsymbol{G}} & \boldsymbol{G}_{\Delta\widetilde{\mu}}\end{bmatrix}\begin{bmatrix}\boldsymbol{y} \\ \Delta\widetilde{\boldsymbol{\mu}}\end{bmatrix} = \boldsymbol{g} \Rightarrow \widetilde{\boldsymbol{G}}\boldsymbol{y} = \widetilde{\boldsymbol{g}}(\Delta\widetilde{\boldsymbol{\mu}}) \tag{9.2-16}$$

where $\Delta\mu_j$ is the control perturbation corresponding to the selected element j. $\widetilde{\boldsymbol{G}}$ is a full-rank square matrix of dimension $2N$ obtained from $\boldsymbol{G}$ by substituting its column corresponding to $\Delta\phi_j$ with the column of $\boldsymbol{G}_{\Delta\mu}$ corresponding to $\Delta\mu_j$. Similarly, $\boldsymbol{G}_{\Delta\widetilde{\mu}}$ has dimension $2N$ by $N-1$ and is obtained from $G_{\Delta\mu}$ by eliminating the column that corresponds to $\Delta\mu_j$. The matrix $\boldsymbol{g}(\Delta\widetilde{\boldsymbol{\mu}})$ is linearly dependent on $\Delta\widetilde{\boldsymbol{\mu}}$. The steady state $\boldsymbol{y}$ is therefore expressed as a function of the $N-1$ independent control variables $\Delta\widetilde{\boldsymbol{\mu}}$. Following Georgiadis et al. [153], the equality constraint of Eq. (9.2-16) is used to eliminate the $2N$ optimization variables included in $\boldsymbol{y}$ and formulate Eq. (9.2-15) in terms of only the independent $N-1$ control variables $\Delta\widetilde{\boldsymbol{\mu}}$

$$\begin{gathered}\min_{\Delta\widetilde{\mu}} t \\ \text{subject to } \left\|\widetilde{\boldsymbol{G}}^{-1}\widetilde{\boldsymbol{g}}(\Delta\widetilde{\boldsymbol{\mu}})\right\| + \|\Delta\widetilde{\boldsymbol{\mu}}\| \le t \\ \boldsymbol{f}_{\Delta\widetilde{\mu}l} \le \boldsymbol{F}_{\Delta\widetilde{\mu}}\Delta\widetilde{\boldsymbol{\mu}} \le \boldsymbol{f}_{\Delta\widetilde{\mu}h}\end{gathered} \tag{9.2-17}$$

where the equality constraint is now eliminated and the inequality constraints on the array factor have been appropriately reformulated in terms of the independent control variables. As an example, let us consider the five-element coupled oscillator array of Section 8.4, assuming that each oscillator output is connected to an antenna and that the antenna elements are placed a half free-space wavelength apart ($kd = \pi$). The free-running oscillator steady state corresponds to an amplitude of $V_0 = 0.442$ V (the output power is 2.9 dBm) and frequency $f_0 = 9.892$ GHz obtained for a control voltage of $\mu_0 = 10$ V.

The coupling network consists of a transmission line section of 360-deg electrical length at f_0 and two series resistors $R = 150\ \Omega$ (Fig. 8-9). The optimization problem given by Eq. (9.2-17) was solved for the case of main beam direction at $\theta_0 = 0$ deg (broadside) and an additional null constraint at $\theta_1 = -60$ deg. The outcome of the optimization procedure is shown in Table 9-1. The phase perturbation of the middle array element 3 was arbitrarily set to zero. The steady-state vector y consisted of the five oscillator amplitude perturbations; the four phase perturbations of oscillators 1,2,4, and 5; and the control voltage of the middle oscillator 3. Correspondingly, the optimization variables were the control voltages of elements 1, 2, 4, and 5.

The resulting array factor is shown in Fig. 9-3. In addition to the result of the optimization problem given by Eq. (9.2-17), the array factor corresponding to the solution of problem given by Eq. (9.2-6) (which does not contain the array steady-state constraint) was also included for comparison, as well as the array factor corresponding to uniform excitation without a null constraint. It can be verified that the null is successfully imposed in the array factor at the expense of higher side-lobe levels and a small shift in the main lobe direction. For this particular case, the solutions of Eqs. (9.2-17) and (9.2-6) overlap, which indicates that there exists a steady-state solution for the coupled-oscillator array that satisfies the pattern constraints given by Eq. (9.2-6).

The optimization problem given by Eq. (9.2-17) was then solved for different values of the coupling resistor R, and the solution stability was examined by calculating the eigenvalues of the linear variational system of differential equations corresponding to the array steady state. The critical eigenvalue having the largest real part (spectral abscissa) is shown in Fig. 9-4 for different values of R. It is seen that, as coupling becomes weaker, the solution eventually becomes unstable. The change of stability occurs for a coupling resistor value of 178 Ω.

Table 9-1. Pattern nulling optimization of Eq. (9.2-17) applied in a five-element linear coupled-oscillator array[a].

Element	Amplitude, ΔV (Volt)	Phase, $\Delta\phi$ (°)	Control, $\Delta\mu$ (Volt)
1	0.0026	–16.257	–0.093
2	–0.0031	5.292	0.119
3	–0.0001	0	0.004
4	0.0033	–6.178	–0.127
5	–0.0027	16.331	0.097

[a] The main beam direction is $\theta_0 = 0$ deg (broadside). A null in the array factor is imposed at $\theta_1 = -60$ deg.

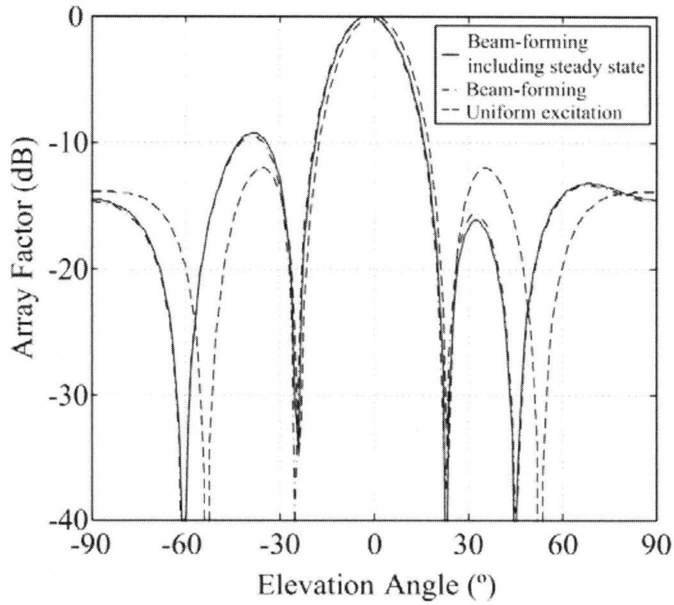

Fig. 9-3. Beamforming capabilities of coupled oscillator arrays. Array factor of five-element array of Section 8.4 for main beam at broadside and one null constraint at $\theta_1 = -60$ deg. The coupling resistor is $R = 150\ \Omega$.

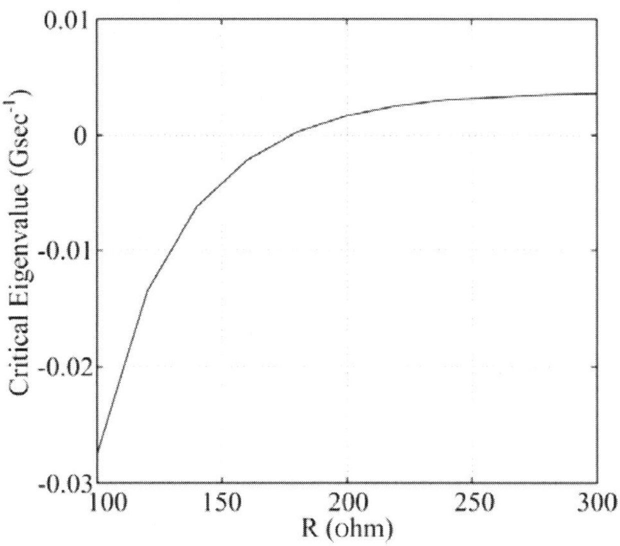

Fig. 9-4. Critical eigenvalue of the solution of the optimization problem of Eq. (9.2-17) versus the coupling resistor R.

9.3 Stability Optimization of the Coupled-Oscillator Steady-State Solution

The stability of the coupled oscillator steady-state solution is verified by examining the linear variational equations corresponding to the system of nonlinear differential equations describing its dynamics. In Section 7.7, a procedure was described to remove the zero eigenvalue that appears due to the free-running nature of the oscillator array. The resulting square matrix $\widetilde{\boldsymbol{K}}$ of dimension $2N - 1$ was derived in Section 7.7, where N is the size of the array. The linear variational equation for $\boldsymbol{\delta}\widetilde{\boldsymbol{x}} = [\boldsymbol{\delta V}^{\boldsymbol{T}} \quad \boldsymbol{\delta}\widetilde{\boldsymbol{\phi}}^{\boldsymbol{T}}]^{\boldsymbol{T}}$ is repeated here for convenience, where $\boldsymbol{\delta}\widetilde{\boldsymbol{\phi}}$ contains $N - 1$ phase differences with respect to an arbitrarily selected oscillator j as a reference

$$\boldsymbol{\delta}\dot{\widetilde{\boldsymbol{x}}} = \widetilde{\boldsymbol{K}}\boldsymbol{\delta}\widetilde{\boldsymbol{x}} \tag{9.3-1}$$

By definition, a steady-state solution is stable if the spectral abscissa of $\widetilde{\boldsymbol{K}}$ is negative. The decay rate of $\widetilde{\boldsymbol{K}}$ is the negative of the spectral abscissa [156], and a steady state is stable if $\widetilde{\boldsymbol{K}}$ has a positive decay rate. Maximizing the decay rate corresponds to a more robust steady-state solution, less likely to lose its stability due to the presence of noise or other perturbations. The matrix $\widetilde{\boldsymbol{K}}(\Delta\widetilde{\boldsymbol{\mu}})$ depends on the steady state defined by the perturbation vector $\boldsymbol{x} = [\boldsymbol{\Delta V}^{\boldsymbol{T}} \quad \boldsymbol{\Delta\phi}^{\boldsymbol{T}}]^{\boldsymbol{T}}$ of the array, which, following Eq. (9.2-16), is determined by the (perturbation) vector of $N - 1$ control voltages $\Delta\widetilde{\boldsymbol{\mu}}$. $\widetilde{\boldsymbol{K}}$ does not depend linearly on $\Delta\widetilde{\boldsymbol{\mu}}$ due to the matrix inversion involved in its derivation and due to the fact that the phase terms appear in exponential terms. This can be easily verified following the formulation that leads to the definition of $\widetilde{\boldsymbol{K}}$ in Section 7.7. However, due to the fundamental assumption that $\boldsymbol{x}$ and $\Delta\widetilde{\boldsymbol{\mu}}$ are small, we may consider the first-order expansion $\widetilde{\boldsymbol{K}}_L(\Delta\widetilde{\boldsymbol{\mu}})$ of $\widetilde{\boldsymbol{K}}(\Delta\widetilde{\boldsymbol{\mu}})$. The derivation of $\widetilde{\boldsymbol{K}}_L(\Delta\widetilde{\boldsymbol{\mu}})$ is straightforward.

A lower bound on the decay rate can be obtained using Lyapunov theory as the maximum q that solves [156]

$$hV_K(\boldsymbol{\delta x}) < -2qV_K(\boldsymbol{\delta x}) \tag{9.3-2}$$

for any δx, where $V_K(\boldsymbol{x})$ is a scalar quadratic potential function defined by a 0real symmetrix matrix P with dimension $2N - 1$ such that

$$V_K(\boldsymbol{\delta x}) = \boldsymbol{\delta x}^{\boldsymbol{T}}\boldsymbol{P}\boldsymbol{\delta x} \tag{9.3-3}$$

Using Eqs.(9.3-1) and (9.3-3) in Eq. (9.3-2), one obtains a matrix inequality

$$\boldsymbol{P}\widetilde{\boldsymbol{K}}_L + \widetilde{\boldsymbol{K}}_L^T\boldsymbol{P} + 2q\boldsymbol{P} < 0 \tag{9.3-4}$$

For a given steady state $\Delta\tilde{\mu}$, finding the symmetric positive definite matrix $\boldsymbol{P}$ that maximizes the decay rate q is a generalized eigenvalue optimization problem. Conversely, given a specific matrix $\boldsymbol{P}$ finding the steady state $\Delta\widetilde{\boldsymbol{\mu}}$ that maximizes q is an eigenvalue optimization problem. As noted in Section 9.1, both such problems can be efficiently solved using convex optimization algorithms. However, finding the steady state $\Delta\widetilde{\boldsymbol{\mu}}$ and matrix $\boldsymbol{P}$ that maximize q is not a convex optimization problem due to the multiplicative terms that appear between the elements of $\Delta\widetilde{\boldsymbol{\mu}}$ and $\boldsymbol{P}$.

It is possible to introduce the decay-rate optimization constraint in the coupled-oscillator array beamforming optimization algorithm following the approach by Georgiadis and Slavakis [153], which is given below. The optimization problem including the stability constraint is written as follows:

$$L(t, q) = \min_{\Delta\widetilde{\boldsymbol{\mu}},\boldsymbol{P}}(t - q)$$

$$\text{subject to (i)} \ \left\|\widetilde{\boldsymbol{G}}^{-1}\widetilde{\boldsymbol{g}}(\Delta\widetilde{\boldsymbol{\mu}})\right\| + \|\Delta\widetilde{\boldsymbol{\mu}}\| \leq t$$

$$\text{(ii)} \ \boldsymbol{F}_{\Delta\widetilde{\mu}}\Delta\widetilde{\boldsymbol{\mu}} \leq \boldsymbol{f}_{\Delta\widetilde{\mu}\boldsymbol{h}} \qquad (9.3\text{-}5)$$

$$\text{(iii)} \ \boldsymbol{P}\widetilde{\boldsymbol{K}}_L(\Delta\widetilde{\boldsymbol{\mu}}) + \widetilde{\boldsymbol{K}}_L^T(\Delta\widetilde{\boldsymbol{\mu}})\boldsymbol{P} + 2q\boldsymbol{P} < \boldsymbol{0}$$

$$\text{(iv)} \ \boldsymbol{P} > 0$$

This is not a convex optimization problem, and its resolution is not straightforward. In Ref. 153, an algorithm was proposed to obtain a solution to the above problem by alternative minimization of two subproblems, an eigenvalue and a generalized eigenvalue problem. The algorithm proceeds as follows

Step 1: Let $e > 0$ be the algorithm termination tolerance and $k = 0$ be the iteration number. Find $\Delta\widetilde{\boldsymbol{\mu}}_{\boldsymbol{k}} = \Delta\widetilde{\boldsymbol{\mu}}_{\boldsymbol{0}}$, the vector of control variables that minimizes the perturbation vector norm $t = t_0$, subject to (i) and (ii). This is the original convex optimization problem of Eqs. (9.2-17) and (9.2-15) that does not include a stability constraint. Obtain the decay rate q_0 corresponding to $\widetilde{\boldsymbol{K}}_L(\Delta\widetilde{\boldsymbol{\mu}}_{\boldsymbol{0}})$ by evaluating its eigenvalues.

Step 2: Repeat {

P1: Find the real symmetric square matrix $\boldsymbol{P}_{\boldsymbol{k}+\boldsymbol{1}}$ that minimizes $L(0, q) = -q$, subject to (iii) and (iv) for a given $\Delta\widetilde{\boldsymbol{\mu}}_{\boldsymbol{k}}$. This is a generalized eigenvalue optimization problem. The optimization objective provides a value of the decay rate $q_{1,k+1}$.

P2: Find the control vector $\Delta\widetilde{\boldsymbol{\mu}}_{\boldsymbol{k}+\boldsymbol{1}}$ that minimizes $L(t, q) = t - q$ using as input the matrix $\boldsymbol{P}_{\boldsymbol{k}+\boldsymbol{1}}$ obtained from the previous step. This is an eigenvalue optimization problem. Additional outputs of this step are the norm of the perturbation vector t_{k+1} and the decay rate $q_{2,k+1}$.

P3: $k = k + 1$

$$\} \text{ until } \left(q_{2,k+1} - q_{1,k+1}\right) < e$$

The algorithm is demonstrated for the case of the five-element coupled-oscillator array considered in Section 9.2, where the main beam was directed broadside ($\theta_0 = 0$ deg) and a null in the array factor was placed at $\theta_1 = -60$ deg. For a coupling resistor $R = 150\ \Omega$ the beamforming optimization problem in Eq. (9.2-17) obtained the stable solution indicated in Table 9-1. The decay rate of this solution was 3.83 million inverse seconds (Ms^{-1}), as shown in Fig. 9-4. Using this solution as a starting point the above algorithm was run in order to find a solution of Eq. (9.3-5) with an optimum decay rate. The algorithm converged after 11 iterations using a termination tolerance of 10^{-5}. The result from P1 and P2 for the various steps of the algorithm is plotted in Fig. 9-5. The final solution of the algorithm had a decay rate of 13.2 Ms^{-1} (which is more than three times the initial value).

The perturbation vector of this solution is indicated in Table 9-2, where one can verify that it is only slightly increased from the original solution of Table 9-1 ($t = 0.484$ compared to the original perturbation of $t = 0.480$). Finally, the radiation pattern of the final solution is almost identical to the radiation pattern of the original starting point solution, shown in Fig. 9-3.

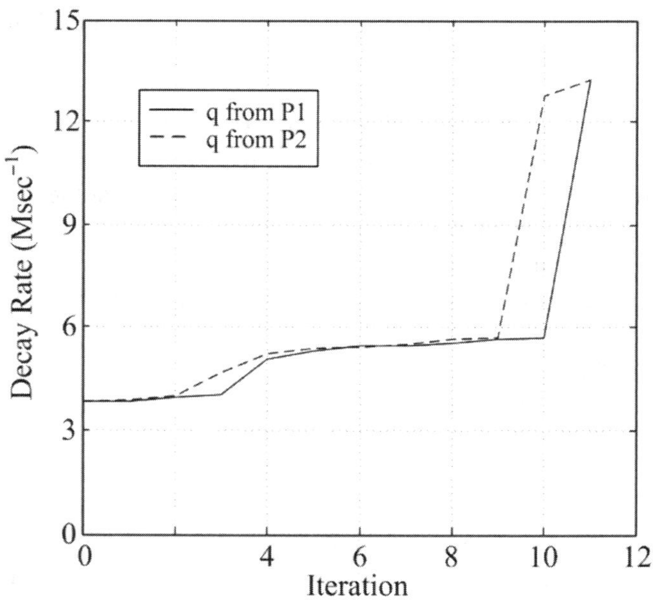

Fig. 9-5. Decay rate of the optimization problem Eq. (9.3-5) versus the iteration number.

Table 9-2. Pattern nulling and stability optimization in Eq. (9.3-5) applied in a five-element linear coupled-oscillator array[a].

Element	Amplitude ΔV (Volt)	Phase $\Delta\phi$ (deg)	Control $\Delta\mu$ (V)
1	0.0025	–17.870	–0.089
2	–0.0027	2.868	0.105
3	–0.0002	0	0.007
4	0.0030	–4.354	–0.114
5	–0.0026	17.033	0.092

[a] The main beam direction is $\theta_0 = 0$ deg (broadside). A null in the array factor is imposed at $\theta_1 = -60$ deg.

9.4 Multi-Beam Pattern Generation Using Coupled-Oscillator Arrays

The synthesis of antenna radiation patterns was formulated as a convex optimization problem by Lebret and Boyd [158]. Considering a uniform linear array for simplicity, its array factor is given by

$$F(\theta) = \sum_{n=1}^{N} v_n e^{jnkd \sin\theta} = \boldsymbol{S}(\theta)^H \boldsymbol{v} \tag{9.4-1}$$

where the vector $\boldsymbol{v} = [v_n] = \left[V_n e^{j\phi_n}\right]$ contains the complex excitations of each element, the element distance is d, and θ is measured from broadside. This formulation is slightly different from the one used in the previous section in order to emphasize the fact that the array factor is a linear function of the complex element excitations.

The pattern-synthesis convex optimization problem is written as [151,158]

$$\begin{aligned} &\min_{v} t \\ &\text{subject to } |F(\theta_i)| < t, \ \forall\, i \in 1, \dots, M \\ &|F(\theta_k)| < U_k, \ \forall\, k \in 1, \dots, P \\ &\left|F(\theta_q)\right| = 1, \ \forall\, q \in 1, \dots, L \end{aligned} \tag{9.4-2}$$

The above formulation contains L equality constraints corresponding to L array factor maxima at angular directions θ_q. Moreover, there exist P maximum level U_k constraints and M array factor minimization constraints. As a result, it is possible to efficiently obtain the complex excitations required to synthesize arbitrary patterns, such as ones having multiple beams and other beam-shaping requirements. Furthermore, the number of the optimization variables may be minimized by using the equality constraints to solve for and eliminate L-dependent variables [158].

Due to the fact that an arbitrary pattern synthesis problem cannot be considered as a perturbation of some initial reference pattern, such as for example the one corresponding to uniform in-phase excitation, the linear constraint for the coupled-oscillator steady state given by Eq. (9.2-14) or (9.2-16) cannot only be used, as the desired steady state may require a large perturbation vector $\boldsymbol{x}$ or $\boldsymbol{y}$, especially in terms of the oscillator phase differences. As a result, the general pattern synthesis problem applied to coupled-oscillator arrays may be approached in three steps. First one obtains the required complex excitations by solving the convex optimization problem Eq. (9.4-2). Second, once the desired amplitude and phase values are found, one uses the steady-state equations corresponding to the coupled-oscillator models in Eq. (7.7-4) or (7.7-12) (which do not assume a linear perturbation for the phase terms) in order to find the coupled-oscillator steady state that closely matches the required amplitude and phase distribution. For example, when using Eq. (7.7-12), one may substitute the phase values obtained by solving Eq. (9.4-2) in the previous step and solve Eq. (7.7-12) for the amplitude and control variables. Alternatively, a nonlinear simulator (such as harmonic-balance optimization) can be used, where the phase values are imposed and fixed, and the amplitude values obtained from convex optimization are used only to initialize the oscillator amplitudes in the simulation and are allowed to be optimized together with the control parameters in order to obtain the steady state. Third, once a coupled-oscillator steady state has been selected, the stability of the solution must be verified, for example by calculating the eigenvalues corresponding to the linear variational system of the array dynamics around the steady state. In fact, as will be seen in the following examples, in this step the designer synthesizes the coupling network in order to guarantee the stability of the steady-state solution.

Difference pattern generation using coupled-oscillator arrays was demonstrated by Heath in Ref. 31. Heath considered a linear coupled oscillator array and, using the generalized phase model to describe its dynamics, extended the application of the beam-steering model initially introduced by York [111] to difference pattern generation and steering. He showed that a stable difference pattern may be generated by a simple modification in the coupling network, that is, by introducing a 180-deg phase shift in the coupling between the central elements of the array, while maintaining a 0-deg phase shift between all remaining elements. In order to steer the difference beam pattern, the following phase distribution should be applied to the array elements [31].

$$\phi_k = \phi_0 + (k-1)\Delta\phi + h_k \tag{9.4-3}$$

where ϕ_0 is an arbitrary phase reference common to all elements, and $\Delta\phi$ is the necessary phase shift that must be imposed to steer the main (difference) beam of the array at an angle $\theta_0 = kd \sin \Delta\phi$. The additional phase h_k should be applied only to half of the array elements

$$h_k = \begin{cases} \pi, & k > N/2 \\ 0, & k < N/2 \end{cases} \qquad (9.4\text{-}4)$$

Assuming a linear coupled-oscillator array with adjacent element coupling, this is easily achieved by using an inter-element coupling network with coupling phase 0 deg between all elements except for the center elements where the coupling network phase is 180 deg.

More importantly, it was further shown by Heath [31] that the difference pattern can be scanned by simply detuning the free-running frequencies of the edge array elements, in the same manner that the sum pattern is scanned (Fig. 9-6).

In Ref. 151, Georgiadis and Collado applied the pattern synthesis optimization algorithm described in this section in a seven-element linear array, in order to synthesize a dual-beam pattern with beam directions at 15 deg and

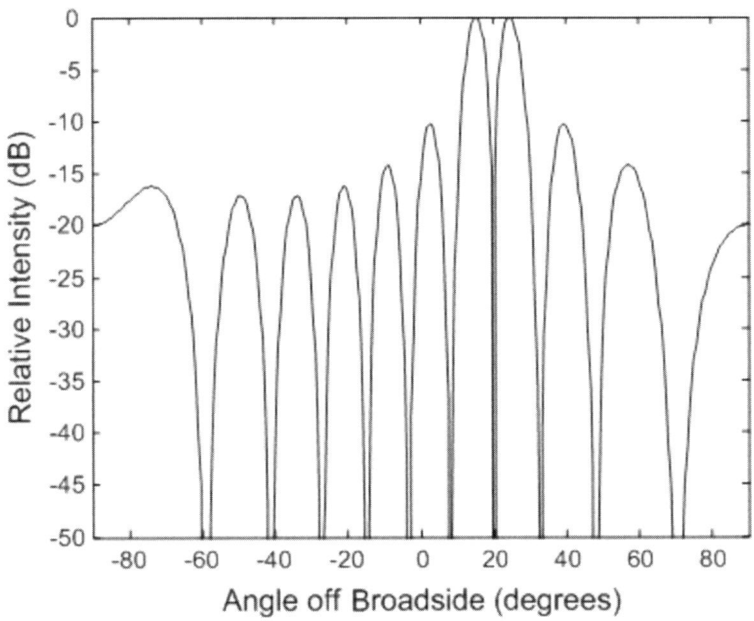

Fig. 9-6. Difference-pattern demonstration using a $N = 20$ element coupled-oscillator array. The beam is steered at $\theta_0 = 20$ deg according to Eq. (9.4-3). (Reprinted with permission from Ref. 31, ©2001 IEEE.)

–25 deg from broadside while imposing a maximum level constraint of –20 decibels relative to the carrier (dBc) between the two beams. The array factor is minimized in the remaining angular directions, imposing the first constraint given by Eq. (9.4-2). The resulting necessary complex element excitations are shown in Fig. 9-7.

The desired phase differences are imposed on the coupled-oscillator array by allowing the free-running frequencies of all oscillators to be tuned. Amplitude control, however, is imposed externally to the oscillator elements by employing variable attenuators (or variable gain amplifiers) at the oscillator outputs. The required excitations are introduced in a harmonic-balance simulator as follows. An ideal probe is connected to the output of each oscillator element, and the phase information is imposed in the probes. Harmonic balance optimization is then used to find the steady-state oscillator amplitudes and control voltages that correspond to the imposed phase distribution. Once the steady state is found, the oscillator output amplitudes are adjusted using attenuators so that the desired amplitude distribution is obtained.

It should be noted that after examination of the required excitation phases obtained from the optimization algorithm (Fig. 9-7), a coupling network was designed such that the coupling phase between elements 2 and 3, and 5 and 6, is 180 deg while the coupling phase of the remaining elements is 0 deg. The rationale behind this choice was that when the coupling network phase is 360 deg, a stable solution with phase difference in the range [–90 deg, +90 deg]

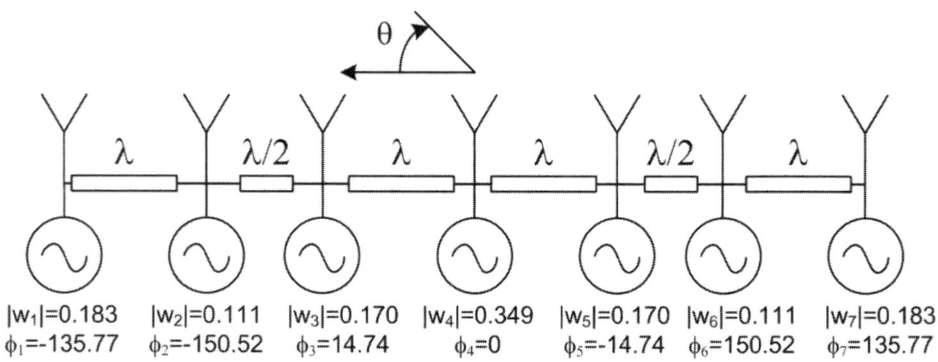

Fig. 9-7. Multi-beam pattern generation using a seven-element coupled oscillator array. Element excitations required to synthesize two main beams at 15 and –25 deg from broadside, with a maximum level constraint of –20 dBc between the two beams. The required coupling network phase shift to ensure a stable solution is indicated. Taken from Ref. 151; copyright EurAAP 2009; used with permission.

can be obtained, whereas if the coupling phase is 180 deg, the stable solutions can have a phase difference in the range of [+90 deg, +270 deg]. As the required phase difference among elements 2 and 3, and 5 and 6, is more than 90 deg (see Fig. 9-7), a coupling network with phase of 180 deg was selected in order to ensure the existence of a stable solution.

The array factor corresponding to the excitations resulting from the solution of the optimization problem given by Eq. (9.4-2) is shown in Fig. 9-8, where it is marked as the ideal pattern. The coupled-oscillator array amplitudes found after the application of the above solution in a harmonic-balance simulator, as described in the previous paragraphs, were used to compute the coupled-oscillator array radiation pattern. The resulting pattern shows an excellent agreement with the ideal pattern. Finally, the array factor corresponding to uniform amplitude excitation and application of only the phase excitation values from Eq. (9.4-2) is superimposed in Fig. 9-8, showing that by imposing the phase condition only it is possible to successfully obtain the two desired main lobes, but it is not sufficient to maintain the sidelobe levels at a sufficiently low value.

Furthermore, it was verified that the two beam patterns can be scanned while maintaining their angular distance of 40 deg by detuning only the free-running frequencies of the end elements. The result of the harmonic-balance simulation is shown in Fig. 9-9. This last example may be viewed as a generalization of the difference pattern synthesis work of Heath [31], in the sense that once a desired phase and amplitude distribution among the array elements is obtained, thus synthesizing a desired array factor, a progressive constant-phase shift distribution may be superimposed by detuning only the end array elements, thereby permitting one to scan the synthesized pattern accordingly.

9.5 Control of the Amplitude Dynamics

Oscillator amplitude control provides an additional degree of freedom in order to synthesize more complex radiation patterns with improved performance capabilities, such as reduced sidelobes. The possibility of controlling the oscillator free-running amplitudes in order to synthesize a desired pattern was investigated by Heath [159]. Furthermore, in the works of Georgiadis et al. [118,150,153] the oscillator amplitude dynamics are included in the beamforming problem formulation. Recently, control of the amplitude dynamics of the coupled oscillator array was also addressed by Jiang and Penno [160], where the generation of triangular amplitude distributions in linear coupled oscillator arrays was demonstrated.

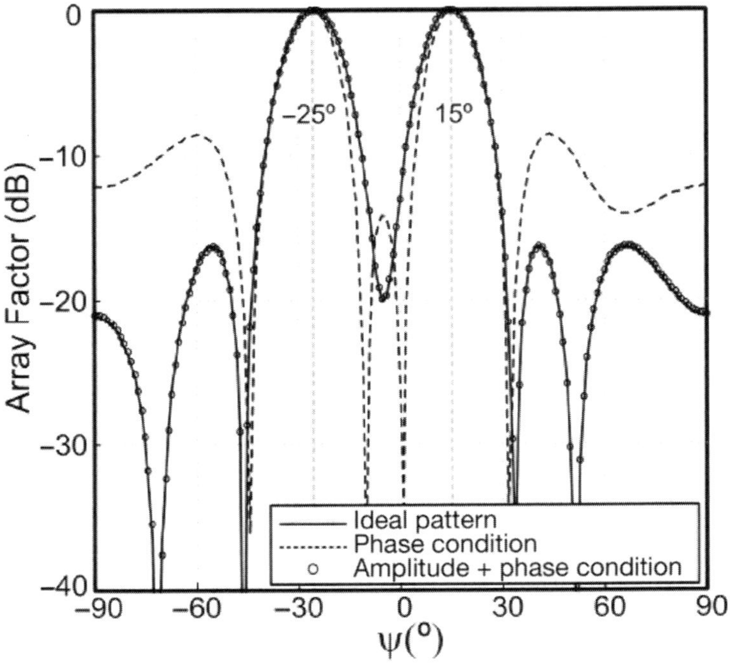

Fig. 9-8. Dual-beam pattern generation using an $N = 7$ element coupled-oscillator array. The solution of the optimization problem (ideal pattern) is compared with the final solution for the array using harmonic-balance optimization (amplitude and phase condition) and with a pattern obtained imposing the phase excitation and uniform magnitude excitation (phase condition). Taken from Ref. 151; copyright EurAAP 2009; used with permission.

Using a complex notation, the oscillator dynamics are described using either of the two models presented in Sections 7.4 and 7.6. The formulation of Heath [159] using the model of Section 7.4, is presented here

$$\dot{a}_m = j\Delta\omega_m a_m + \mu(A_{0m}^2 - |a_m|^2)a_m + \sum_{i=1}^{N} \kappa_{mi} a_i \tag{9.5-1}$$

with $a_m = A_m e^{j\phi_m}$. The periodic steady-state solution is obtained by setting $\dot{A}_m = 0$ and $\dot{\phi}_m = c$ with c an arbitrary constant, resulting in

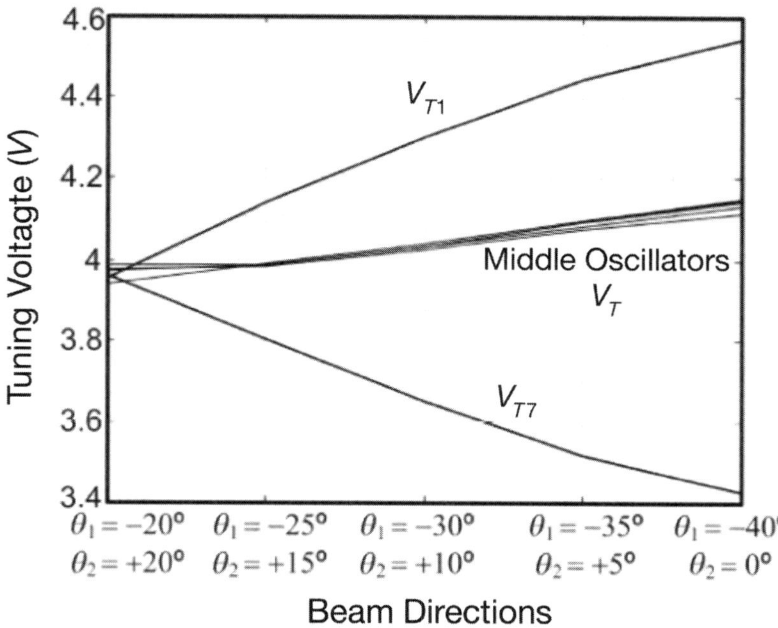

Fig. 9-9. Dual-beam pattern steering using an $N = 7$ element coupled-oscillator array. Oscillator control voltages for different scanning angles. Taken from Ref. 151; copyright EurAAP 2009, used with permission.

$$\rho_m = p_m + j\Delta\omega_m = jc + \mu|a_m|^2 - \sum_{i=1}^{N} \kappa_{mi} \frac{a_i}{a_m} \tag{9.5-2}$$

with $p_m = \mu A_{0m}^2$. Parameters p_m and $\Delta\omega_m$ allow one to independently tune the free-running frequency and free-running amplitude of the oscillator elements in order to synthesize a desired pattern. In Ref. 159, near-neighbor coupling was considered simplifying the coupling network matrix κ. Once a desired amplitude and phase distribution a_m is selected, one may separate the above equation into real and imaginary parts and solve for the tuning parameters, p_m and $\Delta\omega_m$. Finally, the stability of the solution must be examined through the eigenvalues of the linear variational equation given by Eq. (9.5-1), as was described in Section 7.4.

9.6 Adaptive Coupled-Oscillator Array Beamformer

In addition to the beamforming capabilities of coupled-oscillator arrays, an adaptive receive beamformer based on a coupled oscillator array was demonstrated by Ikuma et al. [154]. The steady-state expression of the coupled-oscillator array provides a means for controlling the array-element amplitudes and phases by adjusting the free-running oscillator frequencies and the coupling network. Similarly to the previous paragraph, a complex notation for the array dynamics pertaining to either of the two models of Sections 7.4 and 7.6 may be utilized.

The formulation of Section 7.4, also shown in Eq. (9.5-1), was followed in Ref. 154. The periodic steady-state solution is obtained by setting $\dot{A}_m = 0$ and $\dot{\phi}_m = c$ with c an arbitrary constant, resulting in

$$\rho_m a_m + \sum_{i=1}^{N} \kappa_{mi} a_i = b_m \tag{9.6-1}$$

where $m = 1,2, \dots, N$:

$$\begin{aligned} \rho_m &= \mu A_{om}^2 + j\Delta\omega_m \\ b_m &= (\mu |a_m|^2 + jc) a_m \end{aligned} \tag{9.6-2}$$

Finally, in matrix form one has

$$(\boldsymbol{\rho} + \boldsymbol{\kappa})\boldsymbol{a} = \boldsymbol{b}(\boldsymbol{a}) \tag{9.6-3}$$

where $\boldsymbol{\rho}$ is a diagonal matrix with ρ_m in its main diagonal and $\boldsymbol{b}$ is a vector with b_m in its main diagonal. Matrix $\boldsymbol{\rho}$ contains the oscillator parameters, the free-running amplitudes, and the free-running frequency offsets from ω_0. The frequency offsets can be adjusted, whereas the free-running amplitudes are fixed and assumed equal for all oscillators. Amplitude control may also be achieved using, for example, a variable attenuator or variable-gain amplifier at each oscillator output. The matrix κ contains the coupling-network gain and phase, and it may also be tunable. In Ref. 154, nearest-neighbor coupling is assumed, which results in a bi-diagonal matrix $\boldsymbol{\kappa}$.

There are many possible combinations of $\boldsymbol{\rho}$ and $\boldsymbol{\kappa}$ that can lead to a desired complex amplitude vector $\boldsymbol{a}$. Ikuma et al. [154] considered a reconfigurable coupling network $\boldsymbol{\kappa}$ and identical oscillators without frequency tuning, leading to a fixed $\boldsymbol{\rho}$ matrix. As a result, the coupling matrix $\boldsymbol{\kappa}$ is used to generate the desired amplitude distributions a.

The proposed adaptive receiver of Ikuma et al. [154] is shown in Fig. 9-10. Assuming a receiving uniform linear-antenna array of N elements, the received signal vector from all antennas is $\boldsymbol{r}(t)$.

The received signal is split into two signal paths. The signal in the first path is mixed with a reference oscillator $z_r(t)$; and after passing through a low-pass filter to remove unwanted mixing products, it provides the reference vector

$\boldsymbol{r_a}(t)$. In the second path, the received signal vector is mixed with the coupled-oscillator array vector $\boldsymbol{z}(t)$; and after low-pass filtering, it provides the demodulated scalar output signal $y_a(t)$ of the beamformer. The fixed oscillator is phase locked to the middle element of the coupled-oscillator array. It should be noted that in the block diagram of Fig. 9-10, the analytic representation [149] of the various signals is indicated. As an example, the analytic signal of the reference oscillator is

$$z_r(t) = w_r e^{j\omega_0 t} \tag{9.6-4}$$

with ω_o the reference oscillator frequency and w_r its complex amplitude.

The coupled-oscillator array complex amplitudes are adaptively controlled based on a least-mean-square (LMS) algorithm given by

$$\dot{\boldsymbol{a}} = -\mu \boldsymbol{M} \boldsymbol{r_a}(t) {y_a}^*(t) \tag{9.6-5}$$

in order to minimize the effect of unwanted interfering signals present in the received array signal $\boldsymbol{r}(t)$. The operator ()* denotes the complex conjugate. Matrix $\boldsymbol{M}$ depends on the desired fixed constraints of the beamformer—in other words, on a set of specified array-factor levels at a number of angular directions

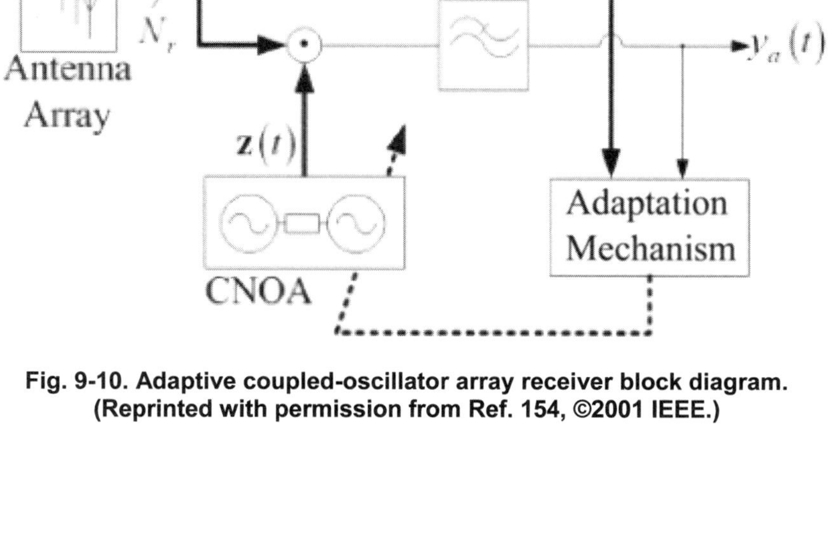

Fig. 9-10. Adaptive coupled-oscillator array receiver block diagram. (Reprinted with permission from Ref. 154, ©2001 IEEE.)

including the direction of arrival of the main beam [154]. Finally, the parameter μ controls the convergence speed of the beamformer. The proposed beamformer operation was verified by computer simulation.

9.7 Conclusion

In this chapter we introduced several optimization problems, demonstrating the beamforming capabilities of coupled oscillator arrays. The beamforming problem has been formulated as a convex optimization problem, which includes the array steady state as a linear constraint. The results of Chapter 7 have been used to provide an expression for the steady state of the coupled-oscillator array. Additionally, the capability of generating and scanning multiple beams has been verified. Furthermore, a nonconvex optimization algorithm, which optimizes the stability of the steady-state solution, has been introduced, and an adaptive beamformer based on coupled-oscillator arrays has been demonstrated. The combination of optimization and signal-processing techniques (together with the rich dynamical properties of coupled-oscillator arrays) reveal the potential and numerous applications of such arrays, which have yet to be explored.

Chapter 10
Overall Conclusions and Possible Future Directions

Active integrated antennas have found numerous applications as phased arrays, retro-directive arrays, and spatial power combiners. Coupled-oscillator antenna arrays represent a very exciting subset of active integrated antennas both from an application point of view and from a research and analysis point of view, due to some very attractive properties, such as their ability to produce arbitrary phase shift distributions, as well as their capabilities of frequency conversion and frequency generation. In addition, they inherit the practical advantages of active integrated antennas, which are compact low-profile circuit implementations that are compatible with low-cost fabrication technologies (such as microstrip and coplanar waveguide), using single and multilayer printed circuit boards.

As we have seen, however, the design of coupled-oscillator antenna arrays is far from trivial due to their highly nonlinear nature, which results in a dynamical behavior that is difficult to simulate and predict accurately and, in effect, increases the difficulty of designing coupled-oscillator arrays demonstrating a robust performance. Nonetheless, the progress of nonlinear circuit simulation and optimization techniques and the increase in computational power of low-cost personal computers has made possible the accurate analysis of coupled-oscillator arrays with as many as a few tens of elements via combining sophisticated nonlinear models for the active devices and electromagnetic analysis for the antenna, transmission lines, and interconnects, using the various methods described in Chapter 8. Efficient analysis of large coupled-oscillator arrays requires the use of approximate perturbation models, infinite-array approximations, and the continuum model that enable understanding of the exhibited behavior at a level permitting design of functional systems. The intuitive understanding of the array behavior and the gain in computational efficiency resulting from application of such methods makes them indispensable tools, complementary to the fully nonlinear simulators. The description and use of such methods has been the focus of Chapter 1 through Chapter 7.

Interest in low-cost, high-performance radio-frequency systems with reconfigurable properties in terms of transmitted beam direction or polarization makes coupled oscillator arrays a strong candidate for many applications including radar, phased arrays, and imaging in the microwave and millimeter wave frequencies. There are still numerous challenges to be addressed and many areas where improvements in coupled-oscillator array technology are desirable. Among these we specifically note the application of new implementation technologies in the design of coupled-oscillator arrays, such as substrate integrated waveguide (SIW) technology and the creation of conformal coupled-oscillator arrays using flexible substrate materials such as paper and liquid crystal polymers (LCP). Preliminary results concerning coupled-oscillator arrays using SIW technology were discussed in Chapter 6, demonstrating the possibility of low-cost single substrate array implementation. On the other hand, fully integrated coupled-oscillator arrays in the millimeter-wave frequencies have also appeared in the literature [161], paving the way for the introduction of coupled-oscillator arrays in millimeter-wave phased-array sensing and communication applications.

We further note that successful demonstrations of only small arrays have been reported in the literature to date, and large arrays employing hundreds of elements remain to be seen. In such large arrays, perimeter oscillator control of the radiated beam will be particularly beneficial. Furthermore, the demonstration of coupled-oscillator arrays using signal processing and optimization techniques in beamforming, and more importantly adaptive beamforming, is an area that should be further exploited. Finally, a number of challenging analysis problems remain to be addressed, such as quantifying the effect of phase noise on the locking range of the array and more detailed study of mode locking for pulsed operation of coupled oscillators in the microwave frequencies.

Despite the progress in the theory and design of coupled-oscillator arrays during the past two decades, active antenna arrays based on coupled oscillators have not yet found widespread practical application, although there have been notable achievements in array implementations such as the ones shown in Chapter 6. It is hoped that the material presented in this book demonstrates the potential of coupled-oscillator arrays and motivates designers to apply them in microwave and millimeter-wave array antennas.

We have endeavored to provide the reader with the understanding and the tools for such application through description of the research to date and mention of a few areas for further study and technological development. The references to the archival literature will, of course, provide more detail than could be included here without rendering the presentation far too cumbersome for the casual reader. However, the literature sometimes presupposes significant familiarity with the approaches currently in vogue. Thus, in parts the present treatment is an overview of the research work while in other parts it provides a tutorial facilitating access to the literature. We hope to have struck a balance between these two styles of presentation, resulting in a book of somewhat wider utility in this field than would be the case for either style alone.

References

[1] K. D. Stephan, "Inter-Injection-Locked Oscillators for Power Combining and Phased Arrays," *IEEE Transactions on Microwave Theory and Techniques*, vol. MTT-34, pp. 1017–1025, October 1986.

[2] K. D. Stephan and W. A. Morgan, "Analysis of Interinjection-Locked Oscillators for Integrated Phased Arrays," *IEEE Transactions on Antennas and Propagation*, vol. AP-35, no. 7, pp. 771–781, July 1987.

[3] K. D. Stephan and S.-L. Young, "Mode Stability of Radiation-Coupled Interinjection-Locked Oscillators for Integrated Phased Arrays," *IEEE Transactions on Microwave Theory and Techniques*, vol. MTT-36, no. 5, pp. 921–924, May 1988.

[4] R. A. York and R. C. Compton, "Quasi-Optical Power-Combining Using Mutually Synchronized Oscillator Arrays," *IEEE Transactions on Microwave Theory and Techniques*, vol. MTT-39, pp. 1000–1009, June 1991.

[5] R. A. York and R. C. Compton, "Mode-Locked Oscillator Arrays," *IEEE Microwave Guided Wave Letters*, vol. 1, no. 8, pp. 215–218, August 1991.

[6] R. A. York and R. C. Compton, "Experimental Observation and Simulation of Mode-Locking in Coupled-Oscillator Arrays," *Journal of Applied Physics*, vol. 71, no. 6, pp. 2959–2965, March 1992.

[7] R. A. York and R. C. Compton, "Measurement and Modeling of Radiative Coupling in Oscillator Arrays," *IEEE Transactions on Microwave Theory and Techniques*, vol. MTT-41, pp. 438–444, March 1993.

[8] J. W. Mink, "Quasi-Optical Power Combining of Solid-State Millimeter-Wave Sources," *Microwave Theory and Techniques*, vol. MTT-34, pp. 273–279, February 1986.

[9] B. van der Pol, "The Nonlinear Theory of Electric Oscillators," *Proceedings of the Institute of Radio Engineers*, vol. 22, pp. 1051–1086, September 1934.

[10] R. Adler, "A Study of Locking Phenomena in Oscillators," *IEEE Proceedings*, vol. 61, pp. 1380–1385, 1973. (Reprinted from *Proceedings of the Institute of Radio Engineers,* vol. 34, pp. 351–357, June 1946.)

[11] D. A. Linkens, "Analytical Solution of Large Numbers of Mutually Coupled Nearly Sinusoidal Oscillators," *IEEE Transactions on Circuits and Systems*, vol. CAS-21, no. 2, pp. 294–300, March 1974.

[12] T. Endo and S. Mori, "Mode Analysis of a Ring of a Large Number of Mutually Coupled van der Pol Oscillators," *IEEE Transactions on Circuits and Systems*, vol. CAS-25, no. 1, pp. 7–18, January 1978.

[13] R. H. Rand, A. H. Cohen, and P. J. Holmes, Ch. 9, "Systems of Coupled Oscillators as Models of Central Pattern Generators," in *Neural Control of Rythmic Movements in Vertebrates*. New York: John Wiley & Sons, 1988.

[14] A. H. Cohen, P. J. Holmes, and R. H. Rand, "The Nature of the Coupling Between Segmental Oscillators of the Lamprey Spinal Generator for Locomotion: A Mathematical Model," *Journal of Mathematical Biology*, vol. 13, pp. 345–369, 1982.

[15] K. Kurokawa, "Injection Locking of Microwave Solid-State Oscillators," *Proceedings of the IEEE*, vol. 61, pp. 1386–1410, October 1973.

[16] H. C. Chang, E. S. Shapiro, and R. A. York, "Influence of the Oscillator Equivalent Circuit on the Stable Modes of Parallel-Coupled Oscillators," *IEEE Transactions on Microwave Theory and Techniques*, vol. MTT-45, no. 8, pp. 1232–1239, 1997.

[17] S. Nogi, J. Lin, and T. Itoh, "Mode Analysis and Stabilization of a Spatial Power Combining Array with Strongly Coupled Oscillators," *IEEE Transactions on Microwave Theory and Techniques*, vol. MTT-41, pp. 1827–1837, October 1993.

[18] B. K. Meadows, T. H. Heath, J. D. Neff, E. A. Brown, D. W. Fogliati, and M. Gabbay, V. In P. Hasler, S. P. Deweeth, and W. L. Ditto, "Nonlinear Antenna Technology," *Proceedings of the IEEE*, vol. 90, no. 5, pp. 882–897, May 2002.

[19] V. Seetharam and L. W. Pearson, "Analysis of Strong Coupling in Coupled Oscillator Arrays," *IEEE Transactions on Antennas and Propagation*, vol. AP-58, no. 4, pp. 1067–1075, April 2010.

[20] M. Armand, "On the Output Spectrum of Unlocked Driven Oscillators," *IEEE Proceedings*, vol. 57, pp. 798–799, 1969.

[21] S. Sancheti and V. F. Fusco, "Active Antenna Top-Cover Frequency Pulling Effects," *IEE Proceedings on Microwave Antennas and Propagation*, vol. 141, no. 5, pp. 374–376, October 1994.

[22] S. Sancheti and V. F. Fusco, "Modeling of Active Antenna Array Coupling Effects—A Load Variation Method," *IEEE Transactions on Microwave Theory and Techniques*, vol. MTT-43, no. 8, pp. 1805–1808, August 1995.

[23] D. E. J. Humphrey and V. F. Fusco, "Nonlinear Two Element Active Antenna Array Stability Analysis," *Electronics Letters*, vol. 32, no. 9, pp. 788–789, April 1996.

[24] D. E. J. Humphrey and V. F. Fusco, "Mode Change Analysis of a Two Element Oscillator-Type Antenna Array," *IEE Proceedings on Microwave Antennas and Propagation*, vol. 144, no. 6, pp. 483–487, December 1997.

[25] D. E. J. Humphrey, V. F. Fusco, and S. Drew, "Active Antenna Array Behavior," *IEEE Transactions on on Microwave Theory and Techniques*, vol. MTT-43, no. 8, pp. 1819–1824, August 1995.

[26] C. A. Lee and G. C. Dalman, "Pulse-Time Modulation of Mutually Coupled Arrays of Oscillators," *IEEE Transactions on Microwave Theory and Techniques*, vol. MTT-47, no. 1, pp. 34–41, January 1999.

[27] V. Seetharam, J. Shen, and L. W. L. W. Pearson, "Effect of Coupling Phase on Mutual Injection Locking Range in Coupled Oscillator Arrays," *IEEE Transactions on Antennas and Propagation*, vol. AP-57, no. 5, pp. 1391–1398, May 2009.

[28] P. Liao and R. A. York, "A New Phase Shifterless Beam-Scanning Technique Using Arrays of Coupled Oscillators," *IEEE Transactions on Microwave Theory and Techniques*, vol. MTT-41, pp. 1810–1815, October 1993.

[29] T. Heath, K. Wiesenfeld, and R. A. York, "Manipulated Synchronization: Beam Steering in Phased Arrays," *International Journal of Bifurcation and Chaos*, vol. 10, no. 11, pp. 2619–2627, 2000.

[30] J-H. Hwang and N-H. Myung, "A New Beam Scanning Technique by Controlling the Coupling Angle in a Coupled Oscillator Array," *IEEE Microwave and Guided Wave Letters*, vol. 8, no. 5, pp. 191–193, May 1998.

[31] T. Heath, "Difference Pattern Beam Steering of Coupled, Nonlinear Oscillator Arrays," *IEEE Microwave and Wireless Components Letters*, vol. 11, no. 8, pp. 343–345, August 2001.

[32] J. J. Lynch and R. A. York, "Synchronization of Oscillators Coupled Through Narrow-Band Coupling Networks," *IEEE Transactions on Microwave Theory and Techniques*, vol. MTT-49, no. 2, pp. 237–249, February 2001.

[33] R. A. York, P. Liao, and J. J. Lynch, "Oscillator Array Dynamics with Broadband N-port Coupling Networks," *IEEE Transactions on Microwave Theory and Techniques*, vol. MTT-42, pp. 2040–2045, November 1994.

[34] R. J. Pogorzelski, "On the Design of Coupling Networks for Coupled Oscillator Arrays," *IEEE Transactions on Antennas and Propagation*, vol. AP-51, pp. 794–801, April 2003.

[35] N. Amitay, V. Galindo, and C. P. Wu, *Theory and Analysis of Phased Array Antennas*. New York: Wiley-Interscience, 1972.

[36] D. E. J. Humphrey and V. F. Fusco, "Capacitively Coupled Oscillator Array Behaviour," *IEE Proceedings—Circuits Devices Systems*, vol. 143, no. 3, pp. 167–170, June 1996.

[37] M. Abramowitz and I. Stegun, *Handbook of Mathematical Functions*. Washington, District of Columbia: National Bureau of Standards, 1964.

[38] R. J. Pogorzelski, P. F. Maccarini, and R. A. York, "A Continuum Model of the Dynamics of Coupled Oscillator Arrays for Phase-Shifterless Beam Scanning," *IEEE Transactions on Microwave Theory and Techniques*, vol. 47, no. 4, pp. 463–470, April 1999, Correction: MTT-48, p. 1422, August 2000.

[39] R. J. Pogorzelski, P. F. Maccarini, and R. A. York, "Continuum Modeling of the Dynamics of Externally Injection-Locked Coupled Oscillator Arrayas," *IEEE Transactions on Microwave Theory and Techniques*, vol. MTT-47, no. 4, pp. 471–478, April 1999, Corrections: MTT-47, p. 1147, July 1999 and MTT-48, p. 1422, August 2000.

[40] R. A. York and T. Itoh, "Injection- and Phase-locking Techniques for Beam Control," *IEEE Transactions on Microwave Theory and Techniques*, vol. MTT-46, pp. 1920–1929, November 1998.

[41] J.-H. Yoon and N.-H. Myung, "Continuous Dynamics for a Linear Coupled-Oscillator Array with Peripheral Coupling-Phase Adjustments," *Microwave and Optical Technology Letters*, vol. 42, no. 5, pp. 422–425, September 2004.

[42] R. J. Pogorzelski, "On the Dynamics of Two-Dimensional Array Beam Scanning via Perimeter Detuning of Coupled Oscillator Arrays," *IEEE Transactions on Antennas and Propagation*, vol. AP-49, no. 2, pp. 234–242, February 2001.

[43] R. J. Pogorzelski, "Two-Dimensional Array Beam Scanning via Externally and Mutually Injection-Locked Couppled Oscillators," *IEEE Transactions on Antennas and Propagation*, vol. AP-49, no. 2, pp. 243–249, February 2001.

[44] R. J. Pogorzelski, "Phased Arrays Based on Oscillators Coupled on Triangular and Hexagonal Lattices," *IEEE Transactions on Antennas and Propagation*, vol. AP-52, no. 3, pp. 790–800, March 2004, Correction: AP-52, 2196, August 2004.

[45] R. J. Pogorzelski, "On the Steady State Phase Distribution in a Hexagonally Coupled Oscillator Array," *IEEE Transactions on Antennas and Propagation*, vol. AP-53, pp. 4058–4064, December 2005.

[46] W. H. Press, B. P. Flannery, S. A. Teukolsky, and W. T. Vetterling, *Numerical Recipies—The Art of Scientific Computing*. Cambridge, United Kingdom: Cambridge University Press, 1986.

[47] R. J. Pogorzelski, "Generalization of the Linearized Discrete Model of Coupled Oscillator Arrays to Account for Coupling Delay," *Radio Science*, vol. 43, no. RS4070, doi:10.1029/2007RS003727, 2008.

[48] R. J. Pogorzelski, "Continuum Modeling of Coupled Oscillator Arrays with Coupling Delay," *Radio Science*, vol. 43, no. RS4S04, doi:10.1029/2007RS003741, 2008.

[49] R. J. Pogorzelski, "Beam Steering Dynamics in Coupled Oscillator Based Phased Arrays with Coupling Delay," *Radio Science*, vol. 44, no. RS2019, doi:10.1029/2008RS004056, 2008.

[50] X. Wang and L. W. Pearson, "Design of Coupled-Oscillator Arrays without a Posteriori Tuning," *IEEE Transactions on Microwate Theory and Techniques*, vol. MTT-53, no. 1, pp. 410–413, January 2005.

[51] J. Shen and L. W. Pearson, "The Phase Error and Beam-Pointing Error in Coupled Oscillator Beam-Steering Arrays," *IEEE Transactions on Antennas and Propagation*, vol. AP-53, no. 1, pp. 386–393, January 2005.

[52] P. S. Hall and P. M. Haskins, "Microstrip Active Patch Array with Beam Scanning," *Electronics Letters*, vol. 28, no. 22, pp. 2056–2057, October 1992.

[53] R. A. York, "Nonlinear Analysis of Phase Relationships in Quasi-Optical Oscillator Arrays," *IEEE Transactions on Microwave Theory and Techniques*, vol. MTT-41, no. 10, pp. 1799–1809, October 1993.

[54] P. Liao and R. A. York, "A Six-Element Beam Scanning Array," *IEEE Microwave and Guided Wave Letters*, vol. 4, no. 1, pp. 20–22, January 1994.

[55] R. Ispir, S. Nogi, M. Sanagi, and K. Fukui, "Transmission-Line Coupling of Active Microstrip Antennas for One- and Two-Dimensional Phased Arrays," *IEICE Transactions on Electronics*, vol. E80-C, no. 9, pp. 1211-1220, September 1997.

[56] T. Kagawa, S. Nogi, and M. Sanagi, "Spatial Power Combining and Phased-Array Behavior of Circularly Polarized Active Patch Antennas," *IEICE Transactions on Electronics*, vol. E82-C, no. 7, pp. 1182–1187, July 1999.

[57] R. Ispir, S. Nogi, and M. Sanagi, "Wide Angle Phase-Shifterless Beam Scanning in Unilaterally Coupled Active Antenna Arrays," *IEICE Transactions on Electronics*, vol. E82-C, no. 7, pp. 1188–1194, July 1999.

[58] J. J. Lynch, H.-C. Chang, and R. A. York, "Coupled-Oscillator Arrays and Scanning Techniques," in *Active and Quasi-Optical Arrays for Solid-State Power Combining*. New York: John Wiley & Sons, 1997, pp. 135–183.

[59] H.-C. Chang, X. Cao, M. J. Vaughn, U. K. Mishra, and R. A. York, "Phase Noise in Externally Injection Locked Oscillator Arrays," *IEEE Transactions on on Microwave Theory and Techniques*, vol. MTT-45, no. 11, pp. 2035–2042, November 1997.

[60] R. J. Pogorzelski, R. P. Scaramastra, J. Huang, R. J. Beckon, S. M. Petree, and C. Chavez, "A Seven Element S-Band Coupled Oscillator Controlled Agile Beam Phased Array," *IEEE Transactions on Microwave Theory and Techniques*, vol. MTT-48, no. 8, pp. 1375–1384, August 2000.

[61] R. J. Pogorzelski, "A Two-Dimensional Coupled Oscillator Array," *IEEE Microwave and Guided Wave Letters*, vol. 10, pp. 478–480, November 2000.

[62] R. J. Pogorzelski, "Coupled Oscillator Based Agile Beam Transmitters and Receivers: A Review of Work at JPL," *Proceedings of the 2006 IEEE Aerospace Conference (Big Sky, Montana, March 14–16, 2006)*, 0-7803-9546-8/06, 2006.

[63] R. J. Pogorzelski, "Experimental Confirmation of the Dynamics of Coupled Oscillator Arrays," *IEEE Tranactions on Microwave Theory and Techniques*, vol. MTT-50, no. 1, pp. 143–149, January 2002.

[64] F. Shigeki, "Waveguide Line (in Japanese)," Japan Patent 06-053711, February 25, 1994.

[65] F. Giuppi, A. Georgiadis, M. Bozzi, A. Collado, and L. Perregrini, "Active Antenna Oscillator Systems in Substrate Integrated Waveguide (SIW) Technology," presented at *Proceedings of the Fourth European Conference on Antennas and Propagation (EuCAP)*, Barcelona, Spain, 2010.

[66] M. Bozzi, L. Perregrini, K. Wu, and P. Arcioni, "Current and Future Research Trends in Susbtrate Integrated Waveguide Technology," *Radioengineering*, vol. 18, no. 2, pp. 201–209, June 2009.

[67] N. Karmakar, "Investigations into a Cavity-Backed Circular Patch Antenna," *IEEE Transactions on Antennas and Propagation*, vol. 50, no. 12, pp. 1706–1715, December 2002.

[68] F. Giuppi, A. Georgiadis, A. Collado, M. Bozzi, and L. Perregrini, "Tunable SIW Cavity Backed Active Antenna Oscillator," *The Institution of Engineering and Technology (IET) Electronics Letters*, vol. 46, no. 5, pp. 1053–1055, July 22, 2010.

[69] R. J. Pogorzelski, "A Frequency Tripled Two-Dimensional Coupled Oscillator Array," *IEEE Transactions on Antennas and Propagation*, vol. AP-52, no. 3, pp. 907–910, March 2004.

[70] R. J. Pogorzelski, "A Five by Five Element Two-Dimensional Coupled Oscillator Array," *IEEE Transactions on Antennas and Propagation*, vol. AP-53, no. 4, pp. 1337–1345, April 2005.

[71] T. Heath, R. R. Kerr, and G. D. Hopkins, "Two Dimensional Nonlinear Oscillator Array Antenna," *Proceedings IEEE Aerospace Conference*, 2005.

[72] S.-C. Yen and T.-H. Chu, "A Beam-Scanning and Polarization-Agile Antenna Array Using Mutually Coupled Oscillating Doublers," *IEEE Transactions on Antennas and Propagation*, vol. 53, no. 12, pp. 4051–4057, December 2005.

[73] X. Cao and R. A. York, "Coupled Oscillator Scanning Techniques for Receiver Applications," *IEEE Antennas And Propagation Symposium Digest*, pp. 1311–1314, 1995.

[74] R. J. Pogorzelski and F. Chiha, "A Demonstration of the Coupled Oscillator Based Agile Beam Receiver Concept," *IEEE Transactions on Antennas and Propagation*, vol. AP-53, no. 11, pp. 3584–3588, DOI: 10.1109/TAP.2005.858835, November 2005.

[75] M. A. Kott, *Phased Array with Wide Null, U.S. Patent 5,343,211*, August 30 1994.

[76] R. J. Pogorzelski, "On a Simple Method of Obtaining Sidelobe Reduction over a Wide Angular Range in One and Two Dimensions," *IEEE Transactions on Antennas and Propagation*, vol. AP-49, pp. 475–482, March 2001.

[77] G. S. Shiroma, R. Y. Miyamoto, and W. A. Shiroma, "A 16-Element Two-Dimensional Active Self-Steering Array Using Self-Oscillating Mixers," *IEEE Transactions on Microwave theory and Techniques*, vol. 51, no. 12, pp. 2476–2482, December 2003.

[78] M. Sanagi, J. Fujiwara, K. Fujimori, and S. Nogi, "Beam Control in Unilaterally Coupled Active Antennas with Self-Oscillating Harmonic Mixers," *IEICE Transactions on Electronics*, vol. E88-C, no. 7, pp. 1375–1381, July 2005.

[79] S. Ver Hoeye, C. Vazquez, M. Fernandez, F. L. Herran, and F. Las-Heras, "Receiving Phased Antenna Array Based on Injection-Locked Harmonic Self-Oscillating Mixers," *IEEE Transactions on Antennas and Propagation*, vol. 57, pp. 645–651, Mar. 2009.

[80] S. Ver Hoeye, L. F. Herran, M. Fernandez, and F.L. Heras, "Design and Analysis of a Microwave Large-Range Variable Phase-Shifter Based on an Injection-Locked Harmonic Self-Oscillating Mixer," *IEEE Microwave and Wireless Components Letters*, vol. 16, no. 6, pp. 342–344, June 2006.

[81] H.-C. Chang, A. Borgioli, P. Yeh, and R. A. York, "Analysis of Oscillators with External Feedback Loop for Improved Locking Range and Noise Reduction," *IEEE Transactions on Microwave Theory and Techniques*, vol. MTT-47, no. 8, pp. 1535–1543, August 1999.

[82] M. Zheng, P. Gardener, P. S. Hall, Y. Hao, Q. Chen, and V. F. Fusco, "Cavity Control of Active Integrated Antenna Oscillators," *IEE Proceedings on Microwave Antennas and Propagation*, vol. 148, no. 1, pp. 15–20, February 2001.

[83] M. Colwell and L. W. Pearson, "Enhancement of Locking Range Through Reactive Loading on the Feedback Path," *IEEE Microwave and Wireless Components Letters*, vol. 13, no. 9, pp. 417–419, September 2003.

[84] H.-C., Chang, X. Cao, U. K. Mishra, and R. A. York, "Phase Noise in Coupled Oscillators: Theory and Experiment," *IEEE Transactions on Microwave Theory and Techniques*, vol. MTT-45, no. 5, pp. 604–615, May 1997.

[85] M. Nick, A. Banai, and F. Farzaneh, "Phase Noise Measurement Using Two Inter-Injection Locked Microwave Oscillators," *IEEE Transactions on Microwave Theory and Techniques*, vol. MTT-54, no. 7, pp. 2993–3000, July 2006.

[86] L. Dussopt and J.-M. Laheurte, "Coupled Oscillator Array Generating Circular Polarization," *IEEE Microwave and Guided Wave Letters*, vol. 94, pp. 160–162, April 1999.

[87] L. Dussopt and J.-M. Laheurte, "Phase Noise Improvement in a Loop Configuration of Oscillating Antennas," *IEEE Microwave and Guided Wave Letters*, vol. 10, no. 3, pp. 111–113, March 2000.

[88] S. Yang, V. F. Fusco, and D. E. J. Humphrey, "Ring-Coupled-Oscillator Sequentially Rotated Active Antenna," *IEEE Transactions on Microwave Theory and Techniques*, vol. MTT-49, no. 8, pp. 1492–1497, August 2001.

[89] J. J. Lynch and R. A. York, "Stability of Mode Locked States of Coupled Oscillator Arrays," *IEEE Transactions on Circuits and Systems -I: Fundamental Theory and Applications*, vol. 42, no. 8, pp. 413–418, August 1995.

[90] J. J. Lynch and R. A. York, "An Analysis of Mode-Locked Arrays of Automatic Level Control Oscillators," *IEEE Transactions on Circuits and Systems I: Fundamental Theory and Applications*, vol. 41, no. 12, pp. 859–865, December 1994.

[91] R. R. Ram, R. Sporer, H.-R. Blank, and R. A. York, "Chaotic Dynamics in Coupled Microwave Oscillators," *IEEE Transactions on Microwave Theory and Techniques*, vol. MTT-48, no. 11, pp. 1909–1916, November 2000.

[92] R. D. Martinez and R. C. Compton, "Electronic Beamsteering of Active Arrays With Phase-Locked Loops," *IEEE Microwave and Guided Wave Letters*, vol. 4, no. 6, pp. 166–168, June 1994.

[93] J. J. Lynch and R. A. York, "A Mode Locked Array of Coupled Phase Locked Loops," *IEEE Microwave and Guided Wave Letters*, vol. 5, no. 7, pp. 213–215, July 1995.

[94] D. W. Jordan and P. Smith, *Nonlinear Ordinary Differential Equations: An Introduction to Dynamical Systems,* 3rd ed. Oxford, United Kingdom: Oxford University Press, 1999.

[95] R. A. York, P. Liao, and J. J. Lynch, "Oscillator Array Dynamics with Broadband N-Port Coupling Networks," *IEEE Transactions on Microwave Theory and Techniques*, pp. 2040–2045, November 1994.

[96] T. Heath, "Simultaneous Beam Steering and Formation with Coupled, Nonlinear Oscillator Arrays," *IEEE Transactions on Antennas and Propagation*, vol. 53, no. 6, pp. 2031–2035, June 2005.

[97] H.-C. Chang, X. Cao, U. K. Mishra, and R.A. York, "Phase Noise in Coupled Oscillators: Theory and Experiment," *IEEE Transactions on Microwave Theory and Techniques*, pp. 604–615, May 1997.

[98] R. J. Pogorzelski, "Experimental Confirmation of the Dynamics of Coupled-Oscillator Arrays and Implications for Angle-Based Modulation," *IEEE Transactions on Microwave Theory and Techniques*, vol. 50, no. 1, pp. 143–149, January 2002.

[99] C. Kykkotis, P.S. Hall, and H. Ghafouri-Shiraz, "Performance of Active Antenna Oscillator Arrays under Modulation for Communication Systems," *IEE Proceedings on Microwaves, Antennas and Propagation*, vol. 145, no. 4, pp. 313–320, 1998.

[100] J. Guckenheimer and P. Holmes, *Nonlinear Oscillations, Dynamical Systems, and Bifurcations of Vector Fields* (*Applied Mathematical Sciences*, vol. 42). New York: Springer-Verlag, 1983.

[101] T. S. Parker and L. O. Chua, *Practical Numerical Algorithms for Chaotic Systems*. New York: Springer-Verlag, 1989.

[102] Y. A. Kuznetsov, *Elements of Applied Bifurcation Theory*, 2nd ed. New York: Verlag, Springer, 1998.

[103] J. A. Sanders and F. Velhurst, *Averaging Methods in Nonlinear Dynamical Systems*. New York: Springer Verlag, 1985.

[104] N. M. Krylov and N. N. Bogoliubov, *New Methods of Nonlinear Mechanics in Their Application to the Investigation of the Operation of Electronic Generators*. Moscow, Russia: United Scientific and Technical Press, 1934.

[105] K. Kurokawa, "Noise in Synchronized Oscillators," *IEEE Transactions on Microwave Theory and Techniques*, vol. 16, no. 4, pp. 234–240, April 1968.

[106] A. C. Scott, "Distributed Multimode Oscillators of One and Two Spatial Dimensions," *IEEE Transactions on Circuit Theory*, vol. 17, no. 1, pp. 55–60, February 1970.

[107] H. M. Aumann, "Standing Waves on a Multimode Ladder Oscillator," *IEEE Transactions on Circuits and Systems*, vol. 21, no. 3, pp. 461–462, May 1974.

[108] T. Endo and S. Mori, "Mode Analysis in a Multimode Ladder Oscillator," *IEEE Transactions on Circuits and Systems*, vol. 23, no. 2, pp. 100–113, February 1976.

[109] T. Endo and S. Mori, "Mode Analysis of a Two-Dimensional Low-Pass Multimode Oscillator," *IEEE Transactions on Circuits and Systems*, vol. 23, no. 9, pp. 517–530, September 1976.

[110] R. A. York and R. C. Compton, "Quasi-Optical Power-Combining Using Mutually Synchronized Oscillator Arrays," *IEEE Transactions on Microwave Theory and Techniques*, vol. MTT-39, pp. 1000–1009, June 1991.

[111] R. A. York, "Nonlinear Analysis of Phase Relationships in Quasi-Optical Coupled Oscillator Arrays," *IEEE Transactions on Microwave Theory and Techniques*, vol. 43, no. 10, pp. 1799–1809, October 1993.

[112] T. Heath, "Beam Steering of Nonlinear Oscillator Arrays Through Manipulation of Coupling Phases," *IEEE Transactions on Antennas and Propagation*, vol. 52, no. 7, pp. 1833–1842, July 2004.

[113] Y. Kuramoto, *Chemical Oscillations, Waves and Turbulence*. Berlin, Germany: Springer Verlag, 1984.

[114] D.E.J. Humphrey, V.F. Fusco, and S. Drew, "Active Antenna Array Behavior," *IEEE Transactions on Microwave Theory and Techniques*, vol. 43, no. 8, pp. 1819–1925, August 1995.

[115] K. Y. Chen, P. D. Biernacki, A. Lahrichi, and A. Mickelson, "Analysis of an Experimental Technique for Determining Van der Pol Parameters of a Transistor Oscillator," *IEEE Transactions on Microwave Theory and Techniques*, vol. 46, no. 7, pp. 914–922, July 1998.

[116] A. Georgiadis, A. Collado, and A. Suarez, "New Techniques for the Analysis and Design of Coupled Oscillator Systems," *IEEE Transactions on Microwave Theory and Techniques*, vol. 54, no. 11, pp. 3864–3877, November 2006.

[117] J.J. Lynch and R.A. York, "Synchronization of Oscillators Coupled Through Narrow-Band Networks," *IEEE Transactions on Microwave Theory and Techniques*, pp. 237–249, February 2001.

[118] A. Georgiadis, A. Collado, and A. Suarez, "Pattern Nulling in Coupled Oscillator Antenna Arrays," *IEEE Transactions on Antennas and Propagation*, vol. 55, no. 5, pp. 1267–1274, May 2007.

[119] A. S. Lewis and M. L. Overton, "Eigenvalue Optimization," *Acta Numerica*, vol. 5, pp. 149–190, 1996.

[120] A. Georgiadis and A. Collado, "Injection Locked Coupled Oscillator Arrays," *IEEE Microwave and Wireless Components Letters*, pp. 900–902, December 2007.

[121] C. Kykkotis, P. S. Hall, H. Ghafouri-Shiraz, and D. Wake, "Modulation Effects in Active Integrated Locked Antenna Oscillator Arrays," in *IEE Tenth International Conference on Antennas and Propagation,* (Conf. Pub. No. 436), 1997, pp. 510–513.

[122] D. T. Auckland, J. Lilly, and R. A. York, "Analysis of Beam Scanning and Data Rate Transmission Performance of a Coupled Oscillator Phased Array," in *IEE Tenth International Conference on Antennas and Propagation* (Conf. Pub. No. 436), 1997, pp. 245–249.

[123] H.-C. Chang, X. Cao, M. J. Vaughan, U.K. Mishra, and R. A. York, "Phase Noise in Externally Injection-Locked Oscillator Arrays," *IEEE Transactions on Microwave Theory and Techniques*, vol. 45, no. 11, pp. 2035–2042, November 1997.

[124] A. Collado and A. Georgiadis, "Performance of Coupled Oscillator Arrays with Angle-Modulated Injection Signals," *IEEE Transactions on Circuits and Systems I*, vol. 57, no. 9, pp. 2343–2352, 2010.

[125] R. J. Pogorzelski, S. Stride, J. Venkatesan, and M. Zawadzki, "A Seven Element X-Band Agile Beam Receiver," in *IEEE International Symposium on Antennas and Propagation*, Honolulu, Hawaii (June 9–15, 2007), 2007, pp. 1925–1928.

[126] R. D. Martinez and R. C. Compton, "Electronic Beamsteering of Active Arrays with Phase-Locked Loops," *IEEE Microwave and Guided Wave Letters*, vol. 4, no. 6, pp. 166–169, June 1994.

[127] J. F. Buckwalter, T. H. Heath, and R. A. York, "Synchronization Design of a Coupled Phase-Locked Loop," *IEEE Transactions on Microwave Theory and Techniques*, vol. 51, no. 3, pp. 952–960, March 2003.

[128] H.-C. Chang, "Analysis of Coupled Phase-Locked Loops with Independent Oscillators for Beam Control Active Phased Arrays," *IEEE Transactions on Microwave Theory and Techniques*, vol. 52, no. 3, pp. 1059–1066, March 2004.

[129] K. Kurokawa, "Some Basic Characteristics of Negative Resistance Oscillator Circuits," *Bell System Technical Journal*, vol. 61, pp. 1937–1955, July–August 1969.

[130] X. Lai and J. Roychowdhury, "Capturing Oscillator Injection Locking via Nonlinear Phase-Domain Macromodels," *IEEE Transactions on Microwave Theory and Techniques*, vol. 52, no. 9, pp. 2251–2261, September 2004.

[131] F. Giannini and G. Leuzzi, *Nonlinear Microwave Circuit Design*. Hoboken, New Jersey: John Wiley and Sons, 2004.

[132] K. S. Kundert, J. K. White, and A. Sangiovanni-Vincentelli, *Steady-State Methods For Simulating Analog and Microwave Circuits*. Norwell, Massachussetts: Kluwer Academic Publishers, 1990.

[133] S. Boyd and L. Vandenberghe, *Convex Optimization*. Cambridge, UK: Cambridge University Press, 2004.

[134] V. Rizzoli and A. Neri, "State of the Art and Present Trends in Nonlinear Microwave CAD Techniques," *IEEE Transactions on Microwave Theory and Techniques*, vol. 36, no. 2, pp. 343–365, February 1988.

[135] D. Sharrit, "Method for Simulating a Circuit," U.S. Patent 5588142, December 24, 1996.

[136] E. Ngoya and R. Larcheveque, "Envelope Transient Analysis: A New Method for the Transient and Steady State Analysis of Microwave Communication Circuits and Systemms," in *IEEE MTT-S International Microwave Symposium*, San Francisco, California, 1996, pp. 1365–1368.

[137] L. O. Chua, "A Switching Parameters Algorithm for Finding Multiple Solutions of Non-Linear Resistive Circuits," *International Journal of Circuit Theory and Applications*, vol. 4, pp. 215–239, July 1976.

[138] Ch.-R. Chang, M. B. Steer, S. Martin, and E. Reese Jr., "Computer-Aided Analysis of Free-Running Microwave Oscillators," *IEEE Transactions on Microwave Theory and Techniques*, vol. 39, no. 10, pp. 1735–1745, October 1991.

[139] R. Quere, E. Ngoya, M. Camiade, A. Suarez, M. Hessane, and J. Obregon, "Large Signal Design of Broadband Monolithic Microwave Frequency Dividers and Phase-Locked Oscillators," *IEEE Transactions on Microwave Theory and Techniques*, vol. 41, no. 11, pp. 1928–1938, November 1993.

[140] A. Suarez and S. Sancho, "Application of the Envelope-Transient Method to the Analysis and Design of Autonomous Circuits," *International Journal of RF and Microwave Computer-Aided Engineering*, vol. 15, no. 6, pp. 523–535, 2005.

[141] J. M. T. Thompson and H. B. Stewart, *Nonlinear Dynamics and Chaos*, 2nd ed. Chichester, West Sussex, United Kingdom: John Wiley and Sons, 2002.

[142] P. Liao and R. A. York, "A New Phase Shifterless Beam-Scanning Technique Using Arrays of Coupled Oscillators," *IEEE Transactions on Microwave Theory and Techniques*, vol. MTT-41, pp. 1810–1815, October 1993.

[143] A. Georgiadis, "A Model for Stability, Noise and Angle Modulation Analysis of Injection-Locked Oscillators," *IEEE Transactions on Microwave Theory and Techniques*, vol. 56, no. 4, pp. 826–834, April 2008.

[144] R. J. Pogorzelski, P. F. Maccarini, and R. A. York, "Continuum Modeling of the Dynamics of Externally Injection-Locked Coupled Oscillator Arrays," *IEEE Transactions on Microwave Theory and Techniques*, vol. 47, no. 4, pp. 471–478, April 1999.

[145] A. Alexanian, H.-C. Chang, and R. A. York, "Enhanced Scanning Range of Coupled Oscillator Arrays Utilizing Frequency Multipliers," in *IEEE International Symposium on Antennas and Propagation (AP-S) Symposium Digest*, Newport Beach, California, 1995, pp. 1308–13010.

[146] M. Sanagi, K. Yano, K. Fujimori, and S. Nogi, "Active Phased Array Antenna Radiating Second Harmonic Output Wave," *Electronics and Communications in Japan, Part 2*, vol. 89, no. 4, pp. 39–50, DOI: 10.1002/ecjb.20168, April 2006.

[147] A. Georgiadis, "Design of Coupled Oscillator Arrays for Second Harmonic Radiation," in *IEEE MTT-S International Microwave Symposium Digest*, Honolulu, Hawaii, June 2007, pp. 1727–1730.

[148] L. F. Herran, S. V. Hoeye, and F. Las Heras, "Nonlinear Optimization Tools for the Design of Microwave High-Conversion Gain Harmonic Self-Oscillating Mixers," *IEEE Microwave and Wireless Components Letters*, vol. 16, no. 1, pp. 16–18, January 2006.

[149] J. G. Proakis and M. Salehi, *Digital Communications*, 5th ed. New York, New York: The McGraw-Hill Companies, 2008.

[150] A. Georgiadis and K. Slavakis, "A Convex Optimization Method for Constrained Beam-Steering in Planar (2-D) Coupled Oscillator Antenna Arrays," *IEEE Transactions on Antennas and Propagation*, vol. 55, no. 10, pp. 2925–2928, October 2007.

[151] A. Georgiadis and A. Collado, "Multibeam Pattern Generation Using Coupled Oscillator Arrays," in *Proceedings of 3rd European Conference on Antennas and Propagation* (EuCAP, March 23–27, 2009), Berlin, Germany, 2009.

[152] T. Heath, "Difference Pattern Beam Steering of Coupled, Nonlinear Oscillator Arrays," *IEEE Microwave and Wireless Components Letters*, vol. 11, no. 8, pp. 343–345, November 2001.

[153] A. Georgiadis and K. Slavakis, "Stability Optimization of the Coupled Oscillator Array Steady State Solution," *IEEE Transactions on Antennas and Propagation*, vol. 58, no. 2, pp. 608–612, February 2010.

[154] T. Ikuma, A. A. Beeks, J. R. Zeidler, and B. K. Meadows, "Adaptive Interference Mitigation with a Coupled Nonlinear Oscillator Array Beamformer," in *IEEE Aerospace Conference*, Big Sky, Montana, 2006, pp. 1–13.

[155] I. Nesterov, A. Nemirovskii, and Y. Nesterov, *Interior Point Polynomial Methods in Convex Programming,* vol. 13, *Studies in Applied Mathematics*. Philadelphia, Pennsylvania: Society for Industrial and Applied Mathematics, 1994.

[156] S. Boyd, L. El Ghaoui, E. Feron, and V. Balakrishnan, *Linear Matrix Inequalities in System and Control Theory*. Philadelphia, USA: SIAM Studies in Applied Mathematics, vol. 15, 1994.

[157] H. Steyskal, "Simple Method for Pattern Nulling by Phase Perturbation," *IEEE Transactions on Antennas and Propagation*, vol. AP-31, no. 1, pp. 163–166, January 1983.

[158] H. Lebret and S. Boyd, "Antenna Array Pattern Synthesis via Convex Optimization," *IEEE Transactions on Signal Processing*, vol. 45, no. 3, pp. 526–532, March 1997.

[159] T. Heath, "Beam Shaping of Coupled Nonlinear-Oscillator Arrays," in *Proceedings of the SPIE*, vol. 4374, 2001, pp. 114–122.

[160] H. Jiang and R. Penno, "Effects of Amplitude Dynamics on Beam Steering and Shaping in Coupled Oscillator Array," *IEEE Antennas and Wireless Propagation Letters*, vol. 9, pp. 474–477, 2010.

[161] J. F. Buckwalter, A. Babakhani, A. Komijani, and A. Hajimiri, "An Integrated Subharmonic Coupled-Oscillator Scheme for a 60-GHz Phased Array Transmitter," *IEEE Transactions on Microwave Theory and Techniques*, vol. 54, Part 2, no. 12, pp. 4271–4280, December 2006.

Acronyms and Abbreviations

AACS	attitude and articulation control subsystem
ABET, Inc.	Accreditation Board for Engineering and Technology
ACORDE	Advanced Communications Research and Development, S.A. (Santander, Spain)
AG	auxiliary generator
AM	amplitude modulation
BMDO	U. S. Ballistic Missile Defense Organization
CMOS	complementary metal oxide semiconductor
COA	coupled-oscillator array
CPG	central pattern generator
CPM	constant-phase modulation
CTTC	Centre Tecnologic de Telecomunicacions de Catalunya (Barcelona, Spain)
dBc	decibels relative to the carrier
dBm	decibels referenced to milliwatts

dc, DC	direct current
DC bias	bias voltage associated with a direct-current system
d.e.	differential equation
deg	degree
E-plane	plane containing the electric field intensity vector
Eq.	Equation
EuCAP	European Conference on Antennas and Propagation
FET	field effect transistor
FFT	fast Fourier transform
GCD	Global Communications Devices (North Andover, Massachusetts)
GHz	gigahertz
GMSK	Gaussian minimum-shift keying
GSM	Global System for Mobile Communications
GTRI	Georgia Tech Research Institute
H-plane	plane containing the magnetic field intensity vector
HSOM	harmonic self-oscillating mixer
IEE	Institution of Electrical Engineers
IEEE	Institute of Electrical and Electronics Engineers
IET	Institution of Engineering and Technology
IRE	Institute of Radio Engineers
IF	intermediate frequency
ILR	inverse locking range

JPL	Jet Propulsion Laboratory
L-band	frequency range between 390 and 1550 MHz
LCP	liquid-crystal polymers
LMDS	local multipoint distribution service
LMS	least mean square
MESFET	metal semiconductor field-effect transistor
MHz	megahertz
MMIC	monolithic microwave integrated circuit
mr	milliradian
Ms^{-1}	mega inverse seconds
NARRA	Novel Architectures for Reconfigurable Reflectarrays and Phased Array Antennas
NASA	National Aeronautics and Space Administration
nH	nanohenry
ns	nanosecond
pF	picofarad
pHEMT	pseudomorphic high electron mobility transistor
PLL	phase-locked loop
PM	phase modulation
R.A.C.	Radio Antenna Communications (Milan, Italy)
Ref.	reference
RF	radio frequency
RFCSET	RF/Microwave Communication Subsystems for Emerging Wireless Technologies
RFID	radio-frequency identification

RLC resistance, inductance, capacitance

RoM reduced order model

S-band RF frequencies 1550–5200 MHz

SIW substrate integrated waveguide

SOM self-oscillating mixer

SWAP (Marie Curie Industry-Academia Pathways and Partnerships project) Symbiotic Wireless Autonomous Powered system

USNC/URSI U.S. National Committee (USNC) of the Union Radio Scientifique Internationale

UWB ultra-wideband

V volt

VCO voltage- controlled oscillator

VHF very high frequency

VNA vector network analyzer

VSA vector signal analyzer

Index